Development and Application of Low-Cost Environmental Sensors for the Evaluation of Ambient Air Quality and Related Emission Sources

Development and Application of Low-Cost Environmental Sensors for the Evaluation of Ambient Air Quality and Related Emission Sources

Guest Editor

László Bencs

MDPI

Basel • Beijing • Wuhan • Barcelona • Belgrade • Novi Sad • Cluj • Manchester

Guest Editor
László Bencs
HUN-REN Wigner Research
Centre for Physics
Budapest
Hungary

Editorial Office
MDPI AG
Grosspeteranlage 5
4052 Basel, Switzerland

This is a reprint of the Special Issue, published open access by the journal *Atmosphere* (ISSN 2073-4433), freely accessible at: https://www.mdpi.com/journal/atmosphere/special_issues/L6N991E81S.

For citation purposes, cite each article independently as indicated on the article page online and as indicated below:

Lastname, A.A.; Lastname, B.B. Article Title. *Journal Name* **Year**, *Volume Number*, Page Range.

ISBN 978-3-7258-5655-8 (Hbk)
ISBN 978-3-7258-5656-5 (PDF)
https://doi.org/10.3390/books978-3-7258-5656-5

Contents

About the Editor

László Bencs

László Bencs started his career in 1992 at the Research Laboratory for Crystal Physics of the Hungarian Academy of Sciences. Since 2003, he has been the head of the analytical spectroscopy laboratory of the latter institute's legal successors: the Research Institute for Solid State Physics and Optics, and the Wigner Research Centre for Physics (Wigner RCP). His main research interests encompass preparative chemical and analytical instrumental methods in water, food, and advanced material (optical crystal) analysis, using spectrochemical techniques. Between 2001 and 2011, he worked as a postdoctoral researcher for six years in the group of Prof. René Van Grieken at the Department of Chemistry, University of Antwerp (UA). During this time, his research focused on the investigation of particulate matter (PM) in indoor/outdoor air in various regions of Northern Belgium, including the North Sea. He is currently working at Wigner RCP on topics such as environmental and advanced materials analysis using spectroscopy and sensor techniques.

Preface

It is my pleasure to announce to the scientific community the launch of a new Special Issue of Atmosphere, connected to the emerging research area of low-cost environmental sensors (LCSs). These pocket-sized, lightweight devices are currently receiving growing interest and usage in atmospheric research, due to their ease of use in field applications.

This Special Issue incorporates ten research papers, written by 37 authors working in various countries (Belgium, Bosnia and Herzegovina, Brazil, Cuba, Hungary, Japan, Thailand, Uganda, and the United States), located on five continents (Africa, Asia, Europe, North America, and South America). This geographical broadness also indicates worldwide interest in research on LCSs in regions with various urban environments and resources.

The topics discussed in this collection of articles address a couple of interesting and important scientific issues, including high temporal resolution monitoring of aerosols (PM_1, $PM_{2.5}$, PM_{10}, black carbon) and a variety of gaseous air pollutants (CH_2O, CO_2, NO_2, O_3, SO_2, and VOC), low-cost and high-end laboratory and field calibration methods, the effects of environmental variables, such as ambient T and RH on accuracy, assessment of the measurement error, correction procedures for bias, drone-assisted sampling for mobile emission sources, development and use of LCSs in city-scale networks for real-time spatial pollutant mapping utilizing user-friendly interfaces, and the monitoring of indoor and outdoor air quality to identify emission sources in undisturbed and construction-related urban environments. Some of the research papers demonstrate fruitful international cooperation between developed and developing regions. There is no doubt that studies with LCSs are among the flourishing areas of atmospheric research and will likely remain so in the future. I am pleased to recommend this Special Issue to any scientist or engineer devoted to the R&D of environmental sensors. I would like to take this opportunity to express my sincere gratitude to the authors for their valuable scientific contributions, and to my co-workers (Anne Hu and Phoebe Ma) for their assistance at the journal *Atmosphere*.

László Bencs
Guest Editor

atmosphere

MDPI

Article

A Low-Cost Calibration Method for Temperature, Relative Humidity, and Carbon Dioxide Sensors Used in Air Quality Monitoring Systems

Rosa Amalia González Rivero [1], Luis Ernesto Morera Hernández [1], Olivier Schalm [2,*], Erik Hernández Rodríguez [1], Daniellys Alejo Sánchez [1], Mayra C. Morales Pérez [1], Vladimir Nuñez Caraballo [3], Werner Jacobs [2] and Alain Martinez Laguardia [1]

1 Department of Chemistry, Universidad Central "Marta Abreu" de Las Villas, Road to Camajuaní Km 5.5, Santa Clara 54830, Villa Clara, Cuba
2 Antwerp Maritime Academy, Noordkasteel Oost 6, 2030 Antwerpen, Belgium
3 Meteorological Center, Marta Abreu Street, Santa Clara 50100, Villa Clara, Cuba
* Correspondence: olivier.schalm@hzs.be

Abstract: Low-cost sensors provide an affordable alternative to monitor environmental parameters with acceptable performance. There is a substantial amount of literature where low-cost sensors are compared with high-end reference measurements. However, not all organizations have access to such reference infrastructure. We propose low-cost calibration methods for temperature, relative humidity, and CO_2 to allow them to collect their own reliable data. These methods are based on simple techniques and procedures that allow temperature calibration to be achieved in the range of 0 to 50 °C, relative humidity from 0 to 90%, and CO_2 between 0 and 1100 ppm. The materials used to create the calibration setups can be purchased online, at hardware stores, and in pharmacies. The reliability of the calibration methods was evaluated using several indicators, such as the airtightness of the calibration box, similarity with the factory calibration, similarity with the reference, similarity between different sensors, replicability of the calibration method, and the similarity with a golden standard. In addition, the results of the low-cost calibration methods were compared with the more advanced calibration methods. A short measurement campaign in the city of Santa Clara, Cuba, demonstrated that such calibrations transform in-house developed monitoring systems into valid low-cost scientific instruments for decision-making. This work creates opportunities for institutions and researchers hosted in low- and mid-income countries to build and validate their own equipment to reliably solve local problems.

Keywords: low-cost sensors; low-cost calibration; temperature; relative humidity; carbon dioxide

1. Introduction

Due to the increase in anthropogenic emissions of pollutants into the atmosphere, the evaluation of ambient air quality has become an important environmental concern during the last decades [1–5]. Many countries have a network of reference monitoring stations that measure air quality in real time. However, the high installation and maintenance costs of these stations mean that low- and middle-income countries are less covered by such stations. In those regions, alternative technologies such as low-cost monitoring systems can be used instead [6–10]. Some define a low-cost sensor as <$100 and a low-cost monitor consisting of one or more sensors and communication/data components as <$1000 [11]. The need for low-cost monitoring devices also exists in specific sectors in high-income countries. For example, temperature (T) and relative humidity (RH) monitoring in museums, archives, or churches are critical to assess the aggressiveness of ambient air toward heritage objects [12–15]. Unfortunately, investments in measuring instruments are frequently limited by the available budgets [16]. Another example is the reduction of the risk of losing expensive instruments by attaching low-cost devices (e.g., the Ozonesonde) to a weather balloon for measuring environmental parameters at high altitudes [17,18]. With the low-cost open hardware technology that exists on the market, citizens and pupils can build their own monitoring systems to measure the local air quality of their interest [19]. Several citizen science platforms such as Sensor. Community or PurpleAir are currently used in research [20–23].

Scientific literature regarding the use of low-cost sensors is growing. A search in the database of ScienceDirect resulted in around 500 hits for 2021 when using the keyword "low-cost sensor", including several reviews [24–29]. It is striking that quite a few publications emphasize the (calibration) limitations and the performance problems of such sensors [30–32]. These studies demonstrate that such sensors do not achieve the same data quality as obtained by reference methods, which is to be expected. In contrast to this negative way of evaluating low-cost sensors, one can also evaluate the possibilities of such technology and show how they can be used to build low-cost scientific instruments that are good enough to evaluate air quality [33]. Sensors and calibrations are complementary and equally important. The search for literature in ScienceDirect using the term "low-cost calibration" resulted in only four hits for 2021, and only one was related to air quality [34]. This suggests that the popularity of comparing low-cost sensors with state-of-the-art technology is higher than the development of low-cost calibration methods for low-cost sensors. This work aims to develop low-cost calibration methods so that institutes or areas without access to high-end equipment can perform their own reliable measurement campaigns.

The calibration of low-cost sensors with state-of-the-art reference instruments and the calibration of low-cost sensors with low-cost calibration methods correspond to the distinction between maximizers and satisficers [35]. A maximizer is someone who wants only the very best and who strives to make a choice that will give them the maximum benefit later on. One evaluates all possible options before one makes a decision. A satisficer is someone who makes a decision or takes action as soon as an option meets his criteria. Once one has found the option that is good enough, one stops his quest for better alternatives. A satisficer does not necessarily settle for mediocrity. His criteria can be very high, but as soon as one finds a low-cost scientific instrument that has the qualities he wants, one is satisfied. What he wants is an air quality monitoring system at a low-cost with high reliability that provides data that can be used by decision-makers who are responsible for improving the air quality. To obtain such a system, the satisficer can enhance the reliability of the monitoring system with a restricted budget by suppressing noise or removing spikes in the acquired data using data processing techniques. Another way to improve the data quality is to calibrate the sensors with low-cost methods. However, such calibration methods have been rarely mentioned in the literature.

A maximizer will only be satisfied when one can perform his analyses with the best possible measuring instruments that are available on the market [36]. When one is obliged to use low-cost sensors, one will strive for sensor calibration using state-of-the-art technology. He will compare sensors with reference measurements in laboratory conditions where environmental parameters are changed in a controlled way [37,38]. An alternative comparative test is to perform co-location experiments near national air quality measuring stations, where pollutants in the air are determined with sufficient accuracy [20,39,40]. Reference instruments can also be built inside a van to make them mobile so that co-location experiments can be performed at any location that is needed [41,42]. Different kinds of calibration procedures are applied to the parallel measurements performed in a laboratory or in co-location [9,28,43]: simple linear regression, multi-linear regression [44], or machine learning [10,45,46] methods. A drawback of the maximizer's methods is that his way of working is impossible when one has no access to high-end equipment. In addition, the advantages of low-cost sensors decrease when expensive calibration methods are required.

This contribution focuses on the satisficer's way of thinking, where data quality obtained by low-cost sensors is improved without the aid of expensive state-of-the art equipment. For example, the T inside a closed box can easily be manipulated using a heater or a Peltier cooler, while a simple glass thermometer acts as the reference device. The RH in a sealed container can be changed in a controlled way using saturated salt solutions [16,47–54]. An alternative method to control the RH is with glycerol solutions [55–60], which appear to be more convenient because one product can generate any kind of RH. For CO_2, methods to generate small but known amounts of gas using disposable medical equipment, such as plastic syringes, have been reported [61]. The setup proposed in this contribution can also be used to generate NO_2 and SO_2 in small but known amounts [62]. For gaseous pollutants, ambient air analyzed with classical methods of wet chemistry analysis, such as titrations [63], turbidity [64], colorimetry [65–68], or fluorimetry [69], can be used as a reference measurement. In addition, several indicators (e.g., airtightness of the closed box) are defined, which gives an insight into the reliability of the calibration methods. Further to the development of low-cost calibration methods for T, RH, and CO_2, the usefulness of the calibration methods is illustrated in a case study where these parameters were monitored at several locations in the city of Santa Clara, Cuba. The measurement campaigns illustrate that in-house developed monitoring systems can be used to generate reliable and useful data without the need for expensive reference measurement devices, even when the quality of the collected data is most probably lower than that of reference instruments.

2. Materials and Methods

2.1. Sensors Used with the Low-Cost Monitoring System

The low-cost monitoring system that was used in the experiments consisted of an Arduino MEGA2560 (provided by Tinytronics, Eindhoven, The Netherlands) as the computer unit and interface board, a real-time clock, and an SD-card for data storage. The total price of the monitoring system was below €200. More information about the monitoring system can be found elsewhere [70–72]. Different sensors were coupled to the monitoring system during the calibration experiments (see Table 1). To distinguish the five different AM2315 sensors (provided by Tinytronics, Eindhoven, The Netherlands) that were used, a number was attributed to them.

Table 1. Overview of the sensors and calibration experiments performed in this work. The sensor numbers (nr.) are used in Figures 7–13.

Nr.	Sensor	Parameter	Calibration Method	Experiment	Location	Calibration Data
1	AM2315	T	Low-cost calibration	Figure 7	Cuba	8 December 2020
2	AM2315	T	Low-cost calibration	Figure 7	Cuba	8 December 2020
3	AM2315	T	Controlled climate chamber	Figure 8	Belgium	28 October 2021–2 November 2021
4	AM2315	T	Controlled climate chamber	Figure 8	Belgium	28 October 2021–2 November 2021
5	AM2315	T	Low-cost calibration, heating	Figure 9	Belgium	22 November 2022
6	BME280	T	Low-cost calibration, heating	Figure 9	Belgium	22 November 2022
1	AM2315	RH	Low-cost calibration	Figure 10	Cuba	21 September 2020–24 September 2020
2	AM2315	RH	Low-cost calibration	Figure 10	Cuba	21 September 2020–24 September 2020
3	AM2315	RH	Controlled climate chamber	Figure 11	Belgium	28 October 2021–2 November 2021
4	AM2315	RH	Controlled climate chamber	Figure 11	Belgium	28 October 2021–2 November 2021

Table 1. *Cont.*

Nr.	Sensor	Parameter	Calibration Method	Experiment	Location	Calibration Data
7	COZIR-A	CO_2	Low-cost calibration	Figure 12	Cuba	28 October 2020
8	COZIR-A	CO_2	Low-cost calibration	Figure 12	Cuba	28 October 2020
9	COZIR-A	CO_2	Low-cost calibration	Figure 13	Cuba	27 April 2022
10	SCD30	CO_2	Low-cost calibration	Figure 13	Cuba	27 April 2022

2.2. Design of a Low-Calibration Laboratory

A low-cost calibration laboratory that can evaluate T, RH, and gas (i.e., CO_2, but other gasses such as NO_2 or SO_2 are also possible with the setup) requires an investment between €1000 and €1500. The prices mentioned in Table 2 are per individual item, but in some cases, the items can only be purchased in larger boxes (e.g., syringes). Other items such as the plastic box can be used in other calibration setups. Medical disposables were also reused several times. It should be noted that the low-cost calibration methods are more labor-intensive. How the items were used to build the calibration setups are shown in Figures 1–4.

Table 2. Components that are needed to build the calibration setups shown in Figures 1–4. Medical disposables were locally purchased from a medical supplier.

Role	Provider	Description	Amount	Price (€)
General purpose	Amazon (München, Germany)	Hot glue gun to close holes and attach components and glue sticks	1	30
	Amazon	Gimlet drill set with a diameter of 2.5–5 mm to make holes in the plastic box	1	10
	VWR (Leuven, Belgium)	Precision balance SE 422 with readability of 0.01 g (inv. nr. 611-3299)	1	350
Temperature calibration	Ikea (Antwerp, Belgium)	Food container with lid, 5.2 l (inv. nr. 692.768.07)	1	8
	Amazon	Digital LED Thermostat Temperature Controller with Sensor Probe, MH1210A Mini (12 VDC)	1	18
	Amazon	Axial fan SUNON 12 VDC, 0.58 W, 60 × 60 × 15 mm (inv. nr. MF60151V31000UA99)	2	7
	RS online (Brussels, Belgium)	RS PRO Mica Heating Pad, 80 W, 230 V AC (inv. nr. 790-4842)	1	34
	VWR	Glass thermometer with range 0–70 °C (inv. nr. 620-0889)	1	56
	Amazon	WiMas Peltier TEC1-12706 Thermoelectric Cooler	1	23
Relative humidity calibration	Sigma Aldrich (Darmstadt, Germany)	Glycerol ReagentPlus, $\geq$99.0%, 1 L (inv. nr. G7757-1L)	1	103
	VWR	Blaubrand Calibrated pycnometer of 10 mL (inv. nr. 614-0702)	1	53
	VWR	Petri dish in glass with a diameter of 150 mm	1	16

Table 2. *Cont.*

Role	Provider	Description	Amount	Price (€)
Gas calibration		Pressure connection tube, 1.5 × 2.8 mm, Luer lock connector	1	1
	Amazon	Polytetrafluoroethylene (PTFE)/Teflon tube with an outer diameter of 6 mm and an inner diameter of 4 mm.	1	13
	Amazon	DEWIN Micro Vacuum Pump, Mini Motor for Air Pump, DC 12 V	1	12
	Mouser, Eindhoven, Netherlands	DIN-Rail Power Supply of Traco power 88%, 12 V, 6.7 A, 80 W, Adjustable (inv. nr. TIB 080-112EX)	1	108
	VWR	Gas washing bottles, Drechsel pattern, VitraPOR® (inv. nr. ROBU40100)	2	82
	VWR	U-shaped drying tube in glass with a length of 100 mm (inv. nr. LENZ05339010)	1	24
	VWR	Supelco silicagel of 1 kg in a glass bottle (inv. nr.: 717185-1KG)	1	93
Gas production	Amazon	Terumo syringe, Luer lock, no needle, volume: 5 mL, box of 100	1	14
		3-Way Blue Sterile Stopcock for Intravenous Drips	2	1
		Minimum volume extension tube, 180 cm, Luer Lock, Dead Space = 1.2 mL	1	3

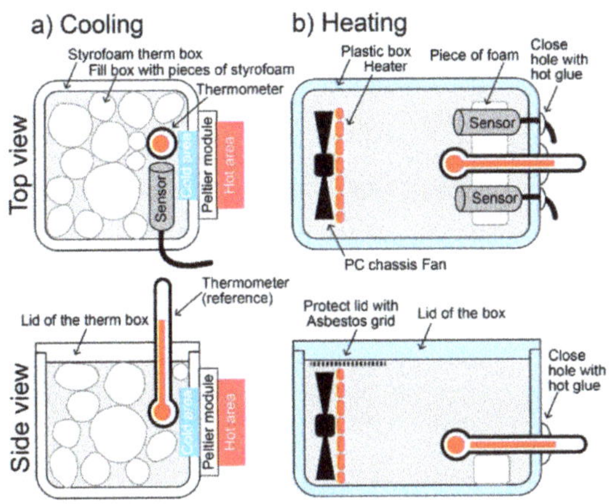

Figure 1. Schematic representation of the setup for the T calibration of the AM2315 sensor. (**a**) Calibration of the lower temperatures relative to room T using a cooling process. (**b**) Calibration of the higher temperatures using a heating process.

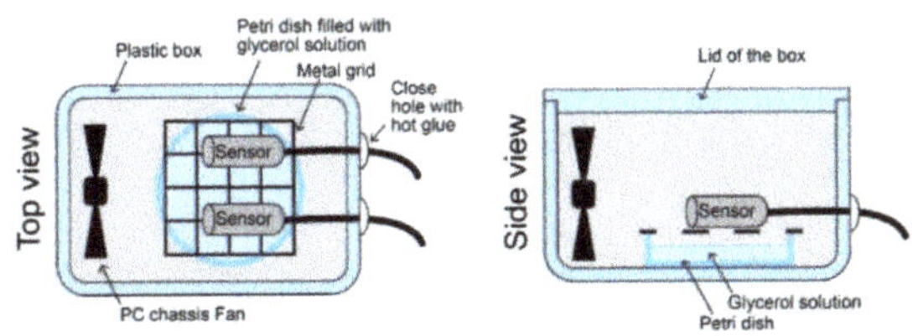

Figure 2. Schematic representation of the calibration setup for RH with AM2315 sensor.

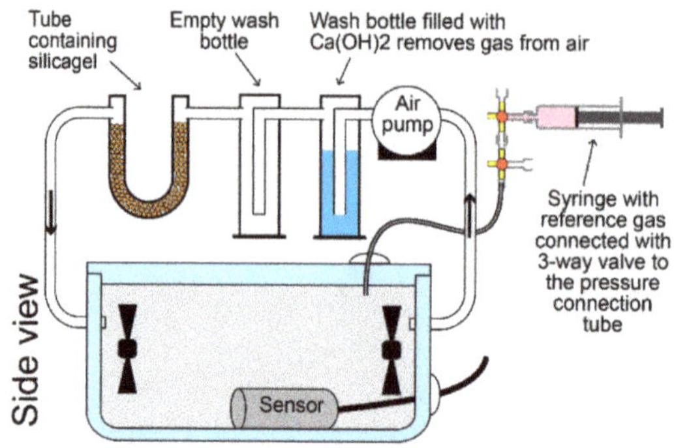

Figure 3. Setup to calibrate a CO_2 sensor and the circuit to remove ambient CO_2 inside the box.

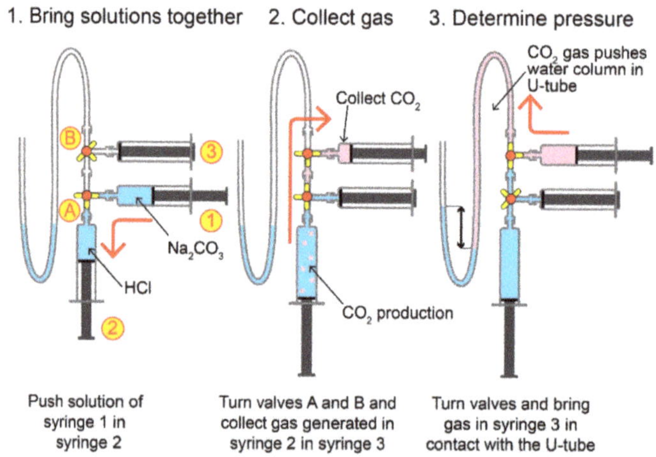

Figure 4. The three critical steps in the production process of CO_2 gas and the position of valves A and B for each step.

2.3. *Temperature Calibration*

Several calibration experiments were performed in a laboratory located in Santa Clara, Cuba. For these experiments, the average air T and RH were 27 °C and 64%. The calibration of the T parameter of the AM2315 sensors was performed in two stages: (1) the cooling of the ambient air inside a Styrofoam box filled with smaller pieces of Styrofoam, leaving only a small volume for the sensors and the reference glass thermometer (Figure 1a) using a Peltier cooler so that the T decreases from room T (26 °C) down to 0 °C; and (2) the heating of ambient air in a closed plastic box (Figure 1b) by placing a mica heater inside the box that is coupled to a thermostat so that the T rises from room T to 50 °C. A glass thermometer acts as a reference device to measure the T inside the box. During the experiment, the sensor readings were registered by the low-cost data logger. Meanwhile, the T of the glass thermometer was visually determined and their values were written down with the corresponding timestamp in a laboratory notebook. Both time series were merged by selecting the sensor reading with the same timestamp as the measured value of the glass thermometer. The calibration curves were determined by processing the data in Microsoft Excel.

2.4. Relative Humidity Calibration

Pure glycerol in a closed box generates a RH of 0%; pure water generates a RH of 100%. The density of the glycerol solution taken from the bottle (Sigma Aldrich, 99%) was determined with a pycnometer (i.e., 1.25960 g/mL) and its density corresponded with a RH of 12.73% (detailed information about the calculations of the corresponding RH can be found in the Supplementary Material). This was the minimum RH in the calibration experiment. Six solutions with glycerol concentrations of 99, 80, 70, 50, 40, and 0 vol% were prepared to perform a series of experiments with increasing RH. The solutions with higher glycerol concentrations resulted in a drying effect of the ambient RH conditions in the calibration box (ambient conditions during the calibration session were about 60–95%); the solution with high water content resulted in humidification. The pure glycerol solution took about 6 h before the RH in the box reached its equilibrium. The time that the calibration box reached equilibrium became shorted with the increasing water content in the mixture.

The experiments were carried out inside a plastic box of 6.27 L closed with a plastic lid. A closed container with a volume larger than 1 L should contain a fan [53]. A Petri dish of 15 cm diameter containing a glycerol solution of 380 mL was placed at the bottom of the box. The Petri dish guarantees a large surface of the solution to allow free diffusion of water. A grid was placed on top of the Petri dish. The sensor was placed on top of that grid. The box also contained two PC chassis fans to assure the homogeneity inside the box and to speed up the moisture exchange between air and glycerol solution so that equilibrium was reached faster (Figure 2). For each experiment, the sensor signal was registered over time and stopped once the signal remained constant for at least 1 h. For the six experiments, the measured and theoretical concentrations determined from the density of the glycerol solutions were brought together in Microsoft Excel to generate the calibration curve.

2.5. Calibration of Gas Sensors

The calibration of the gas sensors was performed inside the same plastic box as used in the RH calibration experiment. The box contained two PC chassis fans to homogenize the air inside the calibration box. For CO_2, the box contained the appropriate sensor that was coupled to the monitoring device. Before the calibration experiments started, the CO_2 concentration in ambient air and also inside the closed box was measured. The box was also coupled to a closed circuit that could remove the CO_2 in the ambient air inside the box (see Figure 3). That circuit contained (1) a pump, (2) a wash bottle with 80 mL of saturated $Ca(OH)_2$ solution plus 20 g of NaOH beads that removed CO_2 from the air according to reaction (1), (3) an empty wash bottle to trap the water droplets present in air, and (4) a U-shaped glass tube containing 10 g of silica gel to dry the air. The CO_2 inside the closed box was removed by switching on the pump. After several minutes, zero air was obtained and the pump was switched off.

$$Ca(OH)_{2\ (aq)} + CO_{2\ (g)} \rightarrow CaCO_{3\ (aq)} + H_2O \quad (1)$$

The concept of gas calibration relies on the generation of small but known amounts of gas using stoichiometric reactions [61,73–76]. The setup used to generate CO_2 gas inside a syringe is shown in Figure 4. It was built with low-cost medical disposables, such as Luer Lock syringes, three-way sterile stopcocks, and an extension tube with Luer Lock to create a U tube manometer (see Table 2). Two syringes were used to mix the two reagents mentioned in reaction (2), while the third was used to collect the generated CO_2 gas. The volume of CO_2 gas can be measured from the tick marks on syringe 3. The T of the gas is equal to the room T and can be measured with a glass thermometer. The pressure of the gas inside the syringe can be measured from the barometric pressure and the differential pressure between the room and the syringe. The differential pressure can be measured from the height difference between the two liquid columns. The calculation of the amount of generated gas in the syringe is described in the Supplementary Material. The obtained amount of gas was in accordance with the reaction stoichiometry.

$$Na_2CO_{3\ (aq)} + 2HCl_{\ (aq)} 2NaCl_{(aq)} + CO_{2\ (g)} + H_2O \quad (2)$$

The syringe filled with CO_2 was closed off with a three-way stopcock. Syringe 3 and stopcock B were disconnected from the setup and attached to the pressure connection tube of 10 cm that was attached to the lid (see Figure 3). Then, the gas in the syringe was transferred to the calibration box. The entire gas volume in the syringe was pushed into the calibration box in consecutive steps to obtain a staircase curve. Since the total amount of CO_2 in the syringe expressed in mole was known (see Supplementary Material), it was possible to calculate the CO_2 concentration inside the box for every step.

2.6. Reliability of Low-Cost Calibration Methods

The calibration curve describes the relationship between the sensor and the reference measurements. Since it must transform the sensor measurements into corrected values, the parameter measured by the sensor has to be the independent variable. The contribution of the random error on the experimental calibration points can be seen as a scattering of the calibration points around the curve. This means that the occurrence of random errors can easily be observed. The occurrence of systematic errors or erroneous trends is harder to identify. However, several indicators were used that gave insight into the presence of such errors. They are listed below. The first five methods in the list give indirect or weak information about constant errors, but their combination is good enough for a satisficer to evaluate the calibration method. A maximizer will only be satisfied with the last method. Unfortunately, that option is only accessible to a limited group of researchers. However, in this contribution, all methods were used to evaluate the low-cost calibration methods.

- **Airtightness of the calibration box:** When the calibration box is not completely airtight, the calibration conditions will gradually degrade over time. To evaluate the importance of leakages, the tracer gas method was used. For this, the background concentration of CO_2 in ambient air was measured first. Then, about 5 mL of CO_2 was introduced into the calibration box using a syringe so that the CO_2 concentration inside the box was substantially higher than outside the box. After some minutes, the sensor reached a stable value. The exponential decay of the tracer gas in the box was monitored for a period of about 20 h. The air exchange rate, which is expressed in air change per hour (h^{-1}), was calculated with the formulas given in the Supplementary Material [77–82]. A calibration box that was sufficiently airtight throughout the calibration experiment (<5% of the total volume) is an indication that the method used was sufficiently valid;
- **Similarity with the factory-calibration:** Some sensors, such as the ones measuring T, RH, or CO_2, are often factory-calibrated. Although their calibration is often not well documented and unreliable [83], the measurements can be expected to have at least some similarity to the reference values. Therefore, the calibration curve is expected to be linear with an intercept and a slope around 0 and 1, respectively. A strong deviation from that expectation is an indication that something is wrong with the calibration (or with the sensor). This test is only valid for new sensors because they age over time and their calibration will change;
- **Similarity with the reference:** For some calibration methods, the parameter to be measured is changed according to a predefined test program. A strong deviation of the sensor from the imposed test program is an indication that something is wrong;
- **Similarity between different sensors:** If identical sensors or a set of factory-calibrated sensors are simultaneously calibrated, then minor differences between the calibration curves are expected;
- **Replicability of the calibration method:** When a calibration method is applied on different sensors of the same type using different calibration boxes, different equipment,

and different regions (i.e., what is the effect of the high RH in a tropical climate?), the calibration curves are expected to be similar. The occurrence of significant differences between the experiments indicate a lack of replicability;

- **Similarity with a golden standard:** The calibration of the T/RH sensors was also performed by VITO, Belgium, using a closed stainless-steel exposure test chamber (Weiss Technik, Germany) with a total internal volume of the chamber at 1.0 m^3. During the performed tests, the exposure test chamber was supplied with a constant and controlled flow of clean humidified air. The environmental conditions inside the chamber were constantly controlled and changed according to a test program to realize a consecutive series of T and RH levels. The homogeneity of the chamber's inner atmosphere was provided by an active mixer installed next to the inlet of the chamber. The T and RH inside the exposure test chamber were continuously monitored and recorded using integrated and calibrated T/RH monitors mounted near the exhaust of the chamber. The reference monitoring devices are yearly calibrated by Weiss. During the development stage of low-cost calibration methods, it is useful to compare low-cost calibrations with reference calibrations that are considered as the golden standard. The more resemblances there are between both calibration curves, the more reliable the low-cost calibration method is.

2.7. Field Study

The air quality in Santa Clara, Cuba, had been already analyzed in previous studies [84]. That expertise was used to select the measuring locations summarized in Table 3. The field study consisted of a short measurement campaign at three different locations (Figure 5): (1) the Yabu site located northwest of the city in a rural region away from roads; (2) the farm site located in a rural area where the wind blows the air towards the city; and (3) the doctor's office site in the center of the city where the air quality is affected by emissions from traffic and industry in the vicinity (e.g., production of pneumatic tires). The devices were placed at a height of 2.5 m above ground level, according to Cuban standard [85]. At the doctor's office and at Yabu, the monitoring systems were hung in the window openings. The window openings were located at least 1 m from buildings, trees, and/or other obstructions. At the farm site, the monitoring system was hung underneath the porch. The calibration of the raw data obtained by the monitoring systems was performed as post-processing using Microsoft Excel. The two-step method had the advantage of allowing others to check the calibration of the collected data and roll back when an error was made in the calculations.

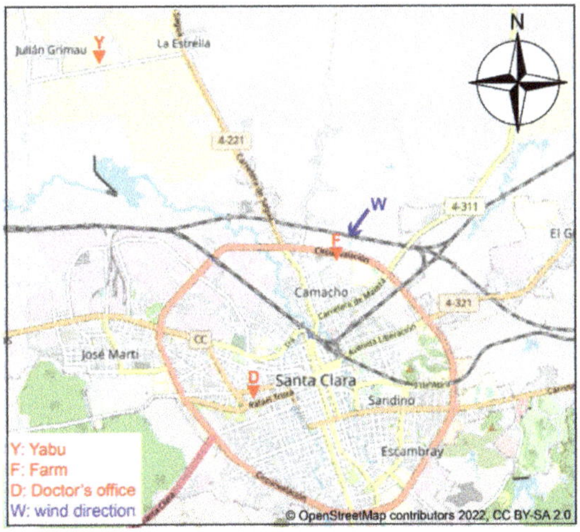

Figure 5. Map showing the three sampling sites in Santa Clara, Cuba.

Table 3. Sampling sites and an organizational overview of the field test.

Code	Site	Position	Sampling Period	Sampling Rate	Variables	Sensor Nr.
Y	Yabu	22°45′58.23″ N, 80°02′21.15″ W	14–29 February 2020	300 s	T (°C), RH (%)	1
F	Farm	22°42′63.74″ N, 79°96′38.64″ W	4–6 March 2020	7 s	T (°C), RH (%) CO_2 (ppm)	2 3
D	Doctor's office	22°40′36.78″ N, 79°97′58.19″ W	4–6 March 2020	7 s	T (°C), RH (%) CO_2 (ppm)	1 4

3. Results and Discussion

3.1. Airtightness of the Calibration Boxes

Before calibration experiments were carried out, it was necessary to check how airtight the calibration box was. The ambient CO_2 concentration before the airtightness test was 376 ± 8 ppm. Then, the CO_2 gas was injected. Once the CO_2 concentration reached its maximum value (i.e., 1267 ppm), the experiment was started. During a period of 17.5 h, the CO_2 concentration exponentially dropped to 935 ppm (see Figure 6). The calibration box had an air exchange rate of 0.02674 h^{-1}. Since the calibration experiments lasted no more than 2 h, a loss of the entire air inside the box of about 5% could be considered as sufficiently small. It can be concluded that the plastic box was sufficiently airtight to use in calibration experiments.

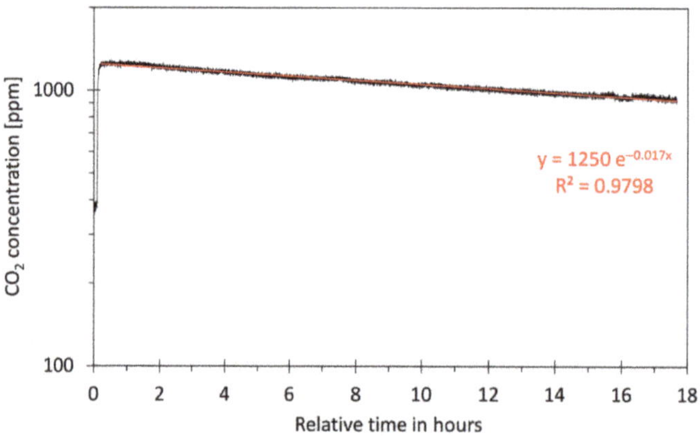

Figure 6. Exponential decay of CO_2 over time due to small leakages in the closed box. The experiment was performed in Cuba (tropical climate). From this experiment, the air exchange rate can be calculated.

3.2. Temperature Calibration

Figure 7 shows the results of two different sensors simultaneously calibrated in the same box. The calibration consisted of a cooling and a heating process. The evolution of the T could not be imposed by a predefined test program, but was monitored with a glass thermometer (i.e., the reference measurement). The cooling process took about 17 min. The heating process was determined by the T set by the thermostat, which was 65 °C. The T raised from room T to the set T in about 5 min. The reference values in Figure 7b are shown in the vertical axis so that regression functions directly give the correct formula to convert measured values into the corrected ones. The following observations could be made:

- **Calibration method:** The uncontrolled test program could be monitored with a glass thermometer. The calibration curve was obtained by plotting the sensor measurements with the corresponding glass thermometer measurements observed at the same moment;

- **Low impact of random errors:** The scattering of the calibration points around the linear regressions is small, suggesting that the method resulted in small random errors;
- **Comparison between sensors:** There is a good similarity between the calibration curves of the two different sensors. Even when the calibration contains a constant error, it is still possible to compensate for calibration differences between sensors;
- **Comparison with factory-calibration:** In the early stage of the lifetime of factory-calibrated sensors, the low-cost calibration and factory-calibration should match. For the cooling process, this was indeed the case, but not for the heating process;
- **Comparison with the reference:** During the heating process, there is no close similarity between sensor and glass thermometer measurements (see Figure 7a). In principle, a polynomial regression with order 3 can be used to describe the entire set of calibration points in Figure 7b covering the total range of 0–63 °C (i.e., correlation coefficient: 0.9979 and 0.9987 for sensors 1 and 2, respectively). However, the indicators used to evaluate the calibration methods suggest that the reliability of the heating process must be checked more closely.

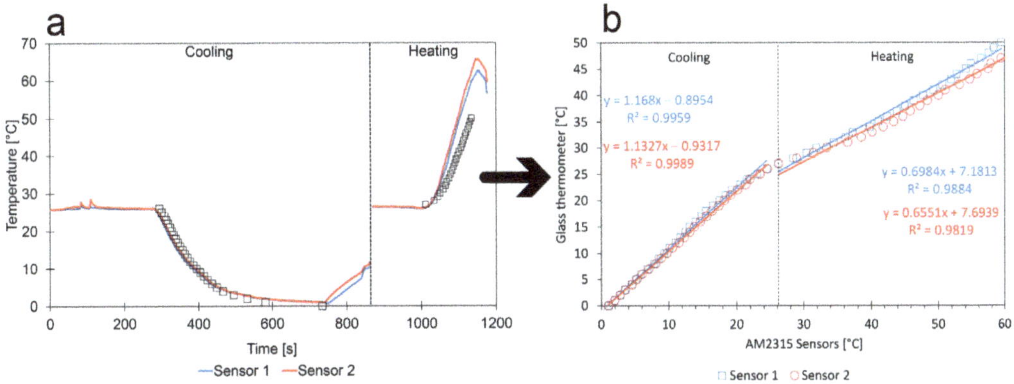

Figure 7. Calibration curves obtained for two AM2315 sensors using a glass thermometer as a reference device. The arrow illustrates the data processing process that converts the raw data into a calibration curve. Visualizing that process helps to regenerate the calibration curve from the same data by others (i.e., reproducibility). The experiments were performed in Cuba (tropical climate); (**a**) raw data of the two sensors obtained during the cooling and heating experiment; (**b**) calibration curves obtained from the raw data.

A similar calibration experiment was performed with a golden standard. Figure 8a shows how the T changed over time according to a predefined program. For every transition between consecutive plateaus (see moments between the black rectangles in Figure 8a), some time was needed before the T reached a stable value. The following observations could be made:

- **Calibration method:** Only a limited number of periods have a stable but known T. These periods are numbered in Figure 8. The average temperatures were calculated for the second half of the plateaus because the chamber and the sensors needed some time to reach equilibrium;
- **Comparison between sensors and with reference:** The behavior of the sensors and the test program as measured by the reference instrument show similar trends over time. In addition, there is a similarity between the calibration curves of the two different sensors (i.e., a linear trend, a slope of around 1). However, the intercepts are clearly different from 0, suggesting that the quality of pre-calibrated sensors needs to be improved before they can be used in the field tests. In addition, it shows that

the calibration during the heating process shown in Figure 7 is incorrect. The other indicators appear to be sensitive enough to identify the calibration error.

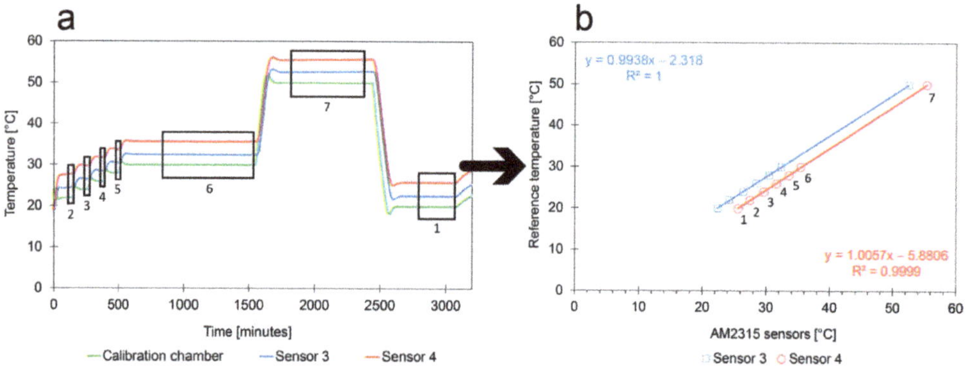

Figure 8. Calibration curves for two different AMS2325 sensors using a state-of-the-art calibration chamber. The numbered rectangles in (**a**) are the areas that were used to calculate the average temperatures shown in (**b**). The reference experiment was performed in Belgium (moderate marine climate); (**a**) raw data as registered by the monitoring system with periods of constant T in the rectangles; (**b**) calibration curve obtained by processing the raw data in the rectangles.

The T calibration experiment using the heating process was repeated at a lower heating rate. In this experiment, the lid did not entirely close the box to allow better dissipation of the heat. In addition, the fan was stopped if the set point was reached. The moment that the T started to drop, the next set point was introduced in the thermostat and the fan was switched on again. As a result, there was first a small drop followed by an increase in T. There was no full control of the staircase function, but the increase in T appeared to be sufficiently slow to obtain a calibration that was similar to the ones obtained with the golden standard.

- **Absolute calibration method:** The calibration relies on the calculation of the average T within the short periods where the T is stable. These periods are numbered in Figure 9. The average temperatures are compared with the glass thermometer measurements;
- **Relative calibration method:** When sensors experience the same conditions during the same calibration experiment, differences between them can be compensated. This compensation is even possible when the low-cost calibration method is unable to accurately determine the absolute values. Consequently, differences in trends from two monitoring campaigns performed in parallel during the field study can be studied in a reliable way;
- **Comparison with the reference:** The experiments suggest that the behavior of the sensors in comparison with the reference show several small differences. The AM2315 follows the values obtained with the glass thermometer with some delay (see Figure 9a). The BME280 sensor appears to be less sensitive to the changes in the calibration box;
- **Comparison between sensors:** The slopes of the calibration curves of the two factory-calibrated sensors were close to 1 and the intercepts had a small deviation from 0. The calibration curves also showed more similarities with the ones obtained with cooling (see Figure 7). This indicates that cooling or heating rates should be sufficiently small to obtain a correct calibration;
- **Replicability of the calibration method:** The same calibration method was used to calibrate a set of low-cost sensors in different regions (Figure 7 in Cuba; Figure 9 in Belgium) using different calibration boxes and with different equipment. Except for the heating process shown in Figure 7, they all gave a linear regression with similarities to the factory calibration. No obvious differences between the low-cost calibration

and the golden standard (Figure 8 performed in Belgium) could be observed. This means that the calibration method resulted in reproducible results and that factory calibrations can be improved.

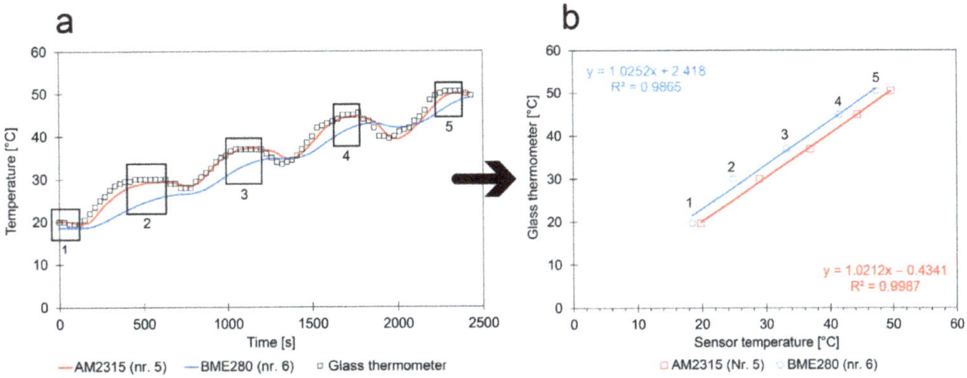

Figure 9. The heating experiment was repeated in Belgium (moderate marine climate) with two different sensors; (**a**) raw data with periods in the rectangles where the glass thermometer shows constant values; (**b**) calibration curve obtained by processing the raw data in the rectangles.

3.3. Relative Humidity Calibration

Figure 10 shows the behavior of two identical AM2315 sensors (the same sensors as in Figure 7) during the RH calibration. To obtain a calibration curve that corrects the RH readings measured by the sensors, a scatter plot was made between the theoretical RH that was determined from the density of the glycerol solution and the sensor measurement after the box reached equilibrium. Each calibration point in Figure 10b is a different experiment that required time before a constant RH was obtained. The different calibration experiments were concatenated into a single time series (see Figure 10a). The trend of series 1 in Figure 10a using the pure glycerol solution was truncated because it took about 14 h to perform that experiment.

- **Calibration method:** The calibration experiments in Figure 10a were stopped once the RH reached a stable value. The average value was calculated from the stable periods shown in the numbered boxes and compared with the theoretical values that were obtained with the glycerol solution. The results show that the calibration performed by the manufacturer can be improved with low-cost calibration experiments (at least in a relative way);
- **Low random error:** The calibration points follow a linear trend, suggesting that the contribution of random errors is low;
- **Comparison between sensors:** The linear calibration functions of both sensors show similarities. The slopes of the calibration curves are close to 1 as expected. The intercepts are similar, but deviate from zero.

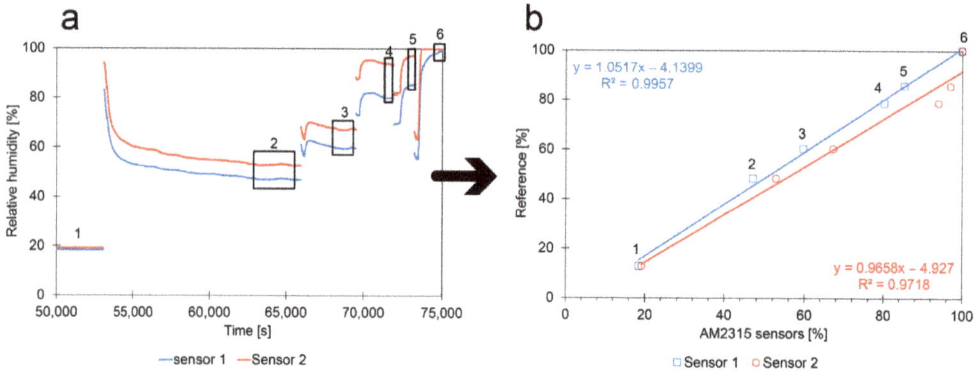

Figure 10. Calibration curves obtained for RH using aqueous glycerol solutions and performed in Cuba (tropical climate); (**a**) the raw data of the different experiments were concatenated into a single time series but the trend of experiment 1 is not entirely shown; (**b**) calibration curve obtained from the experiments.

The calibration experiment performed with a golden standard was also used to evaluate the RH measurements of the same AM2315 sensors as the ones used in Figure 8. Figure 11a shows how the RH changed over time according to a predefined program. Calibration points 1 to 4 were determined at a T of 50 °C to reach the lower values. For points 5 to 9, the constant T is representative for tropical weather.

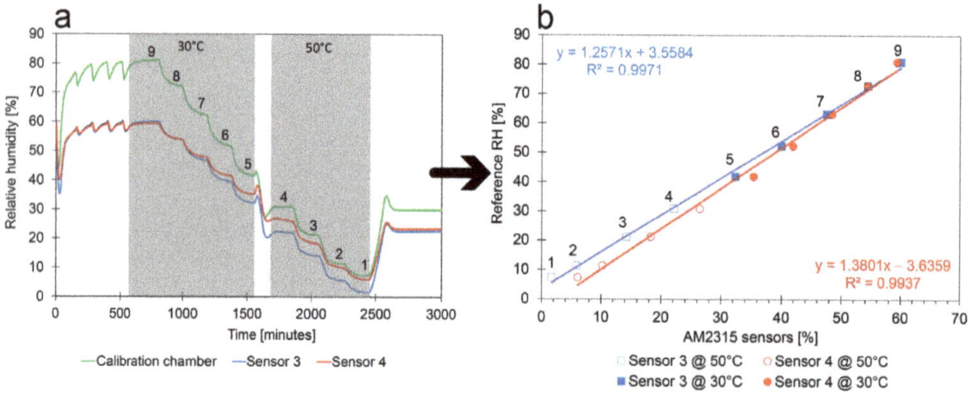

Figure 11. RH calibration of the same AM2315 sensors as in Figure 8 using a state-of-the-art calibration chamber performed in Belgium (moderate marine climate); (**a**) raw data with the moments at the constant T shown in grey; (**b**) calibration curve obtained by processing the numbered areas.

- **Calibration method:** Figure 11b shows the linear calibration of the sensors. The calibration points show the relation between the average values of the numbered plateaus in Figure 11a and the corresponding RH measured by the climate chamber. Only the second half of the plateaus were considered because they were more stable;
- **Comparison between sensors:** The slope and intercept of both sensors are similar but different enough to require additional calibration before they can be used in field measurements. In addition, the slope and intercept are close to the ones determined with the low-cost calibration methods. It should be remarked that the set of calibration points determined at 30 °C and the set of 50 °C do not entirely match the linear regression. There is a small impact of the T on the RH measurements;

- **Replicability of the calibration method:** The calibration method has been tested on several sensors of the same type. The same calibration box and equipment have been used to independently perform the six experiments of Figure 10a. Calibration is also performed with a golden standard. No obvious differences could be observed between the linear regressions. The similarities show that the low-cost RH calibration is sufficiently reliable and that factory calibrations of the sensors can be improved.

3.4. CO_2 Calibration

The behavior of two identical COZIR-A CO_2 sensors enduring the same calibration process is shown in Figure 12a. Once the box was closed, the sensors measured the natural background concentration of CO_2 in ambient air. When the air was pumped through the $Ca(OH)_2$/NaOH solution, the CO_2 concentration decreased below the detection limit of the sensors (i.e., nearly zero). Then, the CO_2 gas in the syringe was introduced in three consecutive steps. At every CO_2 injection, the sensor signal suddenly increased. The sensor readout followed the jumps of a staircase function, while the room T and RH inside the box remained constant (i.e., 27 °C and 65%).

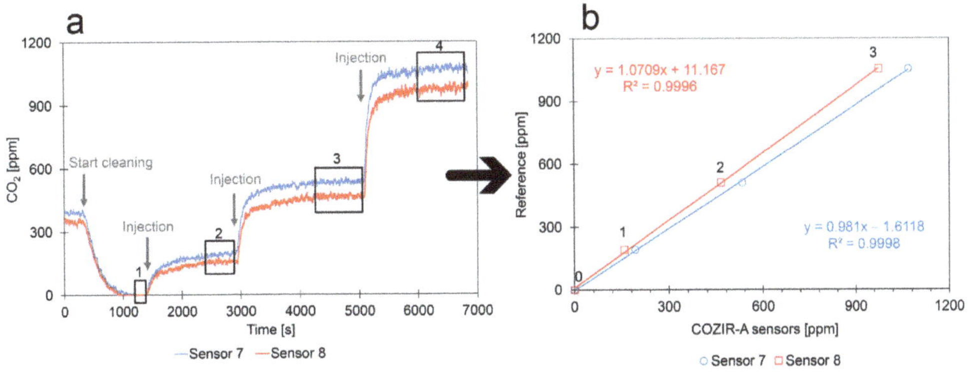

Figure 12. (**a**) Concentration trends of CO_2 during the calibration process for the two COZIR-A sensors performed in Cuba; (**b**) calibration curves obtained for CO_2 sensors.

- **Calibration method:** Because all CO_2 inside the closed box was removed and because known amounts of CO_2 were injected into the box, the staircase function was fully controlled. However, only the concentrations of the plateaus could be calculated from the injected amounts and used for calibration. Consequently, the average values of the numbered rectangles in Figure 12a were used for calibration;
- **Limited random errors:** The mathematical relation of the calibration points can be described by linear regression (see Figure 12b). This suggests that the contribution of random errors during the experiment is low;
- **Comparison between sensors:** Figure 12a shows that both sensors follow the same trend, but that they respond slightly differently to the same environment.

A similar calibration experiment as in Figure 12 was performed without the cleaning of the air. This means that known amounts of CO_2 were introduced inside the closed box that contained air with an unknown amount of CO_2. However, it is known that the background concentration in ambient air is around 412 ppm. The calibration was performed by the standard addition method.

- **Calibration method:** The two factory-calibrated sensors measured a different background concentration (rectangle 0 in Figure 13a) in the closed box (COZIR-A: 476 ppm; Sensirion SCD30: 450 ppm). From the measured background concentrations and the calculated concentrations that were introduced, the reference concentrations of rectangles 1, 2, and 3 could be determined;

- **Comparison between sensors:** Although the intercept of the calibration curves might contain a small but constant error due to an error in the measurement of the background concentration, the slopes can be accurately determined;
- **Replicability of the calibration method:** Different kinds of factory-calibrated sensors have been tested with two different calibration methods. The linear regressions given in Figures 12 and 13 show sufficient similarities to conclude that the low-cost CO_2 calibration methods are sufficiently reliable and that the factory calibration of the sensors can be improved.

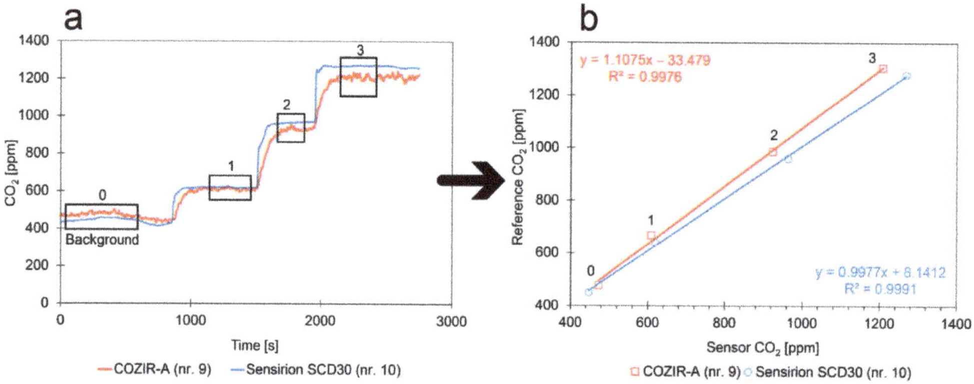

Figure 13. (**a**) Concentration trends of CO_2 when calibration gas is introduced on top of the background concentration performed in Cuba; (**b**) calibration curve obtained by the standard addition method.

3.5. Results of the Field Study

Table 4 presents the descriptive statistics of the time series recorded during the field study. The standard deviations from the two sites measured in parallel, the farm and doctor's office, show that environmental parameters fluctuate more at the farm. In addition, the time series of the three parameters show different signs in the asymmetry coefficients. The values of the coefficient indicate that the distribution of the data around their average is not symmetric (i.e., the positive value means that the right tail is longer; the negative value means a skewness to the left). The statistical overview of Table 4 gives a good overview of the three sites and makes comparison easy, but a substantial amount of information that is present in the dynamics of temporal trends is lost in the descriptive statistics. Therefore, measurement campaigns with reliable low-cost monitoring systems are an important added value. Unfortunately, this way of working was hampered by regular power cuts in the city. In addition, additional efforts are needed in data post-processing to suppress noise and remove spikes. For example, the CO_2 measurements at the farm contained several spikes, with a value of 0 or 2 ppm. They were removed by a simple filter: if the concentration was below 100 ppm, then the data point was replaced by the average value of the previous and next data point.

Figure 14a shows the calibrated results of T and RH during the field experiment at the Yabu site. During that period, there was no precipitation. In this measuring campaign, the trends clearly showed day-night cycles where T and RH were in an opposite phase. For each day, between 5:30 and 7:00, the T dropped down to a minimum of 21–25 °C, while the RH reached a maximum of 80–87%. Around 13:00 and 15:00, the T reached a maximum of 26–28 °C, while the RH during that period was 53–70%. The period of the day-night cycles varied around 24 h but, due to small random changes of the minima and the maxima in the given periods, the cyclic pattern was quasi-periodic. In addition, the local maxima and minima varied. This short measurement campaign suggests that in-house built monitoring systems can be used to generate useful information that is complementary to the statistical data.

Table 4. Descriptive statistics for the time series registered over this field study.

Variable	Measuring Point	Total Data Points	Average	Maximum	Minimum	Standard Deviation	Asymmetry Coefficient
T (°C)	Yabu	4869	27	32.4	12.9	3.1	−0.58
	Farm	99,837	28	43.0	20.6	5.6	0.99
	Doctor's ofice	27,734	33	38.6	28.1	2.9	0.41
RH (%)	Yabu	4869	65	83.5	20.5	9.7	−0.73
	Farm	99,837	43	78.9	21.7	17.3	−0.21
	Doctor's ofice	27,734	62	92.5	31.2	16.0	−0.03
CO_2 (ppm)	Farm	99,837	485	637.0	417.3	36.5	1.12
	Doctor's ofice	27,734	570	671.9	493.1	23.6	−0.44

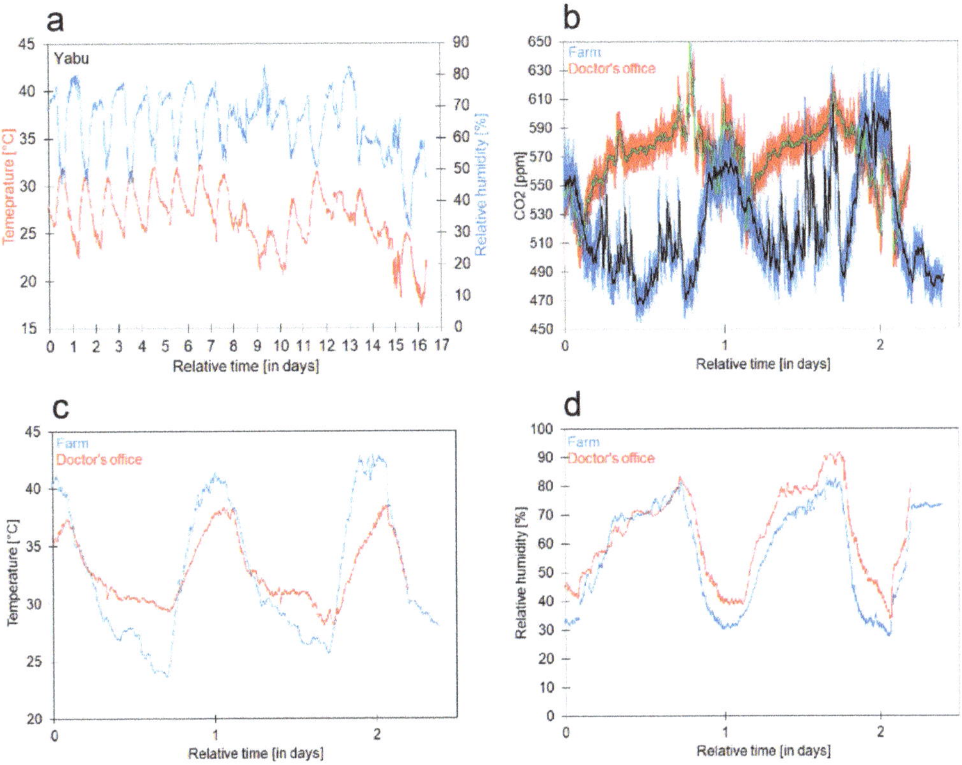

Figure 14. Corrected sensor data for T and RH in the field test at the Yabu site during 14–29 February 2020, and the parallel measuring campaigns at the farm and the doctor's office during 4–6 November 2020; (**a**) results of the campaign in Yabu; (**b**) CO_2 trends at the farm and the doctor's office with a central moving average and a window width of 10 min; (**c**) superimposed temperate trends of the farm and doctor's office; (**d**) superimposed RH trends of the farm and doctor's office.

Figure 14b–d shows the results of CO_2, T, and RH measured in parallel at the farm and the doctor's office during the field tests during 4–6 March 2020. The calibration of both sensors ensures that the differences between locations are only caused by the location and not by measuring errors. At both locations, T, RH, and CO_2 show diurnal cycles. For the farm, the minima and maxima of the T are somewhat more extreme. The RH at both locations shows a similar trend, but the minima observed at the farm are somewhat lower. This suggests that, at the farm, the nights are cooler and less humid. The CO_2 trend at the farm shows similar dynamics to the T, with maxima at around midnight. The CO_2 trend at the doctor's office is more complex. What is striking is the noise in Figure 14b. The noise was suppressed by using a centered moving window with a window size of 10 min. The comparison demonstrates that calibrated low-cost monitoring systems can generate valuable information from which conclusions can be drawn.

Calibration procedures and field tests allowed for reaching the maximum performance reflected in the datasheets of the sensors. Therefore, the system is considered suitable as a measurement instrument, as long as a resolution greater than 0.1 °C for T, 2% for RH, and 30 ppm for CO_2 is not required. Furthermore, when the maximum values exceed 80 °C or 2000 ppm of CO_2, they will cause saturation of the outputs. It is important to mention that the maximum operating temperature for Arduino Mega 2560 computing unit is 85 °C; thus, it is not recommended to exceed this value. The system is not suitable for obtaining measurement resolutions higher than those mentioned or more extreme operating conditions.

4. Conclusions

There is abundant information available to build low-cost monitoring systems that can measure parameters such as temperature, relative humidity, and CO_2. However, applications using such devices rely on the calibration of the manufacturer. This means that the accuracy of such devices is unclear. Some studies compare low-cost monitoring systems with high-end reference instruments. Unfortunately, not everyone has access to such instruments. This contribution proposes low-cost calibration methods for temperature, relative humidity, and CO_2 that can be performed in a laboratory situated in a tropical climate. The calibration methods were used for several sensors using different calibration boxes, equipment, and regions where the laboratory is located. All situations resulted in similar calibration results. The air exchange rates suggested that the airtightness of the calibration box was sufficient. In addition, the calibrations for temperature and relative humidity obtained with the low-cost calibration methods were in line with the golden standard. The experiments showed that factory calibrations can be improved and differences between sensors can be compensated. The field campaigns demonstrated that the calibrated low-cost monitoring systems could be used as a low-cost scientific instrument. The quality of the data generated by that scientific instrument was perhaps not as good as that of state-of-the-art reference instruments, but the field test showed that it was good enough to draw conclusions from the measurement campaigns and that the dynamics in trends gave valuable information that was complementary to a statistical analysis of larger periods. By using the mindset of a satisficer, numerous institutions and researchers from low- and mid-income countries gain access to research infrastructure to solve local problems in a reliable way. This makes scientific research more inclusive. Moreover, technology is systematically improving, so data quality generated by low-cost scientific instruments will increase over time. The satisficer method will become more performant over time.

Supplementary Materials: The following supporting information can be downloaded at: https://www.mdpi.com/article/10.3390/atmos14020191/s1, Figure S1. Schematic overview illustrating the pressure difference between the atmosphere and syringe.; Table S1. Overview of the symbols used in the equations.; Equation (S1)–(S7).

Author Contributions: R.A.G.R.: Processing of results obtained in the air exchange rate determination, calibration of the CO_2 sensors, and manuscript preparation. L.E.M.H.: Processing of results obtained in the calibration of the temperature and relative humidity sensors, manuscript preparation. O.S.: Research conception and design, review and approval of the final version of the manuscript. E.H.R.: Design of the low-cost monitoring system. D.A.S.: Research conception and design, review and approval of the final version of the manuscript. M.C.M.P.: Sampling site selection strategy. Data processing and analysis obtained in the field study. V.N.C.: Sampling site selection strategy. Data processing and analysis obtained in the field study. W.J.: Research conception and design, review and approval of the final version of the manuscript. A.M.L.: Assembly of the low-cost monitoring system, review and approval of the final version of the manuscript. All authors have read and agreed to the published version of the manuscript.

Funding: The present work has been supported by the VLIR-UOS Program as a South Initiative Project SI-2019 nr. CU2019SIN242B124 "AIR@PORT: Low-cost decision support system to evaluate the impact of ships on the air quality in the port city Cienfuegos". Additional financial support has been given by the common Global Minds project BE2017GMHVLHC106 of all Flemish Universities of Applied Sciences and Arts.

Institutional Review Board Statement: Not applicable.

Informed Consent Statement: Not applicable.

Data Availability Statement: All data used in this study are available upon request.

Acknowledgments: The authors thank VLIR-UOS for their financial support of the South Initiative Project SI-2019 nr. CU2019SIN242B124 "AIR@PORT: Low-cost decision support system to evaluate the impact of ships on the air quality in the port city Cienfuegos". They are also grateful for the additional support by the Global Minds project BE2017GMHVLHC106.

Conflicts of Interest: The authors declare no conflict of interest.

References

1. Mannucci, P.; Franchini, M. Health Effects of Ambient Air Pollution in Developing Countries. *IJERPH* **2017**, *14*, 1048. [CrossRef] [PubMed]
2. Chen, B.H.; Hong, C.L.; Pandey, M.R.; Smith, K.R. Indoor Air Pollution in Developing Countries. *World Health Stat. Q.* **1990**, *43*, 127–137. [PubMed]
3. Pinder, R.W.; Klopp, J.M.; Kleiman, G.; Hagler, G.S.W.; Awe, Y.; Terry, S. Opportunities and Challenges for Filling the Air Quality Data Gap in Low- and Middle-Income Countries. *Atmos. Environ.* **2019**, *215*, 116794. [CrossRef] [PubMed]
4. Ibrahim, M.F.; Hod, R.; Nawi, A.M.; Sahani, M. Association between Ambient Air Pollution and Childhood Respiratory Diseases in Low- and Middle-Income Asian Countries: A Systematic Review. *Atmos. Environ.* **2021**, *256*, 118422. [CrossRef]
5. Helmut, M. Air Pollution in Cities. *Atmos. Environ.* **1999**, *33*, 4029–4037.
6. Hodoli, C.G.; Coulon, F.; Mead, M.I. Source Identification with High-Temporal Resolution Data from Low-Cost Sensors Using Bivariate Polar Plots in Urban Areas of Ghana. *Environ. Pollut.* **2023**, *317*, 120448. [CrossRef]
7. deSouza, P.; Nthusi, V.; Klopp, J.; Shaw, B.; Ho, W.; Saffell, J.; Jones, R.; Ratti, C. A Nairobi Experiment in Using Low Cost Air Quality Monitors. *Clean Air J.* **2017**, *27*, 12–42. [CrossRef]
8. Abraham, S.; Li, X. A Cost-Effective Wireless Sensor Network System for Indoor Air Quality Monitoring Applications. *Procedia Comput. Sci.* **2014**, *34*, 165–171. [CrossRef]
9. Aguiar, E.F.K.; Roig, H.L.; Mancini, L.H.; de Carvalho, E.N.C.B. Low-Cost Sensors Calibration for Monitoring Air Quality in the Federal District—Brazil. *JEP* **2015**, *6*, 173–189. [CrossRef]
10. Zimmerman, N.; Presto, A.A.; Kumar, S.P.N.; Gu, J.; Hauryliuk, A.; Robinson, E.S.; Robinson, A.L.; Subramanian, R. A Machine Learning Calibration Model Using Random Forests to Improve Sensor Performance for Lower-Cost Air Quality Monitoring. *Atmos. Meas. Tech.* **2018**, *11*, 291–313. [CrossRef]
11. Morawska, L.; Thai, P.K.; Liu, X.; Asumadu-Sakyi, A.; Ayoko, G.; Bartonova, A.; Bedini, A.; Chai, F.; Christensen, B.; Dunbabin, M.; et al. Applications of Low-Cost Sensing Technologies for Air Quality Monitoring and Exposure Assessment: How Far Have They Gone? *Environ. Int.* **2018**, *116*, 286–299. [CrossRef]
12. Schalm, O.; Cabal, A.; Anaf, W.; Leyva Pernia, D.; Callier, J.; Ortega, N. A Decision Support System for Preventive Conservation: From Measurements towards Decision Making. *Eur. Phys. J. Plus* **2019**, *134*, 74. [CrossRef]
13. Anaf, W.; Leyva Pernia, D.; Schalm, O. Standardized Indoor Air Quality Assessments as a Tool to Prepare Heritage Guardians for Changing Preservation Conditions Due to Climate Change. *Geosci. J.* **2018**, *8*, 276. [CrossRef]
14. Gaudenzi Asinelli, M.; Serra Serra, M.; Molera Marimòn, J.; Serra Espaulella, J. The SmARTS_Museum_V1: An Open Hardware Device for Remote Monitoring of Cultural Heritage Indoor Environments. *HardwareX* **2018**, *4*, e00028. [CrossRef]
15. CEN EN 15757:2010. Conservation of Cultural Property: Specifications for Temperature and Relative Humidity to Limit Climate-Induced Mechanical Damage in Organic Hygroscopic Materials. European Committee for Standardization (CEN): Brussels, Belgium, 2010.
16. Silva, H.E.; Coelho, G.B.A.; Henriques, F.M.A. Climate Monitoring in World Heritage List Buildings with Low-Cost Data Loggers: The Case of the Jerónimos Monastery in Lisbon. *JOBE* **2020**, *28*, 101029. [CrossRef]

17. Tarasick, D.W.; Smit, H.G.J.; Thompson, A.M.; Morris, G.A.; Witte, J.C.; Davies, J.; Nakano, T.; Van Malderen, R.; Stauffer, R.M.; Johnson, B.J.; et al. Improving ECC Ozonesonde Data Quality: Assessment of Current Methods and Outstanding Issues. *Earth Space Sci* **2021**, *8*, e2019EA000914. [CrossRef]
18. Van Malderen, R.; De Muer, D.; De Backer, H.; Poyraz, D.; Verstraeten, W.W.; De Bock, V.; Delcloo, A.W.; Mangold, A.; Laffineur, Q.; Allaart, M.; et al. Fifty Years of Balloon-Borne Ozone Profile Measurements at Uccle, Belgium: A Short History, the Scientific Relevance, and the Achievements in Understanding the Vertical Ozone Distribution. *Atmos. Chem. Phys.* **2021**, *21*, 12385–12411. [CrossRef]
19. Rodriguez-Vasquez, K.A.; Cole, A.M.; Yordanova, D.; Smith, R.; Kidwell, N.M. AIRduino: On-Demand Atmospheric Secondary Organic Aerosol Measurements with a Mobile Arduino Multisensor. *J. Chem. Educ.* **2020**, *97*, 838–844. [CrossRef]
20. Mijling, B.; Jiang, Q.; de Jonge, D.; Bocconi, S. Field Calibration of Electrochemical NO_2 Sensors in a Citizen Science Context. *Atmos. Meas. Tech.* **2018**, *11*, 1297–1312. [CrossRef]
21. Watne, Å.K.; Linden, J.; Willhelmsson, J.; Fridén, H.; Gustafsson, M.; Castell, N. Tackling Data Quality When Using Low-Cost Air Quality Sensors in Citizen Science Projects. *Front. Environ. Sci.* **2021**, *9*, 733634. [CrossRef]
22. Wallace, L.; Bi, J.; Ott, W.R.; Sarnat, J.; Liu, Y. Calibration of Low-Cost PurpleAir Outdoor Monitors Using an Improved Method of Calculating PM. *Atmos. Environ.* **2021**, *256*, 118432. [CrossRef]
23. Karaoghlanian, N.; Noureddine, B.; Saliba, N.; Shihadeh, A.; Lakkis, I. Low Cost Air Quality Sensors "PurpleAir" Calibration and Inter-Calibration Dataset in the Context of Beirut, Lebanon. *Data Br.* **2022**, *41*, 108008. [CrossRef] [PubMed]
24. Karagulian, F.; Barbiere, M.; Kotsev, A.; Spinelle, L.; Gerboles, M.; Lagler, F.; Redon, N.; Crunaire, S.; Borowiak, A. Review of the Performance of Low-Cost Sensors for Air Quality Monitoring. *Atmosphere* **2019**, *10*, 506. [CrossRef]
25. Narayana, M.V.; Jalihal, D.; Nagendra, S.M.S. Establishing A Sustainable Low-Cost Air Quality Monitoring Setup: A Survey of the State-of-the-Art. *Sensors* **2022**, *22*, 394. [CrossRef] [PubMed]
26. Singh, D.; Dahiya, M.; Kumar, R.; Nanda, C. Sensors and Systems for Air Quality Assessment Monitoring and Management: A Review. *J. Environ. Manag.* **2021**, *289*, 112510. [CrossRef] [PubMed]
27. Chojer, H.; Branco, P.T.B.S.; Martins, F.G.; Alvim-Ferraz, M.C.M.; Sousa, S.I.V. Development of Low-Cost Indoor Air Quality Monitoring Devices: Recent Advancements. *Sci. Total Environ.* **2020**, *727*, 138385. [CrossRef] [PubMed]
28. Bigi, A.; Mueller, M.; Grange, S.K.; Ghermandi, G.; Hueglin, C. Performance of NO, NO2 Low Cost Sensors and Three Calibration Approaches within a Real World Application. *Atmos. Meas. Tech.* **2018**, *11*, 3717–3735. [CrossRef]
29. Bean, J.K. Evaluation Methods for Low-Cost Particulate Matter Sensors. *Atmos. Meas. Tech.* **2021**, *14*, 7369–7379. [CrossRef]
30. Concas, F.; Mineraud, J.; Lagerspetz, E.; Varjonen, S.; Liu, X.; Puolamäki, K.; Nurmi, P.; Tarkoma, S. Low-Cost Outdoor Air Quality Monitoring and Sensor Calibration: A Survey and Critical Analysis. *ACM Trans. Sens. Netw. (TOSN)* **2021**, *17*, 1–44. [CrossRef]
31. Venkatraman Jagatha, J.; Klausnitzer, A.; Chacón-Mateos, M.; Laquai, B.; Nieuwkoop, E.; van der Mark, P.; Vogt, U.; Schneider, C. Calibration Method for Particulate Matter Low-Cost Sensors Used in Ambient Air Quality Monitoring and Research. *Sensors* **2021**, *21*, 3960. [CrossRef]
32. Artinano, B.; Narros, A.; Diaz, E.; Gomez, F.J.; Borge, R. Low-Cost Sensors for Urban Air Quality Monitoring: Preliminary Laboratory and in-Field Tests within the TECNAIRE-CM Project. In Proceedings of the 2019 5th Experiment International Conference (exp.at'19), Funchal, Portugal, 12–14 June 2019; pp. 462–466.
33. Schalm, O.; Carro, G.; Lazarov, B.; Jacobs, W.; Stranger, M. Reliability of Lower-Cost Sensors in the Analysis of Indoor Air Quality on Board Ships. *Atmosphere* **2022**, *13*, 1579. [CrossRef]
34. Mao, K.; Xu, J.; Jin, R.; Wang, Y.; Fang, K. A Fast Calibration Algorithm for Non-Dispersive Infrared Single Channel Carbon Dioxide Sensor Based on Deep Learning. *Comput. Commun.* **2021**, *179*, 175–182. [CrossRef]
35. Schwartz, B. *The Paradox of Choice: Why More Is Less*; Harper Perennial: New York, NY, USA, 2005.
36. Villanueva, F.; Ródenas, M.; Ruus, A.; Saffell, J.; Gabriel, M.F. Sampling and Analysis Techniques for Inorganic Air Pollutants in Indoor Air. *Appl. Spectrosc. Rev.* **2022**, *57*, 531–579. [CrossRef]
37. Samad, A.; Obando Nuñez, D.R.; Solis Castillo, G.C.; Laquai, B.; Vogt, U. Effect of Relative Humidity and Air Temperature on the Results Obtained from Low-Cost Gas Sensors for Ambient Air Quality Measurements. *Sensors* **2020**, *20*, 5175. [CrossRef]
38. Wei, P.; Ning, Z.; Ye, S.; Sun, L.; Yang, F.; Wong, K.; Westerdahl, D.; Louie, P. Impact Analysis of Temperature and Humidity Conditions on Electrochemical Sensor Response in Ambient Air Quality Monitoring. *Sensors* **2018**, *18*, 59. [CrossRef]
39. Wahlborg, D.; Björling, M.; Mattsson, M. Evaluation of Field Calibration Methods and Performance of AQMesh, a Low-Cost Air Quality Monitor. *Environ. Monit. Assess.* **2021**, *193*, 251. [CrossRef]
40. Gonzalez, A.; Boies, A.; Swason, J.; Kittelson, D. Field Calibration of Low-Cost Air Pollution Sensors. *Atmos. Meas. Tech. Discuss.* **2019**, *12*, 1–17. [CrossRef]
41. Borrego, C.; Costa, A.M.; Ginja, J.; Amorim, M.; Coutinho, M.; Karatzas, K.; Sioumis, T.; Katsifarakis, N.; Konstantinidis, K.; De Vito, S.; et al. Assessment of Air Quality Microsensors versus Reference Methods: The EuNetAir Joint Exercise. *Atmos. Environ.* **2016**, *147*, 246–263. [CrossRef]
42. Drewnick, F.; Böttger, T.; von der Weiden-Reinmüller, S.-L.; Zorn, S.R.; Klimach, T.; Schneider, J.; Borrmann, S. Design of a Mobile Aerosol Research Laboratory and Data Processing Tools for Effective Stationary and Mobile Field Measurements. *Atmos. Meas. Tech.* **2012**, *5*, 1443–1457. [CrossRef]
43. Arroyo, P.; Gómez-Suárez, J.; Suárez, J.I.; Lozano, J. Low-Cost Air Quality Measurement System Based on Electrochemical and PM Sensors with Cloud Connection. *Sensors* **2021**, *21*, 6228. [CrossRef]

44. Ionascu, M.-E.; Castell, N.; Boncalo, O.; Schneider, P.; Darie, M.; Marcu, M. Calibration of CO, NO_2, and O_3 Using Airify: A Low-Cost Sensor Cluster for Air Quality Monitoring. *Sensors* **2021**, *21*, 7977. [CrossRef] [PubMed]
45. Smith, K.R.; Edwards, P.M.; Ivatt, P.D.; Lee, J.D.; Squires, F.; Dai, C.; Peltier, R.E.; Evans, M.J.; Sun, Y.; Lewis, A.C. An Improved Low-Power Measurement of Ambient NO_2 and O_3 Combining Electrochemical Sensor Clusters and Machine Learning. *Atmos. Meas. Tech.* **2019**, *12*, 1325–1336. [CrossRef]
46. Wijeratne, L.O.H.; Kiv, D.R.; Aker, A.R.; Talebi, S.; Lary, D.J. Using Machine Learning for the Calibration of Airborne Particulate Sensors. *Sensors* **2019**, *20*, 99. [CrossRef] [PubMed]
47. Greenspan, L. Humidity Fixed Points of Binary Saturated Aqueous Solutions. *J. Res. Natl. Bur. Stan. Sect. A* **1977**, *81A*, 89. [CrossRef]
48. Carotenuto, A.; Dell'Isola, M. An Experimental Verification of Saturated Salt Solution-Based Humidity Fixed Points. *Int. J. Thermophys.* **1996**, *17*, 1423–1439. [CrossRef]
49. Lu, T.; Chen, C. Uncertainty Evaluation of Humidity Sensors Calibrated by Saturated Salt Solutions. *Measurement* **2007**, *40*, 591–599. [CrossRef]
50. Dubovikov, N.I.; Podmurnaya, O.A. Measuring the Relative Humidity over Salt Solutions. *Meas. Tech.* **2001**, *44*, 1260–1261. [CrossRef]
51. Young, J.F. Humidity Control in the Laboratory Using Salt Solutions—A Review. *J. Appl. Chem.* **2007**, *17*, 241–245. [CrossRef]
52. Eggert, G. Saturated Salt Solutions in Showcases: Humidity Control and Pollutant Absorption. *Herit. Sci.* **2022**, *10*, 54. [CrossRef]
53. Winston, P.W.; Bates, D.H. Saturated Solutions For the Control of Humidity in Biological Research. *Ecology* **1960**, *41*, 232–237. [CrossRef]
54. ASTM E 104-85. *Standard Practice for Maintaining Constant Relative Humidity by Means of Aqueous Solutions*, ASTM International: West Conshohocken, PA, USA, 1996.
55. Forney, C.F.; Brandl, D.G. Control of Humidity in Small Controlled-Environment Chambers Using Glycerol-Water Solutions. *HortTechnology* **1992**, *2*, 52–54. [CrossRef]
56. Volk, A.; Kähler, C.J. Density Model for Aqueous Glycerol Solutions. *Exp. Fluids* **2018**, *59*, 75. [CrossRef]
57. Glycerine Producers' Association. *Physical Properties of Glycerine and Its Solutions*; Glycerine Producers' Association: New York, NY, USA, 1963.
58. Hook, M.D.; Mayer, M. Miniature Environmental Chambers for Temperature Humidity Bias Testing of Microelectronics. *Rev. Sci. Instrum.* **2017**, *88*, 034707. [CrossRef]
59. *ASTM D 5032-97*, Standard Practice for Maintaining Constant Relative Humidity by Means of Aqueous Glycerin Solutions. American Society for Testing and Materials: West Conshohocken, PA, USA, 1998.
60. Bosart, L.W.; Snoddy, A.O. New Glycerol Tables. *Ind. Eng. Chem.* **1927**, *19*, 506–510. [CrossRef]
61. Alyea, H.N. Syringe Gas Generators. *J. Chem. Educ.* **1992**, *69*, 65. [CrossRef]
62. Gonzalez-Rivero, R.A.; Álvarez-Cruz, A.; Alejo-Sanchez, D.; Morales-Perez, M.C.; Martinez, A.; Schalm, O. *Calibration of a Low-Cost Sensor for SO2 Monitoring in a Rural Area of Cienfuegos*; Cayo Santa Maria, Universidad Central "Marta Abreu" de Las Villas: Las Villas, Cuba, 2022.
63. Crossno, S.K.; Kalbus, L.H.; Kalbus, G.E. Determinations of Carbon Dioxide by Titration: New Experiments for General, Physical, and Quantitative Analysis Courses. *J. Chem. Educ.* **1996**, *73*, 175. [CrossRef]
64. Yim, B.; Sim, Y.-A.; Kim, S.-T. Evaluation of Short-Term CO_2 Passive Sampler for Monitoring Atmospheric CO_2 Levels. *ksccr* **2016**, *7*, 1–8. [CrossRef]
65. Passaretti Filho, J.; da Silveira Petruci, J.F.; Cardoso, A.A. Development of a Simple Method for Determination of NO_2 in Air Using Digital Scanner Images. *Talanta* **2015**, *140*, 73–80. [CrossRef]
66. Ugucione, C.; Gomes Neto, J.d.A.; Cardoso, A.A. Método colorimétrico para determinação de dióxido de nitrogênio atmosférico com preconcentração em coluna de c-18. *Quím. Nova* **2002**, *25*, 352–357. [CrossRef]
67. Felix, E.P.; Cardoso, A.A. Colorimetric Determination of Ambient Ozone Using Indigo Blue Droplet. *J. Braz. Chem. Soc.* **2006**, *17*, 296–301. [CrossRef]
68. Rodrigues, B.A.; Pitombo, L.R.M.; Cardoso, A.A. Study on the Use of Oxidant Scrubbers for Elimination of Interferences Due to Nitrogen Dioxide in Analysis of Atmospheric Dimethylsulfide. *J. Braz. Chem. Soc.* **2000**, *11*, 71–77. [CrossRef]
69. Felix, E.P.; Cardoso, A.A. A Method for Determination of Ammonia in Air Using Oxalic Acid-Impregnated Cellulose Filters and Fluorimetric Detection. *J. Braz. Chem. Soc.* **2012**, *23*, 142–147. [CrossRef]
70. Martinez, A.; Hernandez-Rodriguez, E.; Hernandez, L.; Schalm, O.; Gonzalez-Rivero, R.A.; Alejo-Sanchez, D. Design of a Low-Cost System for the Measurement of Variables Associated with Air Quality. *IEEE Embedded Syst. Lett.* **2022**, *14*, 1. [CrossRef]
71. Hernández Rodríguez, E.; Schalm, O.; Martínez, A. Development of a Low-Cost Measuring System for the Monitoring of Environmental Parameters That Affect Air Quality for Human Health. *ITEGAM-JETIA* **2020**, *6*, 22–27. [CrossRef]
72. Hernandez-Rodriguez, E.; Kairuz-Cabrera, D.; Martinez, A.; Gonzalez-Rivero, R.A.; Schalm, O. *Low-Cost Portable System for the Estimation of Air Quality*; Cuban National Bureau of Standards: La Habana, Cuba, 2022.
73. Jacob, L.A. A Simple and Inexpensive Gas Trap or Gas Inlet for Microscale Work. *J. Chem. Educ.* **1992**, *69*, A313. [CrossRef]
74. Najdoski, M. Gas Chemistry: A Microscale Kipp Apparatus. *Chem. Educ.* **2011**, *5*, 295–298.
75. Najdoski, M.; Stojkovikj, S. A Simple Microscale Gas Generation Apparatus. *J. Sci. Educ.* **2014**, *1*, 49–50.
76. Najdoski, M.; Stojkovikj, S.; Cyril, S. Cost Effective Microscale Gas Generation Apparatus. *Chemistry* **2010**, *19*, 444–449.

77. Thickett, D.; David, F.; Luxford, N. Air Exchange Rate—The Dominant Parameter for Preventive Conservation? *Conservator* **2005**, *29*, 19–34. [CrossRef]
78. Charlesworth, P.S. *Air Exchange Rate and Airtightness Measurement Techniques—An Applications Guide*; Air Infiltration and Ventilation Centre: Coventry, UK, 1988.
79. You, Y.; Niu, C.; Zhou, J.; Liu, Y.; Bai, Z.; Zhang, J.; He, F.; Zhang, N. Measurement of Air Exchange Rates in Different Indoor Environments Using Continuous CO_2 Sensors. *J. Environ. Sci.* **2012**, *24*, 657–664. [CrossRef]
80. Calver, A.; Holbrook, A.; Thickett, D.; Weintraub, S. Simple Methods to Measure Air Exchange Rates and Detect Leaks in Display and Storage Enclosures. In Proceedings of the 14th triennial meeting the Hague Preprints, The Hague, Netherlands, 12–16 September 2005; pp. 597–609.
81. ASTM E741. Standard Test Method for Determining Air Change in a Single Zone by Means of a Tracer Gas Dilution. American Society for Testing and Materials: West Conshohocken, PA, USA, 2006.
82. Tétreault, J.; Hagan, E. *Airtightness Measurement of Display Cases and Other Enclosures—Technical Bulletin 38*; Technical Bulletins; Canadian Conservation Institute: Ottawa, ON, Canada, 2022; p. 91.
83. Báthory, C.; Kiss, M.L.; Trohák, A.; Dobó, Z.; Palotás, Á.B. Preliminary Research for Low-Cost Particulate Matter Sensor Network. *E3S Web Conf.* **2019**, *100*, 00004. [CrossRef]
84. Alejo, D.; Morales, M.C.; de la Torre, J.B.; Grau, R.; Bencs, L.; Van Grieken, R.; Van Espen, P.; Sosa, D.; Nuñez, V. Seasonal Trends of Atmospheric Nitrogen Dioxide and Sulfur Dioxide over North Santa Clara, Cuba. *Environ. Monit. Assess.* **2013**, *185*, 6023–6033. [CrossRef]
85. NC 111: 2004. Calidad Del Aire—Reglas Para la Vigilancia de la Calidad del Aire en Asentamientos Humanos. Cuban National Bureau of Standards: La Habana, Cuba, 2004.

atmosphere

MDPI

Article

Evaluation and Correction of PurpleAir Temperature and Relative Humidity Measurements

Evan Couzo [1,*], Alejandro Valencia [2] and Phoebe Gittis [3]

1 Department of Atmospheric Sciences, University of North Carolina Asheville, Asheville, NC 28804, USA
2 Sonoma Technology, Inc., Petaluma, CA 94954, USA; avalencia@sonomatech.com
3 Department of Environmental Studies, University of North Carolina Asheville, Asheville, NC 28804, USA; pgittis@unca.edu
* Correspondence: ecouzo@unca.edu

Abstract: The PurpleAir PA-II sensor provides low-cost in situ measurements of meteorological variables including temperature and relative humidity (RH), as well as fine particulate matter ($PM_{2.5}$) in real time. The sensors have been used in several studies investigating intracity differences in temperature and $PM_{2.5}$. While the adoption and use of low-cost sensors has many benefits, care must be taken to ensure proper calibration and testing. This is true not only for $PM_{2.5}$ measurements but also for temperature and RH given the synergistic health impacts from extreme heat and air pollution exposure. Here, we compare continuous temperature and RH measurements from a PA-II sensor to measurements from a Campbell Scientific 107 temperature probe and Vaisala HMP45C RH probe. All three instruments were co-located from December 2021 to June 2023 in Asheville, North Carolina. We found that the PA-II has an overall high temperature bias of 2.6 °C and root mean square error (RMSE) of 2.8 °C. Applying a linear regression correction reduces RMSE to 1.0 °C, while applying the constant 4.4 °C correction suggested by PurpleAir reduces RMSE to only 2.2 °C. Our PA-II RH measurements have a low bias of −17.4% and uncorrected RMSE of 18.5%. A linear regression correction improves the RH RMSE to 4.5%. Applying the constant 4% RH correction suggested by PurpleAir reduces RMSE to only 14.8%. We present new correction factors that differ from those suggested by PurpleAir, which overcorrect the high temperature bias and undercorrect the low RH bias. We also show that our correction factors improve estimates of dewpoint temperature (RMSE of 0.6 °C and 0.9 °C) compared to the corrections suggested by PurpleAir.

Keywords: PurpleAir; low-cost sensor; air pollution; temperature; relative humidity

1. Introduction

Air pollution monitoring has enjoyed a notable surge in the popularity of low-cost sensors, marking a shift in data availability and spatial coverage. The proliferation of these sensors has been driven by advancements in technology and a growing need for neighborhood-scale measurement data [1]. Characterized by affordability and ease of use, low-cost sensors are democratizing air quality data collection and empowering communities to actively participate in monitoring efforts [2,3]. Several studies have investigated possible uses from citizen science efforts to personal exposure monitoring [4–7], and sensor data can supplement existing monitoring networks [8]. This is of particular importance for pollutants—such as fine particulate matter ($PM_{2.5}$)—with high spatiotemporal variability [9–12].

Many low-cost sensors used for air pollution monitoring employ optical particle counters that estimate mass concentration based on how particulates in the sample scatter light. Several studies have described this technology in detail along with some of the drawbacks and potential sources of error when compared to federal reference method (FRM) or federal equivalent method (FEM) instruments [13–16]. The South Coast Air

Quality Management District's AQ-SPEC program has tested dozens of commercially available low-cost $PM_{2.5}$ sensors in laboratory and field settings [17]. Their results show a wide range of sensor quality with some sensors, including the PurpleAir PA-II sensor used in our study, exhibiting good precision and accuracy when compared to FRM and FEM monitors. The US EPA, through conversations with multiple stakeholders [18], developed guidance for testing $PM_{2.5}$ sensors in ambient, non-regulatory settings [19], and the agency also created an Air Sensor Toolbox [20] and Guidebook [21], which provides best practices for using low-cost sensors. Additionally, the US EPA incorporates adjusted data from over 10,000 PurpleAir sensors on the AirNow Fire and Smoke map in recognition of that sensor's usefulness and popularity [22].

While it is well known that exposure to air pollutants and extreme heat are individually hazardous, it is becoming clear that the negative health effects of poor air quality are magnified by high temperatures. A study of 17 cities in France found that the risk of mortality associated with a 10 $\mu g/m^3$ increase in PM_{10} was 14.2% greater during a heat wave [23]. Other studies of European cities reported similar findings [24,25]. Rahman et al. [26], investigating the link between extreme heat and $PM_{2.5}$ in California, found that co-exposure had a greater effect on all-cause mortality (21.0% increase in risk) than the sum of the individual effects (6.1% increase in risk for high temperature days and 5.0% increase in risk for high $PM_{2.5}$ days). The effect was magnified when looking specifically at increased risk of cardiovascular mortality and respiratory mortality. Further, recent reviews of the literature on human population health studies found compelling evidence for synergistic effects of extreme heat and $PM_{2.5}$ [27–29].

Populations living in urban areas are particularly vulnerable to extreme heat and elevated $PM_{2.5}$ concentrations. Individual risk is communicated via the well-known heat index (HI), which integrates ambient temperature with relative humidity (RH), and air quality index (AQI), which relies on ambient concentrations of pollutants such as $PM_{2.5}$ and ozone. Yet, there is no consistent method to communicate the combined risk. Fever et al. [30] unified the HI and AQI for $PM_{2.5}$ and ozone in a combined index using several different models. They found that improved mortality predictions were possible with their combined indices in Monterrey, Mexico, but cautioned that their results might not be generalizable to regions with different air pollutant and weather patterns. Steeneveld et al. [31] developed an urban climate index that combined PM_{10}, ozone, and nitrogen dioxide concentrations with urban heat island effects. Of relevance to low-cost sensor networks, they highlight the importance of fine-scale measurement data when assessing the health risks of extreme heat and air pollution. A third study took a different approach to investigating the combined impact of extreme heat and $PM_{2.5}$ exposure on urban communities. Sabrin et al. [32] used neighborhood-scale data to assess the importance of underlying environmental and social parameters and to communicate the risk to vulnerable populations. Each of these investigations sought to integrate information about heat and air pollution exposure, which indicates the need for high-quality measurements of both.

Several studies have examined the performance of PurpleAir sensors with respect to $PM_{2.5}$ [33–37]. To date, these efforts have occurred in several countries and under a variety of ambient and laboratory conditions. Most often, $PM_{2.5}$ measurements from PurpleAir sensors are compared to nearby FRM or FEM instruments. While nearly all attempts to correct the $PM_{2.5}$ concentrations have recognized the importance of factoring in temperature and/or RH values, very few studies have investigated bias and error in these meteorological measurements from the PurpleAir sensors themselves. Holder et al. [38] found strong correlations to reference measurements (r^2 of 0.91 and 0.84 for temperature and RH) over a range of ambient conditions in North Carolina. They reported a mean temperature bias of 5.2 °C and a mean RH bias of −24.3%. Another study in Greece found excellent inter-sensor agreement between eight different PurpleAir sensors and strong correlations with reference temperature and RH [39]. Though specific values were not provided, this study also noted persistent high and low biases for temperature and RH, respectively. These biases are known by PurpleAir, and specific correction factors

(−4.4 °C and +4%) are provided on the company's community forum [40], though without information about how those correction factors were determined. PurpleAir attributes the measurement discrepancy to heat generated from the internal WiFi module and adds a disclaimer that the on-board temperature and RH readings are not meant to reflect actual ambient conditions. A third study reports yet another set of biases but blames the near-infrared laser counters for the elevated temperature and lower RH values [41]. They found a temperature difference of 4.4 °C and an RH difference of −15% without providing any explanation other than to mention a long period of comparison to an indoor monitor. Finally, Malings et al. [15] states that PurpleAir temperature and RH were 2.7 °C above and 9.7% below average, though again, without a description of how the reference values were obtained.

The current study provides the first long-term, systematic investigation of PurpleAir's on-board temperature and RH measurements. We compared data from the low-cost sensor to co-located research-grade meteorological instruments and provide correction factors to estimate ambient temperature and RH from values reported by PurpleAir. While it has been noted elsewhere that reasonably accurate PurpleAir $PM_{2.5}$ corrections are possible even without knowing actual ambient conditions [42], our results enhance the capability of the sensor by allowing the public health community to combine air quality information with temperature and RH data.

2. Materials and Methods

2.1. PurpleAir PA-II Sensor

The PurpleAir PA-II sensor is a commercially available low-cost particulate matter and temperature/RH/pressure measurement platform. It is most often used to obtain ambient $PM_{2.5}$ levels using a pair of Plantower PMS-5003 optical particle counters. Mass concentrations are calculated using a proprietary algorithm. Descriptions of the PA-II's operating principles have been detailed elsewhere (for example, Ardon-Dryer et al. [33]). Meteorological measurements are provided by a Bosch BME280 sensor.

The PA-II's components are housed in a white plastic shell (85 mm × 85 mm × 125 mm) to protect the electronics from the elements. The plastic housing is open on the bottom to allow the sensors to measure ambient air. In addition to the two PMS-5003 particle counters and the BME280 sensor, the PA-II platform contains a wireless data transmitter, microSD card data logger, and programmable circuit board with a battery-powered real time clock. Figure 1 shows the PA-II sensor and its internal components.

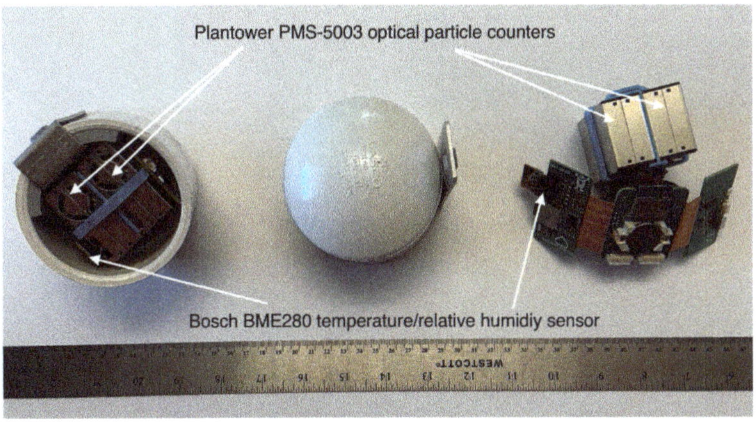

Figure 1. Images of the PurpleAir PA-II sensor labeled with the Plantower PMS-5003 optical particle counters and Bosch BME280 temperature/relative humidity sensor. The lefthand image shows the bottom of the PA-II, and the center image shows the top. The righthand image shows the internal electronic components.

A more recent version of the PurpleAir sensor, the PurpleAir Flex, is also available. The operating principles are identical, though the optical particle counters have been updated to the Plantower PMS-6003 model and the meteorological sensor was updated to the Bosch BME688.

2.2. Measurement Site

All measurements were taken in Asheville, North Carolina, at a weather station maintained by the Department of Atmospheric Sciences at the University of North Carolina Asheville (UNCA). The station, which is part of the North Carolina Environment and Climate Observing Network (ECONet) [43], is 2367 m above sea level and located at 35.62° N, 82.57° W. Temperature data were recorded using a Campbell Scientific 107 probe 2 m above the ground, and RH data were recorded using a Vaisala HMP45C probe also 2 m above the ground. All data are publicly available at https://atms.unca.edu/datarequest/ (accessed on 1 February 2024). North Carolina ECONet stations undergo rigorous quality control and maintenance procedures [44].

Continuous weather station measurements were taken at 1-min intervals for 553 days from 15 December 2021 to 30 June 2023. Both temperature and RH records were more than 99.9% complete. Upon inspection, some temperature data on 12 August 2022 had anomalously low readings and were removed from this analysis. The affected times were 11:12 through 13:14 (local time); RH data were also removed during this interval. After removing the anomalous data points, the 1-min data were aggregated to 5-min averages.

One PA-II sensor was installed on the UNCA weather tower at 3 m and measured temperature and RH. No adjustments were made to the data to account for the 1-m height difference between the PA-II sensor and the UNCA instruments. Continuous measurements were taken at 2-min intervals from 15 December 2021 through 26 July 2022. Measurements beginning on 27 July 2022 were taken at 5-min intervals. All data were aggregated to 5-min averages. Data completeness for the PA-II was 91.4%. Data were unavailable from 14:30 10 January 2023 through 08:35 23 January 2023 (local time), and several data points were removed on 29 October 2022 because the $PM_{2.5}$ measurements were greater than 500 $\mu g/m^3$, indicating a possible data logging malfunction.

2.3. Dewpoint

We used the Python package MetPy version 1.5.1 [45] to calculate dewpoint temperatures (DP) from temperature and RH measurements. While the Campbell scientific 107 probe was used to obtain reference temperatures for comparison to the PA-II's temperature measurements, we relied on temperature from the Vaisala HMP45C probe to calculate DP. It is recommended to use temperature and RH values from the same sensor when calculating DP. Values were calculated for both the UNCA and PA-II data sets using the metpy.calc.dewpoint_from_relative_humidity function.

2.4. Performance Metrics

To evaluate the reliability of the PA-II's meteorological measurements, we calculated mean bias (bias) and root mean square error (RMSE) to compare the UNCA and PA-II data sets. All comparisons were made on time-paired data, so we excluded records where only one data set had a valid measurement. Bias and RMSE were calculated using the following equations:

$$\text{Mean Bias} = \frac{1}{n}\sum_{1}^{n}(x(i) - y(i)) \tag{1}$$

$$\text{RMSE} = \sqrt{\frac{1}{n}\sum_{1}^{n}(x(i) - y(i))^2} \tag{2}$$

where x(i) and y(i) are time-paired measurements from PA-II and UNCA.

3. Results and Discussion

This study compared temperature and RH measurements from one PurpleAir PA-II low-cost sensor to UNCA's research-grade instruments during a 553-day period extending from 15 December 2021 through 30 June 2023. We performed a simple linear regression correction on the PA-II's values using the UNCA data as the reference. Our corrected measurements were then compared to corrections suggested by PurpleAir [40]. Results are presented separately for temperature, RH, and DP.

3.1. Temperature

Measurements from the PA-II are strongly correlated with the UNCA temperature data. The Pearson correlation coefficient (r) between the two data sets is 0.99. Comparisons between the PA-II and UNCA temperatures reveal a consistent and systematic high bias, as shown in Figure 2. The uncorrected PA-II measurements, shown in blue, have a mean bias of 2.6 °C and RMSE of 2.8 °C. A total of 148,230 pairs of 5-min data were analyzed. The differences between the time-paired PA-II and UNCA values ranged from −2.7 °C to 9.1 °C, with 91.1% of the data points exhibiting a high bias. The maximum and minimum temperatures recorded during the study period were 33.8 °C and −19.3 °C for UNCA and 39.2 °C and −17.8 °C for PA-II.

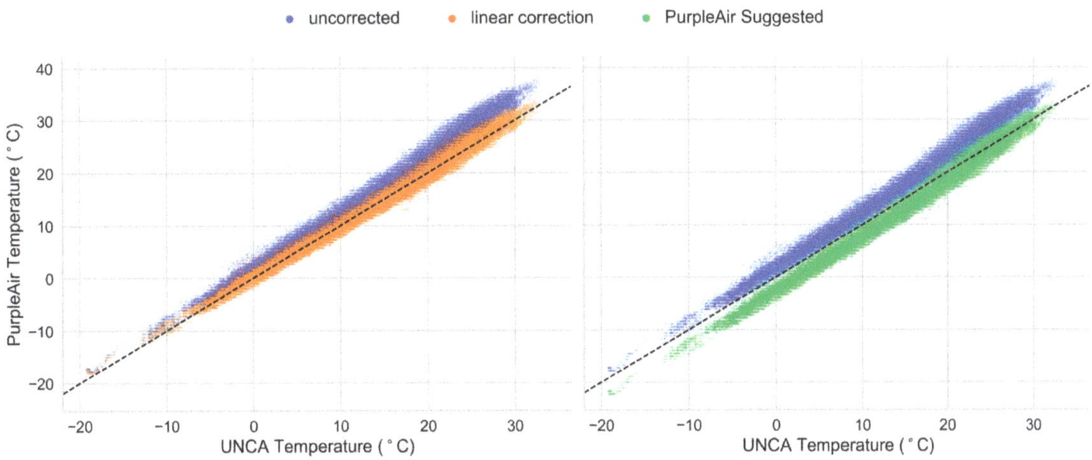

Figure 2. Comparisons of the time-paired 5-min PA-II and UNCA temperature measurements. In both plots, the blue markers show the uncorrected data. Orange markers in the left-hand plot show the PA-II data after applying a simple linear regression correction, and the green markers in the right-hand plot show the PA-II data after applying PurpleAir's suggested correction. The dotted line drawn across each plot is the 1:1 line.

We performed a simple linear regression on the uncorrected PA-II data to remove the high temperature bias. The left-hand side of Figure 2 compares the corrected data (orange) to the uncorrected data (blue). The linear regression correction (slope = 1.07, intercept = 1.60) reduces RMSE to 1.0 °C, and the range in differences between the data sets is now −5.4 °C to 5.3 °C. PurpleAir is aware of the high temperature bias in the PA-II, and they recommend a constant 4.4 °C correction to account for the heat generated by the sensor's WiFi module. We applied this suggestion to the uncorrected PA-II measurements, and the result is shown on the right-hand side of Figure 2 in green. The suggested correction introduces a readily apparent low bias especially for temperatures below 20 °C. Overall mean bias is −1.9 °C and RMSE is 2.2 °C, and the range in differences when comparing PurpleAir's suggested correction to the UNCA data set is −7.2 °C to 4.6 °C, with 84.1% of all data pairs exhibiting a low bias. Thus, while the suggested correction does reduce the magnitude of the bias and RMSE, it performs worse than a simple linear regression correction.

As a next step, we separated the measurement data into bins to determine whether the PA-II's bias is temperature-dependent. We chose five separate temperature bins using UNCA temperature as the reference, as shown in Table 1. Both the bias and RMSE increase monotonically with increasing temperature. Minimum bias (RMSE) is 1.8 °C (2.0 °C) when temperatures are −5 °C or below; maximum bias (RMSE) is 4.2 °C (4.4 °C) when temperatures are greater than 25 °C. Both conditions are rarely met occurring only about 8% of the time. The temperature dependence is reversed when considering the corrections suggested by PurpleAir, which are also provided in Table 1. Here, the bias and RMSE decreases from colder to warmer temperatures.

Table 1. Comparison of PA-II and UNCA temperature measurements.

Correction Type	Temperature Bin (°C)	Mean Bias (°C)	RMSE (°C)	r	Slope	Intercept	n
uncorrected	All	2.6	2.8	0.99	1.07	1.60	148,230
	(−∞, −5]	1.8	2.0	0.98	1.01	1.89	2239
	(−5, 5]	1.9	2.1	0.96	1.03	1.86	26,589
	(5, 15]	2.2	2.3	0.96	1.04	1.71	52,162
	(15, 25]	3.0	3.2	0.94	1.17	−0.31	57,574
	(25, ∞]	4.2	4.4	0.80	1.07	2.36	9666
simple linear regression correction	All	0.0	1.0	0.99	1.00	0.00	148,230
	(−∞, −5]	0.8	1.1	0.98	0.94	0.27	2239
	(−5, 5]	0.2	0.8	0.96	0.96	0.24	26,589
	(5, 15]	−0.2	0.8	0.96	0.97	0.11	52,162
	(15, 25]	0.0	1.0	0.94	1.09	−1.78	57,574
	(25, ∞]	0.6	1.3	0.80	0.99	0.71	9666
PurpleAir suggested correction	All	−1.9	2.2	0.99	1.07	−2.85	148,230
	(−∞, −5]	−2.6	2.8	0.98	1.01	−2.55	2239
	(−5, 5]	−2.6	2.7	0.96	1.03	−2.59	26,589
	(5, 15]	−2.3	2.4	0.96	1.04	−2.73	52,162
	(15, 25]	−1.4	1.9	0.94	1.17	−4.76	57,574
	(25, ∞]	−0.2	1.3	0.80	1.07	−2.08	9666

The high temperature bias presented here is less than previously reported (Table 2). Holder et al. [38] found a bias of 5.2 °C. Their sensors were deployed for nine months in North Carolina. A lower bias of 2.7 °C was found during a March–June field measurement in Pennsylvania [15]. The third reported bias −4.4 °C, which is identical to the correction suggested by PurpleAir—is from an indoor sensor [41].

Table 2. Comparison of temperature and RH bias to previous studies.

Study	T (°C)	RH (%)	Location	Dates
This study	2.6	−17.4	Asheville, NC (USA)	15 Dec 2021–30 June 2023
Malings et al. [15] [1]	2.7	−9.7	Pittsburgh, PA (USA)	30 Mar 2018–4 June 2018
Holder et al. [38] [1]	5.2	−24.3	Research Triangle Park, NC (USA)	10 Aug 2018–30 Apr 2019
Wallace et al. [41] [1,2]	4.4	−15	Santa Rosa, CA (USA)	unknown
PurpleAir [40] [1]	4.4	−4	Unknown	unknown

[1] These studies do not provide a clear methodology for determining the temperature and RH bias. [2] Temperature and RH were measured indoors.

3.2. Relative Humidity

PA-II RH, like temperature, has a consistent and systematic bias. Unlike temperature, however, the RH bias is low as shown in Figure 3. This bias is the result of the excess heat produced by the sensor's electronics. The excess heat increases the local temperature inside the sensor's enclosure thereby lowering RH. Despite the bias, the RH measurements from the PA-II are strongly correlated to the UNCA measurements illustrated by an r-value of 0.98. The overall RH bias and RMSE is −17.4% and 18.5%. The differences between the time-paired PA-II and UNCA measurements ranged from −45.4% to 2.7%, with nearly all measurements exhibiting a low bias (91.4% of all data points). The ranges of measured RH during the study period are 9.6% to 99.6% for UNCA and 8.0% to 90.0% for PA-II.

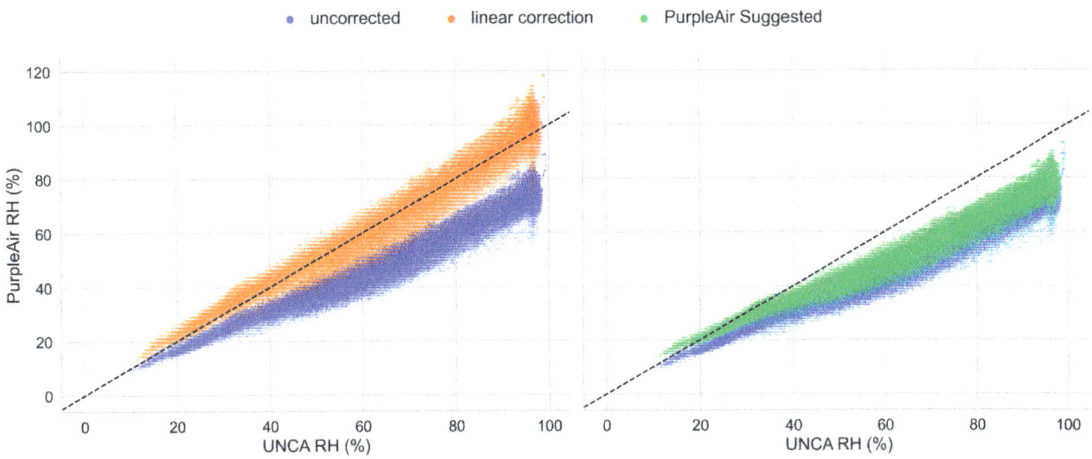

Figure 3. Comparisons of the time-paired 5-min PA-II and UNCA RH measurements. In both plots, the blue markers show the uncorrected data. Orange markers in the left-hand plot show the PA-II data after applying a simple linear regression correction, and the green markers in the right-hand plot show the PA-II data after applying PurpleAir's suggested correction. The dotted line drawn across each plot is the 1:1 line.

Applying a simple linear regression correction (slope = 0.75, intercept = 0.12) on the PA-II removes the high temperature bias. The left-hand side of Figure 3 compares the corrected data in orange to the uncorrected data in blue. RMSE decreases to 4.5% and the range in differences between the sets of measurements is now −28.7% to 20.5%. PurpleAir suggests a 4% correction to RH measurements from the sensor, and we applied this to the uncorrected PA-II data set. The right-hand side of Figure 3 illustrates that the suggested correction is not enough to overcome the persistent low RH bias; 89.0% of all PA-II measurements after the 4% correction are still less than the UNCA values. Overall bias and RMSE for the suggested correction are −13.4% and 14.8%, which is an improvement over the uncorrected data. But, as with temperature, PurpleAir's suggested RH correction has worse performance than the simple linear regression correction.

We partitioned the RH data into separate bins to investigate how the PA-II's bias depends on RH. Table 3 summarizes these results using UNCA RH values as the reference for determining each bin. Bias and RMSE are greater at higher RH. The largest bias (−22.0%) and RMSE (22.3%) was calculated for RH values between 80% and 100. Since nearly half of the data points come from this bin, the overall bias and RMSE are dominated by these values. The uncorrected data in Figure 3 shows a marked increase in data points with a lower bias starting around UNCA RH values of 40%. The picture is similar when considering the data after applying PurpleAir's suggested corrections. Worse RH performance occurs at higher

RH values. Below 40% RH, however, the suggested correction has similar performance to the simple linear regression correction.

Table 3. Comparison of PA-II and UNCA RH measurements.

Correction Type	RH Bin (%)	Mean Bias (%)	RMSE (%)	r	Slope	Intercept	n
uncorrected	All	−17.4	18.5	0.98	0.75	0.12	148,230
	[0, 20]	−2.1	2.5	0.84	0.74	2.21	1323
	(20, 40]	−6.0	6.4	0.92	0.74	2.01	16,506
	(40, 60]	−13.0	13.4	0.79	0.63	6.10	29,269
	(60, 80]	−18.6	19.0	0.79	0.78	−2.95	39,682
	(80, 100]	−22.0	22.3	0.82	0.92	−14.56	61,450
simple linear regression correction	All	0.00	4.5	0.98	1.00	0.00	148,230
	[0, 20]	2.6	3.0	0.84	0.99	2.77	1323
	(20, 40]	2.2	3.1	0.92	0.99	2.51	16,506
	(40, 60]	−0.6	3.8	0.79	0.83	7.96	29,269
	(60, 80]	−1.7	4.9	0.79	1.03	−4.09	39,682
	(80, 100]	0.7	4.8	0.82	1.22	−19.5	61,450
PurpleAir suggested correction	All	−13.4	14.8	0.98	0.75	4.12	148,230
	[0, 20]	1.9	2.3	0.84	0.74	6.21	1323
	(20, 40]	−2.0	3.0	0.92	0.74	6.01	16,506
	(40, 60]	−9.0	9.6	0.79	0.63	10.10	29,269
	(60, 80]	−14.6	15.1	0.79	0.78	1.05	39,682
	(80, 100]	−18.0	18.4	0.82	0.92	−10.56	61,450

The low RH bias in our study is within the bounds of previous investigations (Table 2). Holder et al. [38] reported a bias of −24.3%, which is substantially lower than our overall bias of −17.4%. They also found a larger high temperature bias, which is consistent with a lower RH bias. Our RH bias is lower than the −9.7% bias reported by Malings et al. [15]. However, their temperature bias was slightly higher than ours, which would suggest a lower bias as found by [38]. Given the differences in data collection location, longevity, and time of year, however, some inconsistencies should be expected especially considering the nonlinear relationship between RH and temperature and the importance of local conditions in determining RH. Interestingly, our RH bias was close to the −15% reported by [41], which collected temperature and RH measurements indoors.

3.3. Dewpoint Temperature

DP is a derived variable that is calculated from temperature and RH. It is arguably more relevant to predicting physical discomfort and dangerous heat conditions than the measurements reported by the PA-II. Because DP is determined from temperature and RH, it is subject to the same biases found in the previous sections, though it is not immediately obvious how the high temperature bias and low RH bias will affect bias in DP.

Figure 4 shows the uncorrected DP in blue. Correlation between the PA-II and UNCA values is strong with an r-value of 0.99. The overall DP bias is −1.6 °C and the RMSE is 1.7 °C. Differences between the time-paired PA-II and UNCA measurements ranged from −7.3 °C to 7.7 °C. Nearly all data points (88.9%) exhibited a low bias. The maximum and minimum DP were 24.5 °C and −23.1 °C for UNCA and 22.4 °C and −25.1 °C for PA-II.

Corrected DP values were calculated by using the corrected temperature and RH values obtained from the simple linear regression. The left-hand side of Figure 4 shows that the low bias is largely eliminated. The correction reduces the RMSE to 0.6 °C. The range in differences between the corrected PA-II and UNCA is −6.3 °C to 8.7 °C. We also calculated DP using the suggested PurpleAir corrections to temperature and RH, as shown on the right-hand side of Figure 4. The combination of these suggested corrections leads to a small high bias in DP. Overall bias is 0.6 °C and RMSE is 0.9 °C, and the range in differences when comparing PurpleAir's suggested correction to the UNCA data set is

−5.1 °C to 9.9 °C. As with temperature and RH, the suggested PurpleAir corrections reduce the magnitude of DP bias and RMSE but not to the extent seen by applying the simple linear regression correction.

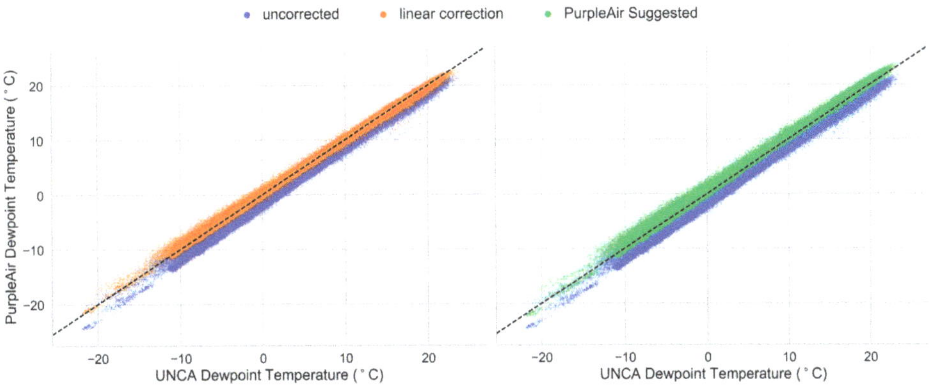

Figure 4. Comparisons of the time-paired 5-min PA-II and UNCA DP values. In both plots, the blue markers show the uncorrected data. Orange markers in the left-hand plot show the PA-II data after applying a simple linear regression correction, and the green markers in the right-hand plot show the PA-II data after applying PurpleAir's suggested correction. The dotted line drawn across each plot is the 1:1 line.

We then separated the DP data into bins (using UNCA DP as the reference) that roughly correspond to physical comfort levels. The results, compiled in Table 4, provide limited evidence for temperature dependence. As DP rises, the uncorrected bias becomes more negative with the lowest bias (−2.6 °C) observed when the DP is above 24 °C. This occurred rarely (n = 3), so care should be taken when considering this result. Temperature dependence is also observed with both corrected sets of DP values. However, the signs of the biases for the two corrected data sets are different. The low bias from the simple linear regression correction means that the actual DP is greater than would be expected for the higher DP bins, while the high bias from PurpleAir's suggested corrections means that the actual DP is lower than would be expected.

Table 4. Comparison of calculated PA-II and UNCA DP values.

Correction Type	DP Bin (°C)	Mean Bias (°C)	RMSE (°C)	r	Slope	Intercept	n
uncorrected	All	−1.6	1.7	0.99	1.00	−1.58	148,236
	[−∞, 15]	−1.6	1.8	0.99	1.00	−1.59	113,091
	(15, 20]	−1.5	1.6	0.94	1.01	−1.73	29,944
	(20, 22]	−1.6	1.7	0.67	0.91	0.33	4813
	(22, 24]	−1.9	2.0	0.43	0.68	5.26	385
	(24, ∞]	−2.6	2.6	0.67	1.59	−17.00	3
simple linear regression correction	All	0.1	0.6	0.99	0.98	0.27	148,236
	[−∞, 15]	0.2	0.7	0.99	0.97	0.27	113,091
	(15, 20]	−0.1	0.6	0.93	0.98	0.23	29,944
	(20, 22]	−0.3	0.7	0.60	0.83	3.16	4813
	(22, 24]	−0.7	1.0	0.40	0.65	7.06	385
	(24, ∞]	−1.5	1.6	0.73	1.48	−13.22	3
PurpleAir suggested correction	All	0.6	0.9	0.99	1.00	0.64	148,236
	[−∞, 15]	0.6	1.0	0.99	1.00	0.63	113,091
	(15, 20]	0.7	0.8	0.94	1.01	0.49	29,944
	(20, 22]	0.6	0.8	0.67	0.91	2.55	4813
	(22, 24]	0.3	0.7	0.43	0.68	7.49	385
	(24, ∞]	−0.4	0.5	0.67	1.59	−14.78	3

4. Conclusions

While the qualitative results of our research were known previously (high temperature bias and low RH bias), this is the first *systematic* investigation of the sensor's bias in the scientific literature. Our study adds a missing piece to the literature in several ways. First, we conducted a long-term systematic investigation of the PurpleAir PA-II sensor's meteorological measurements spanning over 550 days across a variety of environmental conditions. Our results show that the PurpleAir PA-II has a persistent high temperature bias and low RH bias when compared to research-grade meteorological instrumentation. We quantified the correlation, bias, and RMSE between the low-cost sensor and reference measurements for several temperature and RH ranges and found that the biases persist across a wide range of values.

We used a simple linear regression model to calculate correction factors for temperature and RH and compared these to the corrections suggested by PurpleAir. Importantly, we found that PurpleAir's suggested corrections perform substantially worse than our corrections. Their suggested −4.4 °C adjustment over corrects temperature readings and introduces a low bias. Meanwhile, their suggested +4% adjustment to RH is not nearly enough to correct the negative bias. For all variables, we find that our linear regression corrections reduce error compared to the suggested PurpleAir corrections.

Finally, we extended our analysis to dew point calculations. Including corrections for this parameter enhances the usefulness of the PurpleAir PA-II sensor. Reliable correction factors for the sensor's meteorological measurements, provided here for the first time, coupled with reliable correction factors for the sensor's fine particulate matter measurements (provided elsewhere in literature) will allow researchers to investigate the synergistic health effects of exposure to extreme heat and particulate matter.

Other long-term studies of the meteorological capabilities of the PA-II in different climates will strengthen the scientific community's understanding of the sensor's bias and provide more confidence in its measurements. Future studies should also attempt to evaluate multiple PA-II sensors to validate our results. The PA-II sensor's well-documented ability to provide reliable $PM_{2.5}$ data has made it popular among the research, citizen-science, and emergency planning communities. As these sensors become more common for routine air quality monitoring applications, the temperature and RH measurements can also provide valuable information to an increasingly warming and urbanized world.

Author Contributions: Conceptualization, E.C.; methodology, E.C.; formal analysis, E.C. and A.V.; investigation, E.C.; resources, E.C.; data curation, E.C. and A.V.; writing—original draft preparation, E.C. and P.G.; writing—review and editing, E.C. and P.G.; visualization, E.C. and A.V.; supervision, E.C.; project administration, E.C. All authors have read and agreed to the published version of the manuscript.

Funding: This research received no external funding.

Institutional Review Board Statement: Not applicable.

Informed Consent Statement: Not applicable.

Data Availability Statement: The meteorological data from the UNC Asheville weather station are made public by the Department of Atmospheric Sciences (https://atms.unca.edu/datarequest/ (accessed on 1 February 2024)). Data from the PurpleAir PA-II sensor can be accessed through the PurpleAir API (https://api.purpleair.com/) by requesting sensor index 139672.

Conflicts of Interest: The authors declare no conflicts of interest. The author affiliated with the company Sonoma Technology, Inc. declares that there is no conflict of interest associated with the content of this work.

References

1. Kumar, P.; Morawska, L.; Martani, C.; Biskos, G.; Neophytou, M.; Di Sabatino, S.; Bell, M.; Norford, L.; Britter, R. The rise of low-cost sensing for managing air pollution in cities. *Environ. Int.* **2015**, *75*, 199–205. [CrossRef] [PubMed]

2. Commodore, A.; Wilson, S.; Muhammad, O.; Svendsen, E.; Pearce, J. Community-based participatory research for the study of air pollution: A review of motivations, approaches, and outcomes. *Environ. Monit. Assess.* **2017**, *189*, 378. [CrossRef] [PubMed]
3. Morawska, L.; Thai, P.K.; Liu, X.; Asumadu-Sakyi, A.A.; Ayoko, G.; Bartonova, A.; Bedini, A.; Chai, F.; Christensen, B.; Dunbabin, M.; et al. Applications of low-cost sensing technologies for air quality monitoring and exposure assessment: How far have they gone? *Environ. Int.* **2018**, *116*, 286–299. [CrossRef] [PubMed]
4. Ellenburg, J.A.; Williford, C.J.; Rodriguez, S.L.; Andersen, P.C.; Turnipseed, A.A.; Ennis, C.A.; Basman, K.A.; Hatz, J.M.; Prince, J.C.; Meyers, D.H.; et al. Global ozone (GO3) project and AQTreks: Use of evolving technologies by Students and citizen scientists to monitor air pollutants. *Atmos. Environ. X* **2019**, *4*, 100048. [CrossRef]
5. Jerrett, M.; Donaire-Gonzalez, D.; Popoola, O.; Jones, R.; Cohen, R.C.; Almanza, E.; de Nazelle, A.; Mead, I.; Carrasco-Turigas, G.; Cole-Hunter, T.; et al. Validating novel air pollution sensors to improve exposure estimates for epidemiological analyses and citizen science. *Environ. Res.* **2017**, *158*, 286–294. [CrossRef] [PubMed]
6. Miskell, G.; Salmond, J.; Williams, D.E. Low-cost sensors and crowd-sourced data: Observations of siting impacts on a network of air-quality instruments. *Sci. Total Environ.* **2017**, *575*, 1119–1129. [CrossRef] [PubMed]
7. Cao, T.; Thompson, J.E. Personal monitoring of ozone exposure: A fully portable device for under $150 USD cost. *Sens. Actuators B Chem.* **2016**, *224*, 936–943. [CrossRef]
8. Snyder, E.G.; Watkins, T.H.; Solomon, P.A.; Thoma, E.D.; Williams, R.W.; Hagler, G.S.W.; Shelow, D.; Hindin, D.A.; Kilaru, V.J.; Preuss, P.W. The changing paradigm of air pollution monitoring. *Environ. Sci. Technol.* **2013**, *47*, 11369–11377. [CrossRef]
9. Chambliss, S.E.; Preble, C.V.; Caubel, J.J.; Cados, T.; Messier, K.P.; Alvarez, R.A.; LaFranchi, B.; Lunden, M.; Marshall, J.D.; Szpiro, A.A.; et al. Comparison of mobile and fixed-site black carbon measurements for high-resolution urban pollution mapping. *Environ. Sci. Technol.* **2020**, *54*, 7848–7857. [CrossRef]
10. Van den Bossche, J.; Peters, J.; Verwaeren, J.; Botteldooren, D.; Theunis, J.; De Baets, B. Mobile monitoring for mapping spatial variation in urban air quality: Development and validation of a methodology based on an extensive dataset. *Atmos. Environ.* **2015**, *105*, 148–161. [CrossRef]
11. Wu, H.; Reis, S.; Lin, C.; Beverland, I.J.; Heal, M.R. Identifying drivers for the intra-urban spatial variability of airborne particulate matter components and their interrelationships. *Atmos. Environ.* **2015**, *112*, 306–316. [CrossRef]
12. Kaur, S.; Nieuwenhuijsen, M.J.; Colvile, R.N. Fine particulate matter and carbon monoxide exposure concentrations in urban street transport microenvironments. *Atmos. Environ.* **2007**, *41*, 4781–4810. [CrossRef]
13. Badura, M.; Batog, P.; Drzeniecka-Osiadacz, A.; Modzel, P. Evaluation of low-cost sensors for ambient $PM_{2.5}$ monitoring. *J. Sens.* **2018**, *2018*, 5096540. [CrossRef]
14. Kelly, K.E.; Whitaker, J.; Petty, A.; Widmer, C.; Dybwad, A.; Sleeth, D.; Martin, R.; Butterfield, A. Ambient and laboratory evaluation of a low-cost particulate matter sensor. *Environ. Pollut.* **2017**, *221*, 491–500. [CrossRef]
15. Malings, C.; Tanzer, R.; Aliaksei, H.; Saha, P.K.; Robinson, A.L.; Presto, A.A. Fine particle mass monitoring with low-cost sensors: Corrections and long-term performance evaluation. *Aerosol Sci. Technol.* **2020**, *54*, 160–174. [CrossRef]
16. Wang, Y.; Li, J.; Jing, H.; Zhang, Q.; Jiang, J.; Biswas, P. Laboratory evaluation and calibration of three low-cost particle sensors for particulate matter measurement. *Aerosol Sci. Technol.* **2015**, *49*, 1063–1077. [CrossRef]
17. South Coast AQMD. Air Quality Sensor Performance Evaluation Center. Available online: http://www.aqmd.gov/aq-spec (accessed on 1 February 2024).
18. Williams, R.; Duvall, R.; Kilaru, V.; Hagler, G.; Hassinger, L.; Benedict, K.; Rice, J.; Kaufman, A.; Judge, R.; Pierce, G.; et al. Deliberating performance targets workshop: Potential paths for emerging $PM_{2.5}$ and O_3 air sensor progress. *Atmos. Environ. X* **2019**, *2*, 100031. [CrossRef] [PubMed]
19. Duvall, R.M.; Clements, A.L.; Hagler, G.; Kamal, A.; Kilaru, V.; Goodman, L.; Frederick, S.; Barkjohn, K.K.; VonWald, I.; Greene, D.; et al. *Performance Testing Protocols, Metrics, and Target Values for Fine Particulate Matter Sensors*; EPA/600/R-20/280; US Environmental Protection Agency: Research Triangle Park, NC, USA, 2021; p. 27711.
20. US Environmental Protection Agency. Air Sensor Toolbox. Available online: https://www.epa.gov/air-sensor-toolbox (accessed on 1 February 2024).
21. Clements, A.; Duvall, R.; Greene, D.; Dye, T. *The Enhanced Air Sensor Guidebook*; EPA/600/R-22/213; US Environmental Protection Agency: Research Triangle Park, NC, USA, 2022; p. 27711.
22. US Environmental Protection Agency. EPA Research Improves Air Quality Information for the Public on the AirNow Fire and Smoke Map. Available online: https://www.epa.gov/sciencematters/epa-research-improves-air-quality-information-public-airnow-fire-and-smoke-map (accessed on 1 February 2024).
23. Pascal, M.; Wagner, V.; Alari, A.; Corso, C.; Le Tertre, A. Extreme heat and acute air pollution episodes: A need for joint public health warnings? *Atmos. Environ.* **2021**, *249*, 118249. [CrossRef]
24. Burkart, K.; Canário, P.; Breitner, S.; Schneider, A.; Scherber, K.; Andrade, H.; Alcoforado, M.J.; Endlicher, W. Interactive short-term effects of equivalent temperature and air pollution on human mortality in Berlin and Lisbon. *Environ. Pollut.* **2013**, *183*, 54–63. [CrossRef]
25. Chen, K.; Wolf, K.; Breitner, S.; Gasparrini, A.; Stafoggia, M.; Samoli, E.; Andersen, Z.J.; Bero-Bedada, G.; Bellander, T.; Hennig, F.; et al. Two-way effect modifications of air pollution and air temperature on total natural and cardiovascular mortality in eight European urban Areas. *Environ. Int.* **2018**, *116*, 186–196. [CrossRef]

26. Rahman, M.M.; McConnell, R.; Schlaerth, H.; Ko, J.; Silva, S.; Lurmann, F.W.; Palinkas, L.; Johnston, J.; Hurlburt, M.; Yin, H.; et al. The effects of coexposure to extremes of heat and particulate air pollution on mortality in California. *Am. J. Respir. Crit. Care Med.* **2022**, *206*, 1117–1127. [CrossRef] [PubMed]
27. Anenberg, S.C.; Haines, S.; Wang, E.; Nassikas, N.; Kinney, P.L. Synergistic health effects of air pollution, temperature, and pollen exposure: A systematic review of epidemiological evidence. *Environ. Health* **2020**, *19*, 130. [CrossRef]
28. Breitner-Busch, S.; Mücke, H.-G.; Schneider, A.; Hertig, E. Impact of climate change on non-communicable diseases due to increased ambient air pollution. *J. Health Monit.* **2023**, *8*, 103–121. [CrossRef] [PubMed]
29. Mika, J.; Forgo, P.; Lakatos, L.; Olah, A.B.; Rapi, S.; Utasi, Z. Impact of 1.5K global warming on urban air pollution and heat island with outlook on human health effects. *Curr. Opin. Environ. Sustain.* **2018**, *30*, 151–159. [CrossRef]
30. Fever, S.K.; Kahl, J.D.W.; Kalkbrenner, A.E.; Bretón, R.M.C.; Bretón, J.G.C. A new combined air quality and heat index in relation to mortality in Monterrey, Mexico. *Int. J. Environ. Res. Public Health* **2022**, *19*, 3299. [CrossRef] [PubMed]
31. Steeneveld, G.-J.; Klompmaker, J.O.; Groen, R.J.A.; Holtslag, A.A.M. An urban climate assessment and management tool for combined heat and air quality judgements at neighbourhood scales. *Resour., Conserv. Recycl.* **2018**, *132*, 204–217. [CrossRef]
32. Sabrin, S.; Karimi, M.; Nazari, R. Developing vulnerability index to quantify urban heat islands effects coupled with air pollution: A case study of Camden, NJ. *Int. J. Geo-Inf.* **2020**, *9*, 349. [CrossRef]
33. Ardon-Dryer, K.; Dryer, Y.; Williams, J.N.; Moghimi, N. Measurements of $PM_{2.5}$ with PurpleAir under atmospheric conditions. *Atmos. Meas. Technol.* **2020**, *13*, 5441–5458. [CrossRef]
34. Barkjohn, K.K.; Holder, A.L.; Frederick, S.G.; Clements, A.L. Correction and accuracy of PurpleAir $PM_{2.5}$ measurements for extreme wildfire smoke. *Sensors* **2022**, *22*, 9669. [CrossRef]
35. Koehler, K.; Wilks, M.; Green, T.; Rule, A.M.; Zamora, M.L.; Buehler, C.; Datta, A.; Gentner, D.R.; Putcha, N.; Hansel, N.N.; et al. Evaluation of calibration approaches for indoor deployments of PurpleAir monitors. *Atmos. Environ.* **2023**, *310*, 119944. [CrossRef]
36. Robinson, D.L.; Goodman, N.; Vardoulakis, S. Five years of accurate $PM_{2.5}$ measurements demonstrate the value of low-cost PurpleAir monitors in areas affected by woodsmoke. *Int. J. Environ. Res. Public Health* **2023**, *20*, 7127. [CrossRef] [PubMed]
37. Tryner, J.; L'Orange, C.; Mehaffy, J.; Miller-Lionberg, D.; Hofstetter, J.C.; Wilson, A.; Volckens, J. Laboratory evaluation of low-cost PurpleAir PM monitors and in-field correction using co-located portable filter samplers. *Atmos. Environ.* **2020**, *220*, 117067. [CrossRef]
38. Holder, A.L.; Mebust, A.K.; Maghran, L.A.; McGown, M.R.; Stewart, K.E.; Vallano, D.M.; Elleman, R.A.; Baker, K.R. Field evaluation of low-cost particulate matter sensors for measuring wildfire smoke. *Sensors* **2020**, *20*, 4796. [CrossRef] [PubMed]
39. Stavroulas, I.; Grivas, G.; Michalopoulos, P.; Liakakou, E.; Bougiatioti, A.; Kalkavouras, P.; Fameli, K.M.; Hatzianastassiou, N.; Mihalopoulos, N.; Gerasopoulos, E. Field evaluation of low-cost PM sensors (PurpleAir PA-II) under variable urban air quality conditions in Greece. *Atmosphere* **2020**, *11*, 926. [CrossRef]
40. PurpleAir. PurpleAir Sensors Functional Overview. Available online: https://community.purpleair.com/t/purpleair-sensors-functional-overview/150 (accessed on 1 February 2024).
41. Wallace, L.; Bi, J.; Ott, W.R.; Sarnat, J.; Liu, Y. Calibration of low-cost PurpleAir outdoor monitors using an improved method of calculating $PM_{2.5}$. *Atmos. Environ.* **2021**, *256*, 118432. [CrossRef]
42. Magi, B.I.; Cupini, C.; Francis, J.; Green, M.; Hauser, C. Evaluation of $PM_{2.5}$ measured in an urban setting using a low-cost optical particle counter and a federal equivalent method beta attenuation monitor. *Aerosol Sci. Technol.* **2020**, *54*, 147–159. [CrossRef]
43. North Carolina State University. Environment & Climate Observing Network. Available online: https://econet.climate.ncsu.edu/ (accessed on 1 February 2024).
44. Saia, S.M.; Heuser, S.P.; Neill, M.D.; LaForce IV, W.A.; McGuire, J.A.; Dello, K.D. A technical overview of the North Carolina ECONet. *J. Atmos. Ocean Technol.* **2023**, *40*, 701–717. [CrossRef]
45. MetPy. User Guide. Available online: https://unidata.github.io/MetPy/v1.5/userguide/index.html (accessed on 1 February 2024).

atmosphere

MDPI

Article

Relevance and Reliability of Outdoor SO_2 Monitoring in Low-Income Countries Using Low-Cost Sensors

Rosa Amalia González Rivero [1], Olivier Schalm [2], Arianna Alvarez Cruz [1], Erik Hernández Rodríguez [3], Mayra C. Morales Pérez [1], Daniellys Alejo Sánchez [1,*], Alain Martinez Laguardia [3], Werner Jacobs [2] and Luis Hernández Santana [3]

1 Faculty of Chemistry, Universidad Central "Marta Abreu" de Las Villas, Road to Camajuaní Km 5.5, Santa Clara 54830, Villa Clara, Cuba
2 Antwerp Maritime Academy, Noordkasteel Oost 6, 2030 Antwerpen, Belgium
3 Faculty of Electrical Engineering, Universidad Central "Marta Abreu" de Las Villas, Road to Camajuaní Km 5.5, Santa Clara 54830, Villa Clara, Cuba
* Correspondence: daniellysas@uclv.edu.cu

Abstract: In the Western world, the SO_2 concentration in ambient air dropped to low levels, but some emission sources (e.g., merchant ships) and some regions (e.g., low-income countries) still emit substantial amounts of SO_2. At those locations, SO_2 monitoring is critical. However, low-income countries do not have much access to expensive reference instruments. Low-cost gas sensors might be an alternative, but it is unclear how reliable such measurements are. To evaluate the performance of the low-cost alternative, the same SO_2 gas sensor has been subjected to three different calibration methods: (1) low-cost calibration performed in the tropical climate of Cuba; (2) high-end calibration performed in Belgium; (3) a field calibration at an air quality measuring station in Belgium. The first two methods showed similar trends, suggesting that the gas sensor can be calibrated with a low-cost method. The field calibration was hampered by the low SO_2 concentrations. For the monitoring campaign in Cienfuegos, Cuba, the low-cost SO_2 sensor calibrated by the low-cost method appeared to be sufficiently reliable. The reliability of the sensor increases with the increase in SO_2 concentration, so it can be used in Cuba instead of Belgium.

Keywords: pollutants; SO_2; low-cost sensors; low-cost calibration; reliability; Cuba

1. Introduction

Belgium was the first country in continental Europe to follow the UK into the industrial revolution. Heavy industries such as iron foundries or glass production settled in the proximity of rivers and coal mines in the south of Belgium during the 19th century. However, this economic success also had a downside [1–4]. Air pollution and acidifying deposition were suspected to damage vegetation (e.g., oaks, fruit trees, crops) and cause nuisance by the mid-19th century. However, the relationship between air pollution and mortality was only confirmed after the deadly fog in the Meuse Valley on 1–5 December 1930 [5,6]. During the Meuse Valley incident, hundreds of people endured pulmonary attacks, and 63 people suddenly died in Engis, Belgium. This disaster triggered multiple scientific studies and led to the first scientific proof that pollution could cause deaths and diseases. The source of harm appeared to be the combustion of coal. Since fossil fuels such as coal or crude oil contain 1–2% sulphur by weight, they emit smoke and SO_2 when burned. People exposed to the fog inhaled sulphuric acid. Similar disasters occurred at other locations, with the most famous being the Great Smog in London on 5–9 December 1952 [7–10]. At that time, the average SO_2 concentration in London was around 70 ppb, with a peak at 1339 ppb on December 8 [11]. Since then, legislation, abatement strategies, and emission control measures have been implemented [12]. After 1970, global SO_2 emission started to drop in Europe [13]. In Europe and North America, the SO_2 concentrations became so low that acid rain, which caused, among other things, accelerated dissolution of limestone in historical statues and monuments [14], is not a major issue anymore. Today, ambient SO_2 concentrations in Belgium are in the range of 1–2 ppb, with some peaks up to 20 ppb [15,16]. These concentrations are close to the detection limit of low-cost SO_2 gas sensors, and it is unclear if such sensors collect reliable data in such regions. With 'reliable,' we mean some useful information can be extracted from sensor readings to evaluate air quality while operating in a real-world application.

Nevertheless, SO_2 emissions are still significant in certain sectors, such as merchant shipping, where heavy fuel containing sulphur (<0.5%) is used as fuel outside the SECA areas (i.e., regions at sea where sulphur content in fuel must be lower than 0.1%). SO_2 emissions also occur in rapidly growing regions such as East and South Asia. This is also a problem in low-income countries, where the urgency to remove sulphur from crude oil is lower than other issues. For example, Cuba produces mainly heavy crude oil, which contains 7 to 8% sulphur by weight [17–19]. This feature leads to increased SO_2 emissions during combustion. In addition, Cuban oil exhibits several distinct features, including a high asphaltene content, a low API gravity (i.e., American Petroleum Institute gravity), and elevated concentrations of heavy metals such as nickel and vanadium. These metals are known to contribute to the generation of particulate matter (PM) during combustion. According to data from the Central Bank of Cuba, fuel oil production in 2020 was around 3.8 million tons, which represented 41% of the national demand [20]. The rest of the demand was imported from Venezuela, Russia, or Algeria. In addition to these countries, Cuba has also imported fuel oil from other countries such as Mexico, Angola, Brazil, China, and Nigeria, although to a lesser extent. The derivatives, such as diesel oil or fuel oil, have higher fossil-fuel emission factors for SO_2 than the desulfurized equivalents used in the Western world. In addition, the use of old technologies in industry and transportation leads to higher fuel consumption levels. At the same time, the treatment of atmospheric emissions is insufficient. Despite the fact that Cuba's economic situation makes it impossible to replace existing technologies with more efficient and less polluting alternatives in the short term, there is a need to control SO_2. Because of the large number of diesel engines that act as SO_2 sources, there is a particular need for low-cost control systems.

Recently conducted net emission inventories in Cuba showed that the following pollutants can be ranked in decreasing importance: $SO_2 > NO_2 > CO$ [21]. This suggests the importance of SO_2 as a pollutant in Cuba. A Gaussian dispersion modeling system for pollutants suggests that the maximum concentration of SO_2 in rural environments is 30 ppb [22]. Another study performed in Santa Clara City, Cuba reported an average concentration of 4 ppb for the period February to April 2010 [23]. In Havana, a study about air pollution due to vehicular traffic revealed SO_2 average concentrations between 5 and 15 ppb, with peaks of up to 60 ppb [24]. These studies suggest that the average SO_2 concentration in the air is higher in Cuba than in Belgium.

It is known that the reliability of low-cost sensors is lower in comparison with high-end and expensive trace-level UV fluorescence analysers that are considered the gold standard [25–29]. Therefore, one must know the most important calibration problems (e.g., no access to high-end equipment to perform calibrations, the effect of a tropical climate, etc.), the impact of varying environmental parameters, and the interference of other pollutants on the measurements to ensure the best possible reliability of the collected data [30,31]. This contribution will explore the reliability of low-cost SO_2 monitoring. It will be illustrated with a measuring campaign performed in Antwerp, Belgium, on the roof of an air quality monitoring station analysing outdoor air, and a measuring campaign performed in Cienfuegos, Cuba.

2. Background

Electrochemical 4-electrode gas sensors are widely used in air quality monitoring and toxic gas detection applications due to their relative high sensitivity and selectivity [32–35]. A well-known example of such low-cost gas sensors are the ones of the manufacturer

Alphasense. They are provided with calibration information. However, the accurate calculation of pollutant concentrations remains a challenge [36,37]. According to the manufacturer, the sensors register the voltages corresponding to the working electrode (WE) and auxiliary electrode (AE). The WE provides information about the target analyte, while the AE is designed to mimic the WE in the absence of the target gas while considering the response to changes in meteorological conditions such as temperature and relative humidity. Despite the apparent simplicity of the formulas to calculate the concentration from WE and AE, the information about Alphasense is confusing. This leads to different interpretations of the formulas in the literature. Some assume a correction factor n_T equal to 1 [38–40]. Others consider n_T a constant provided in tabular or in graphic form by Alphasense [41,42]. For the Alphasense SO2-A4 sensor, the manufacturer suggests for n_T a value of 0.4 in one document [43] and 1.15 in another document [44]. It is not clear whether these variations are due to a change in the construction of the electrochemical cell. In the 2014 Alphasense Application Note AAN 803-01, "Correcting for background currents in four electrode toxic gas sensors", n_T is defined as the ratio WE_0/AE_0 for zero air at the temperature at which the measurement is performed. This definition suggests a variable ratio. The scientific literature has demonstrated that corrections of WE and AE recommended by the manufacturer do not always lead to pollutant concentration values with acceptable precision [45]. There are several factors that can affect the accuracy of gas concentration measurements, such as cross-sensitivity, drift, calibration, and other environmental factors that can interfere with the sensor signal.

The voltage values of both electrodes for air without pollutants (WE_0 and AE_0) can be obtained from a calibration process using zero air. The ratio allows the conversion of the measured AE-value into the corresponding background baseline $WE_{background}$. The background baseline that needs to be subtracted from WE can be calculated with Equation (1). If one assumes n_T as a variable ratio and the measured WE as the sum of the background base line (i.e., $WE_{background}$) and a contribution that is concentration-dependent (i.e., WE_{gas}), then the formula of Alphasense can be obtained (see Equations (1)–(3)). Consequently, the sensor signal (WE_{gas}), expressed in mV, is directly proportional to the gas concentration.

$$n_T = \frac{WE_0}{AE_0} = \frac{WE_{background}}{AE} \Leftrightarrow WE_{background} = \frac{WE_0}{AE_0} AE, \tag{1}$$

$$WE = WE_{background} + WE_{gas} \Leftrightarrow WE_{gas} = WE - \frac{WE_0}{AE_0} AE, \tag{2}$$

$$WE_{gas} = WE - \frac{WE_0}{AE_0} AE + [n_T AE_0 - WE_0] = (WE - WE_0) - n_T(AE - AE_0). \tag{3}$$

Once the concentration-dependent signal WE_{gas} is calculated, the gas pollutant concentration expressed in ppb is calculated using Equation (4). The gas sensitivity (S_{gas}) is specific to each sensor, and its value is provided by the manufacturer. This means that the calculation of the concentration is a two-step process: (1) correct for the background baseline $WE_{background}$; (2) convert WE_{gas} into the corresponding concentration.

$$c_{gas} = \frac{WE_{gas}}{S_{gas}}. \tag{4}$$

It is usually assumed that n_T of the calibration process is identical to n_T of the measuring campaign. However, the contexts of both measurements are not necessarily the same, and this assumption may lead to an incorrect estimation of the background baseline $WE_{background}$. If there is no in situ zero air calibration available, then WE_0 and AE_0 can be estimated from the collected WE and AE time series. The moment t_{min} is characterized by the lowest WE-value in the time series, which is denoted by $WE_{min}(t = t_{min})$. The corresponding AE-value is given by $AE(t = t_{min})$. That moment is characterized by the minimum concentration, C_{min}. If the calibration curve is known, then C_{min} can be used to calculate the corresponding $WE_{gas}(t_{min})$. There is often sufficient knowledge available about the measuring location to estimate C_{min}. For example, there will be a moment in the measuring campaign of SO_2 in outdoor air in Belgium when it will reach zero concentration. When AE_0 is not recorded at the same moment as WE_0, the average of AE_0 in the time series can, in principle, be used to calculate the sensor signal.

We know that:

$$WE_{gas}(t_{min}) = WE_{min}(t_{min}) - WE_{background}(t_{min}) \tag{5}$$

therefore,

$$WE_{background}(t_{min}) = WE_{min}(t_{min}) - WE_{gas}(t_{min}) \tag{6}$$

$WE_{gas}(t_{min})$ can be expressed by the following formula if we clear it in Equation (4).

$$WE_{gas}(t_{min}) = S_{gas}C_{min}(t_{min}) \tag{7}$$

When $WE_{gas}(t_{min})$ is substituted into Equation (6), the following expression is obtained:

$$WE_{background}(t_{min}) = WE_{min}(t_{min}) - S_{gas}C_{min}(t_{min}) \tag{8}$$

In addition, we know that:

$$WE_{background}(t_{min}) = n_T AE(t_{min}) \tag{9}$$

therefore,

$$n_T = \frac{WE_{background}(t_{min})}{AE(t_{min})} \tag{10}$$

$WE_{background}$ in Equation (10) can be substituted by Equation (9) (see Equation (11)). When $C_{min} = 0$, Equation (11) can be further simplified, showing that n_T is the ratio WE_0/AE_0 for zero air.

$$n_T = \frac{WE_{min}(t_{min}) - S_{gas}C_{min}(t_{min})}{AE(t_{min})} \tag{11}$$

3. Materials and Methods

3.1. Low-Cost Monitoring System

The principles of low-cost monitoring systems have already been described in previous work [46,47]. Here, only the principle of the gas sensor measurements will be given in more detail. The SO_2 sensor type A belongs to one of the gas sensor families from the manufacturer Alphasense. It is inserted in an analog conditioning board (analog front end), designed by Alphasense to accommodate 4 different sensors (part number 810-0023-00). The configuration used includes sensors for NO_2, OX (O_3 + NO_2), CO, and SO_2. The analog front end (AFE) has an internal power supply of 3.3 V and only requires an input voltage between 3.5 and 6.5 V (DC). The conditioning board provides 8 analog outputs (2 for each sensor; more precisely, the "working electrode" and the "auxiliary electrode") that must be wired to analog-to-digital converters (ADC). The sensitivity of the SO_2 sensor is 0.295 mV/ppb. The low-cost monitoring system that was used in the experiments consisted of an Arduino MEGA2560 as the computer unit and interface board, a real-time clock, and an SD card for data storage. The Arduino Mega 2560 contains an ADC with a 10-bit resolution. This ADC, with a reference voltage of 5 V, is capable of detecting voltage variations of about 4 mV. This means that it cannot detect changes of 1 ppb. To resolve this quantization problem, a pair of 4-channel analog-to-digital converters with 16-bit resolution (ADS 1115) have been added to the system. This resolution allows the detection of voltage changes of 0.076 mV, which is sufficient to detect 1 ppb concentration variations. The ADC has 4 analog inputs, hence the need to use 2 converters. The ADS 1115 module is connected to the Arduino Mega through the I2C communication protocol.

3.2. Sensor Calibration

The calibration curve describes the relationship between the sensor signal and the reference measurements. Since that relationship must transform the sensor measurements into concentrations, the sensor signal is considered the independent variable. The same set of A-series Alphasense gas sensors for NO_2, O_3 (i.e., OX sensor), CO, and SO_2 has been subjected to 3 different types of calibration experiments. The experiments are summarized in Table 1 and include the most common calibration methods listed in the literature [48–52]. This work only focuses on the calibration of the SO_2 sensor. Calibration method 1 is a low-cost calibration method performed in Cuba using low-cost electronics purchased via the internet, medical disposables, and common laboratory equipment (e.g., balance, pipettes). This calibration method can be performed with a limited budget in low- or mid-income countries. The calibration setup needed is described elsewhere [53]. In the high-end calibration process number 2, all sensors have been exposed to the 4 different gases during separate calibration experiments. This makes it possible to evaluate the impact of NO_2, O_3, and CO gases on the signal of the SO_2 sensor, that is the cross-interference. This calibration method allows the sensor response to be compared with a reference analyzer. Sensor calibration methods under laboratory conditions are relatively fast. The concentration range of the pollutant and the temperature and relative humidity conditions can also be selected to perform the calibration. However, in these methods, the sensor is not exposed to real-field conditions. An advantage of field calibration is that data provided by low-cost sensors can be compared with official data reported by monitoring stations. This method is considered the most efficient because it allows the correction of meteorological and other pollutant fluctuations in the field environment by complex mathematical models. However, it is applicable only to the site where the calibration was performed. Another advantage of calibration method number 3 is that n_T for the outdoor calibration in ambient conditions must be fairly similar to the measuring location at a different outdoor location in the vicinity. The Alphasense formula with local values for WE_0 and AE_0 was used as well. A comparison of the different calibration processes makes it possible to evaluate the reliability of the low-cost calibration. Table 1 and the list below describe calibration experiments in detail.

Table 1. Overview of the calibration experiments performed on the same set of gas sensors, the period in which that calibration was performed, the average temperature, and relative humidity (RH) during the calibration process.

Nr.	Calibration Method	Location	T and RH Range	Calibration Period
1	Low-cost calibration of the SO_2 sensor by generating SO_2 gas inside a syringe	Santa Clara, Cuba	Tropical climate (23–25 °C and 62–65% RH)	19 November 2021
2	High-end calibration in a climate chamber using NO_2, O_3, CO and SO_2 calibration gases	Mol, Belgium	Moderate marine climate (23 °C and 47–50% RH)	16–19 May 2022
3	Field calibration where the low-cost monitoring device is installed on the roof of the national outdoor air quality measuring station 42R802. The station is close to a crossroad.	Antwerp, Belgium	Moderate marine climate (10–30 °C, 30–94% RH)	30 May 2022–30 June 2022

- **Low-cost calibration in laboratory conditions:** The low-cost calibration setup consists of a closed plastic box that contains the gas sensor. The air inside the calibration box is first cleaned by pumping the air through a solution saturated with $Ca(OH)_2$. The cleaned air can be used to determine WE_0 and AE_0. First, SO_2 gas is generated inside a syringe. The gas is generated in a separate setup using the stoichiometric reaction R1 [54–56]. Graduated syringes with scale intervals of 0.2 mL were used.

One syringe contains 0.2 mL of 2.14 mol/L Na_2SO_3 solution. The second syringe contains 0.2 mL of concentrated 95% (*w*/*w*) H_2SO_4. The reagents are brought together to generate 9 mL of SO_2 gas in a third syringe. The acid solutions did not deteriorate the plastic setup and could be used for several calibration experiments. The amount of SO_2 inside the syringe expressed in moles can be calculated from its volume, the pressure of the gas inside the syringe, and the ambient temperature using the ideal gas law. Then, the syringe filled with SO_2 gas is connected to the calibration box using a pressure connection tube. One experiment entails several injections of small but known volumes of SO_2. The corresponding amounts of SO_2 introduced in the zero air inside the box were 110, 220, 330, 440, and 550 ppb. Each step lasted approximately 1 min. The low-cost calibration is performed by determining the relationship between the concentration-dependent sensor signal WE_{gas} and the reference SO_2 concentration injected into the plastic box.

$$Na_2SO_{3(aq)} + H_2SO_{4(l)} \rightarrow SO_{2(g)} + Na_2SO_{4(aq)} + H_2O \quad \text{(R1)}$$

- **High-end calibration using a high-end climate chamber:** The set of NO_2, OX (i.e., the sum of O_3 and NO_2), CO, and SO_2 sensors is subjected to a calibration experiment at VITO, Belgium, which uses certified and traceable calibration gases to ensure the accuracy and reliability of the measurements. More detailed information on the calibration setup can be found elsewhere [53]. The temperature in the laboratory was controlled and kept at around 21 °C. The entire set of gas sensors was exposed to concentration ranges of 0 to 662.71 ppb, 0 to 145 ppb, 0 to 100 ppb, and 0 to 6 ppm of SO_2, NO_2, O_3, and CO, respectively, in separate calibration experiments. The gas sensors have also been submitted to the calibration experiment of the AM2315 sensor for temperature and relative humidity in the ranges of 20 to 50 °C and 5 to 80% RH. The calibration of the SO_2 sensor was performed in 1 experiment with continuous injections of the reference gas. A staircase profile with 12 SO_2 concentration levels was used. The reference concentrations were 27.44; 97.82; 239.18; 389.02; 532.92; 662.71; 540.00; 402.55; 257.70; 115.62; 43.10 and 4.51 ppb. Each step lasted between 3 and 4 h.
- **Field calibration:** The sensor's performance was also evaluated by a field calibration, where the data provided by the low-cost sensor is compared to a reference measuring station that recorded reliable SO_2 concentrations. The calibration is conducted at station 42R801 of the Flemish Environment Society (Vlaamse Milieu Maatschappij VMM). The station generated a time series with a sampling time of 1 h. The low-cost monitoring systems used a sampling time of 2 min. With the function VLookUp in Microsoft Excel, the sensor data could be resampled so that values are obtained at the same timestamps as the measurements in the reference time series. The resampling allowed the integration of the two-time series into a single database. In this way, the relationship between the SO_2 sensor signal and the reference SO_2 concentration could be determined. Some authors consider that the duration of the field calibration is intrinsically related to the calibration model [57]. An optimal field calibration should cover all environmental conditions; therefore, it should be performed during different seasons and use complex mathematical models. In this study, the results obtained will be compared with calibration methods under laboratory conditions. For this reason, the simplest model (a linear regression model) and a shorter period during the summer were used. This season has the highest similarity to the Cuban climate. The field calibration in Antwerp, Belgium lasted 1 month. The visualization of the data recorded during the field calibration was generated with software developed by [58] in Python 3.

From the measured WE and AE data of the SO_2 gas sensor, the impact of how the concentration-dependent signal (WE_{gas}) is calculated on the calibration results will be discussed. This analysis was performed on calibration methods 1 and 2 from Table 1. The curves were compared with the Alphasense calibration. Alphasense calibration refers to the calculation of gas concentrations using Equations (3) and (4). A simple method is to calculate the difference between the two electrodes (WE–AE). The sensor signal was also calculated by subtracting WE_0 from WE and using Equations (3) and (12). Equation (12) is similar to Equation (3), but the parameter n_T is not considered.

$$WE_{gas} = (WE - WE_0) - (AE - AE_0) \tag{12}$$

3.3. Reliability Testing

The reliability of an electrochemical gas sensor can be assessed through a number of indicators. Several reliability indicators are defined in the list below. They give an insight into the occurrence of calibration or measuring errors. By analyzing these indicators, an insight into the reliability of the measured concentration during the field campaign in Cienfuegos is obtained.

- The coefficient of determination of the linear regression between the sensor signal and the reference concentration is high, suggesting the absence of random errors during the calibration experiment.
- The stability of the SO_2 concentration inside the calibration box should be sufficiently long to perform a stable measurement of the signal at a given concentration.
- There is an agreement between the different types of calibration processes performed on the same sensor.
- The monitoring campaign can be considered reliable when the AE-signal remains constant over time or only gently fluctuates around its average;
- It is possible to remove the background baseline from the measured WE signal and obtain WE_{gas}.
- The detection and quantification limits of the gas sensor are sufficiently low so that the target gas is able to generate a reliable signal.
- The scientific literature gives evidence of the effect of cross-sensitivity in electrochemical sensors [59,60]. The sensor is reliable when the concentration of interfering gases is lower than the signal generated by the target gas or when a correction can be applied.
- Drift in the calibration constants of low-cost sensors can be caused by several factors, such as environmental conditions, lifetime, chemical decomposition, contamination, electromagnetic interference, and others. Measurements will be reliable when the drift in the calibration constants within the period of the calibration experiment and measuring campaign is sufficiently low. Sufficiently low means that the error introduced by drift is not significantly higher than other sources of error.
- The scientific literature demonstrates the influence of temperature and relative humidity on electrochemical sensor measurements [61,62]. Measurements will be sufficiently reliable when the interference of these variables on the measurement of the target gas is sufficiently low (i.e., the error is not significantly higher than other sources of error).

3.4. Measuring Campaign

A measuring campaign has been performed in Cienfuegos, Cuba, from 14 March to 22 April 2022, with a sampling time of 2 min. The urban historic centre of Cienfuegos is a part of UNESCO World Heritage, but Cienfuegos Bay is also an important harbour and industrial centre. The measuring location is situated 9 km south of the city (22°03′55″ N, 80°28′58″ W), more specifically between the Caribbean Sea and Cienfuegos Bay (see Figure 1). The monitoring system has been placed outside at the Centro de Estudios Ambientales (CEAC) close to a window at a height of 2 m above ground level, according to the prescriptions of the Cuban standard NC 111:2014 [63]. The most important stationary pollution sources in the vicinity are the emissions from the Carlos Manuel de Céspedes thermoelectric plant, which uses national crude or fuel oil, and the Camilo Cienfuegos oil refinery. Examples of pollution sources closer to the measuring location are a small diesel power unit and a diesel combustion boiler. In addition, close to the sampling area, wood is burned for charcoal production. Figure 1 shows the location of the low-cost monitoring system and the most important pollution sources in its vicinity. The measuring campaign resulted in a data matrix with measurements arranged in 27,852 rows and 9 parameters arranged in columns. The data matrix contains a column for the WE-value of the SO_2 sensor and another column for the corresponding AE-value. The visualization of the data recorded during the measuring campaign was also generated with the visualization software for environmental data.

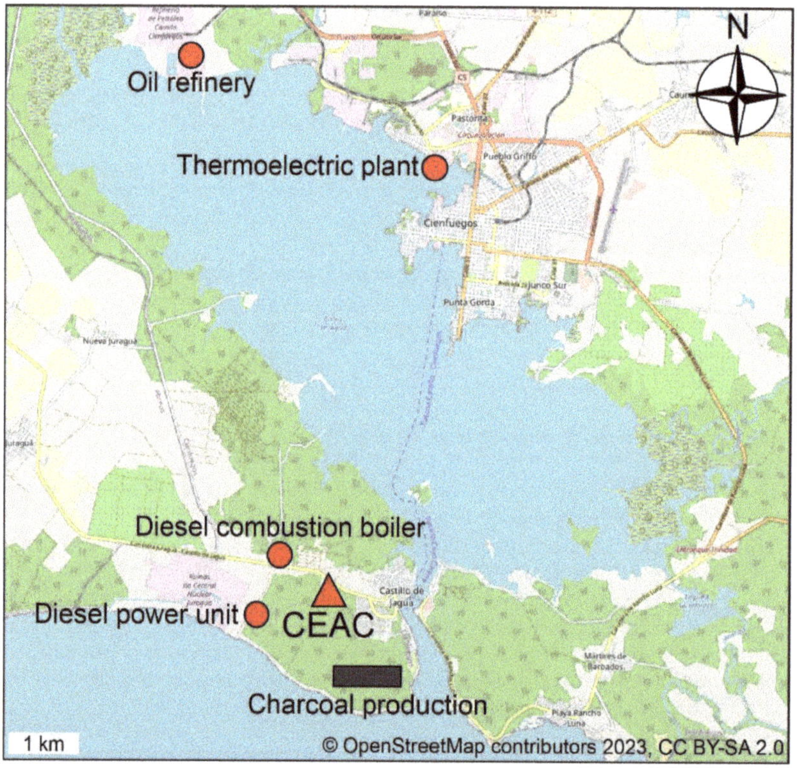

Figure 1. Map showing the sampling location in Cienfuegos, Cuba and the most important pollution sources in its vicinity.

4. Results

4.1. Sensor Calibration

The behaviour of the same SO_2 sensor subjected to different calibration experiments is shown in Figure 2. Each experiment generates raw signals for the WE and AE electrodes. This information is shown in Figure 2a,c,e. From the sensor data, the average concentration dependent signal, WE_{gas} must be calculated. The calculated signal is shown in the X-axes of Figure 2b,d,f where the relationships with the reference concentrations are described by linear regressions. The calibrations performed in laboratory conditions (Figure 2b,d) use elevated SO_2 concentrations and result in linear trends with high coefficients of determination (>0.99). These coefficients of determination can be used as reliability indicators. For the field calibration performed in Belgium, the SO_2 concentration range is smaller, and a fraction of that range falls below the detection limit of the low-cost sensor. The smallest detectable concentration must result in a signal that is higher than the random fluctuations around the average value of WE_0, $<WE_0> + 3s_{WE0}$, where $<WE_0>$ stands for the average signal at zero concentration and s_{WE0} for the corresponding standard deviation. The minimum signal can be determined from point 12 in Figure 2c and compared with the signal at point 1, which appears to be just above the detection limit (i.e., 6.19 ppb). This means that the data from almost the entire measuring campaign are below the detection limit of the sensor. It explains why the linear trend in the field calibration is not reliable. Besides these general observations, several specific conclusions can be drawn from the calibration experiments. They are described below.

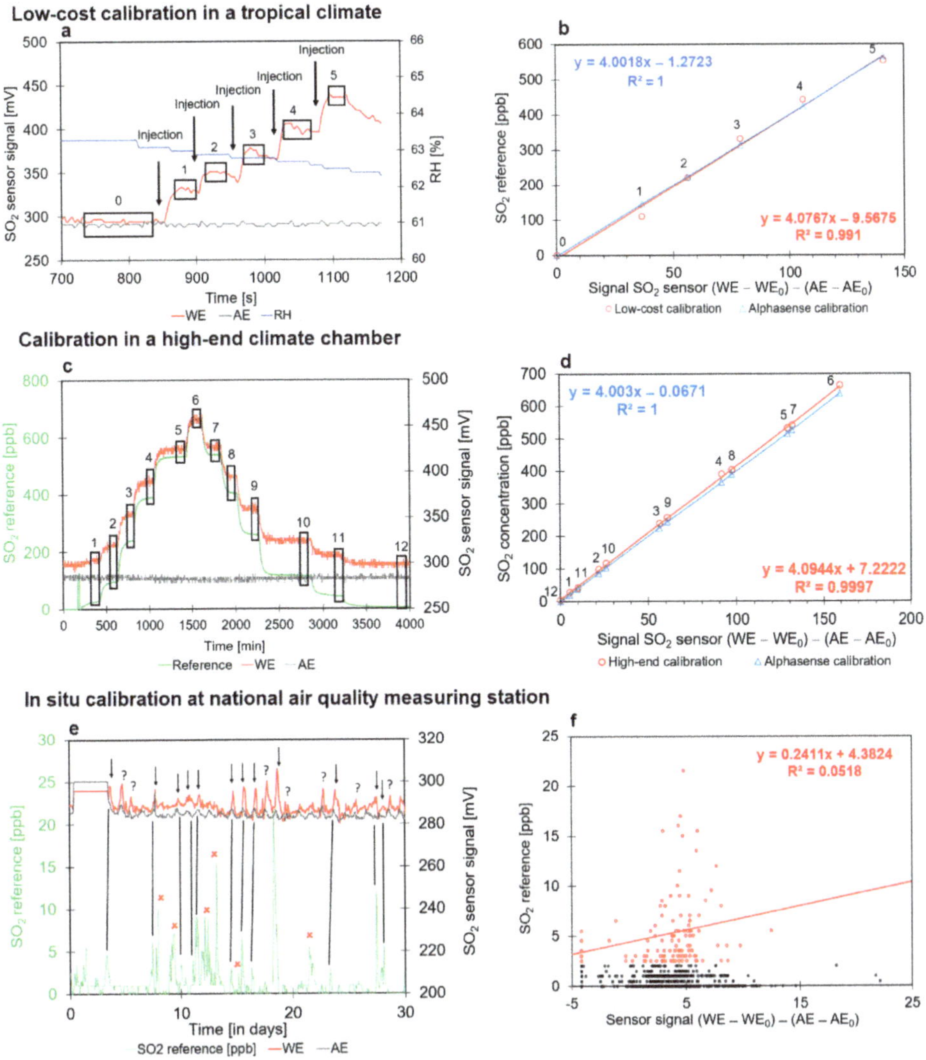

Figure 2. Different calibration processes applied on the same SO_2 sensor; the sensitivity is expressed in ppb/mV and the intercept in ppb. (**a**,**b**) Low-cost calibration under controlled conditions in laboratory performed in a tropical climate in Cuba. (**c**,**d**) High-end calibration under controlled conditions in laboratory performed in a moderate marine climate in Belgium. (**e**,**f**) Field calibration at a VMM air quality monitoring station, where the environmental conditions fluctuate in an uncontrolled way.

- **Low-cost calibration in laboratory:** The WE_0 and AE_0 signals of the SO_2 gas sensor in zero air (i.e., point 0 in Figure 2a) are 294.6 mV and 290.8 mV, respectively, with $n_T = 1.01$. The sensor measures the sudden jumps in the SO_2 concentration when the gas is injected in steps into the calibration box. The average values of the numbered rectangles in Figure 2a are used for calibration. A linear trend between sensor signal and SO_2 concentration is observed (Figure 2b). When using the sensor characteristics determined in zero air, the quantification method of Alphasense gives similar results to the low-cost calibration. The slopes for both methods are similar. However, there is a difference between the intercepts.
- **High-end calibration in laboratory:** The WE_0 and AE_0 signals of the same SO_2 sensor as used in the low-cost calibration in the laboratory in zero air (i.e., point 12 in Figure 2c) are 298.6 mV and 284.2 mV, with $n_T = 1.05$. This means that WE_0 and AE_0 are different from the ones at the low-cost calibration. The moments in Figure 2c marked with black rectangles indicate the stable measurements for each point in the calibration process. These moments are located in the second half of the plateaus because the chamber and sensor need some time to reach equilibrium. Figure 2d shows the linear trend between the average sensor signal from the numbered plateaus and the corresponding reference SO_2 concentration. In addition, Figure 2d also shows the calibration with the Alphasense formula (Equations (3) and (4)) using the characteristics determined at zero air in point 12. Both calibrations show a similar slope but a different intercept.
- **Field calibration:** To determine the local WE_0 and AE_0, the minimum value for WE in the time series has been identified, together with the accompanying AE or average AE value (see Equation (5)). The reference concentrations clearly show that C_{min} within the time series reached zero. This gives 279 mV for WE_0 and 285 mV for AE_0, which corresponds to $n_T = 0.979$. Each calibration experiment appears to be characterized by different WE_0 and AE_0 values. This suggests the need for in situ characterization of the sensor properties. During the measuring campaign, the reference monitoring system registered an average concentration of 0.5 ppb, with peaks of up to 20 ppb. In addition, 87% of the measurements are in the range of 0–2 ppb. Some of the peaks observed in the VMM data are also detected by the gas sensor with some time delay. At the same time, the gas sensor also shows SO_2 peaks that are not detected by VMM, and the VMM data contains peaks that are not detected by the gas sensor. Some peaks may be attributed to a cross-sensitivity effect and not to a significant SO_2 signal because, during the monitoring period, the concentrations of interfering gases such as NO_2, O_3, and CO were higher than that of SO_2. This indicates that at locations with low SO_2 concentrations (e.g., Belgium), the low-cost sensor is not a good alternative for monitoring the pollutant concentration.

Figure 3 shows the different methods used to calculate the sensor signal. Although the calibration method of the manufacturer approached the low-cost calibration of Figure 2b and the high-end calibration of Figure 2d, a difference between both linear trends can be seen in Figure 3a when WE-AE is used as a sensor signal. This difference is not caused by calibration errors but rather by an imperfect subtraction of the background baseline in the sensor signal. This error becomes smaller when the signals are processed differently (see Figure 3b–d). In addition, the experimental values of n_T in calibration methods 1, 2, and 3 are close to 1. For this reason and for the sake of simplicity, n_T is considered 1. Thus, Equation (12) was used to calculate the sensor signal and make the calibration curves shown in Figure 2. When local WE_0 and AE_0 values are used and the background contribution is determined as accurately as possible, both calibration processes result in similar calibration curves. It also means that the low-cost calibration performed on another continent can be considered sufficiently good to calibrate the gas sensor.

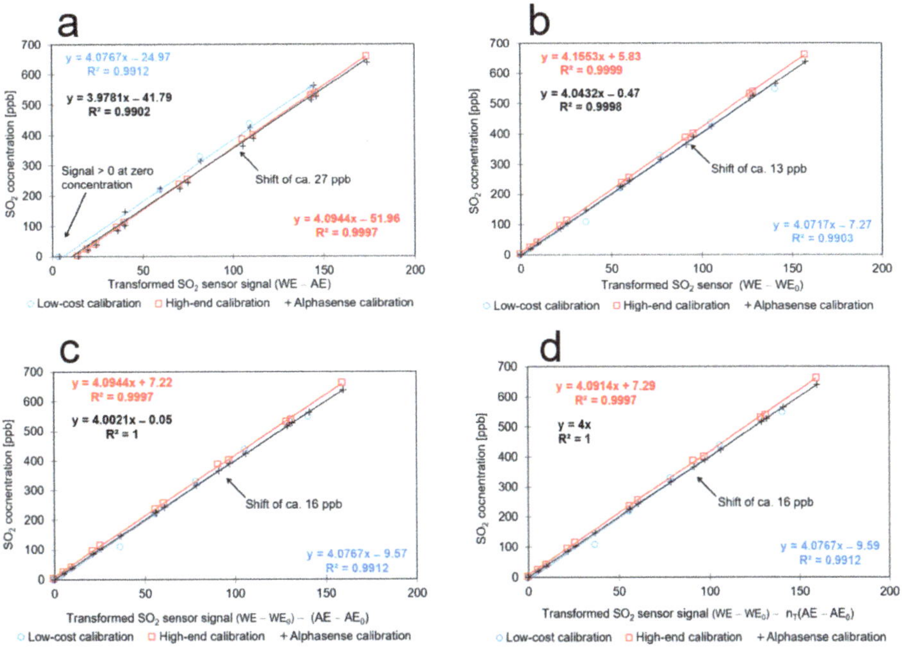

Figure 3. The calibration process of Figure 2a,c, where the concentration-dependent sensor signal has been calculated in different ways in comparison with the Alphasense equation using the appropriate WE_0 and AE_0 for both calibration processes. (**a**) The sensor signal has been calculated using the corresponding voltage values of both electrodes WE-AE; (**b**) The sensor signal has been calculated using the voltage values of the working electrode and the recorded value for zero air; (**c**) The sensor signal has been calculated using the voltage values of the working and auxiliary electrodes and the values recorded for zero air without considering n_T; (**d**) The sensor signal has been calculated using Equation (3) suggested by the manufacturer.

Figure 2 shows that the low-cost calibration is sufficiently reliable under controlled conditions. However, it is unclear if the performance of the sensor is as good in real-life situations. For that reason, several aspects have been studied that give insight into the reliability of SO_2 measurements:

- **Coefficient of determination of the linear regressions:** The low-cost sensor showed a strong linear relationship with the reference concentrations in the calibrations performed under laboratory conditions (0.9912 and 0.9997 for the low-cost and high-end calibrations, respectively). This is an indication that the sensor can be considered reliable. However, the sensor was not reliable in the field calibration, obtaining a coefficient of determination of 0.0518.
- **Agreement between calibration processes:** The calibration curves shown in Figure 2 indicate that the low-cost calibration resulted in similar results as the high-end calibration. The high-end calibration resulted in concentrations that are systematically higher (ca. 17 ppb). However, the slopes are similar. Although the match of both calibration curves is not perfect, this observation gives satisfactory reliability to the low-cost calibration.
- **Stability of the SO_2 concentration inside the calibration box:** The sensor signal did not remain constant in the calibration box. The decreasing trend is most probably not caused by the air exchange rate because it occurs too fast. This behaviour might be attributed to a chemical reaction between SO_2 gas and water vapour. The resulting H_2SO_4 seems to remain invisible to the SO_2 gas sensor so, the concentration drops.

In addition, the RH in the box drops over time and seems to confirm our hypothesis (see Figure 2a). For each step in the low-cost calibration, the zones with the highest concentrations have been used as the most reliable situations.

- **Stability of the signal generated by the AE:** Although the AE signals of the three calibration experiments are different from each other, they appeared to be rather constant during the experiment. However, the AE signal during the field study contained peaks superposed on a slowly decreasing trend. It is not clear if this decreasing trend should be associated with a calibration drift because the background contribution can effectively evolve towards smaller values. It is unclear if this drop is caused by changing values for WE_0 and AE_0. Figure 4a shows the trends of the auxiliary electrode of the SO_2 sensor when the NO_2 (red line), CO (blue line), and O_3 (green line) sensors were exposed to the high-end calibration. This behaviour is shown as a function of relative time in days. The AE trends of the SO_2 sensor show some fluctuations (see Figure 4a). During the NO_2 calibration, a decreasing AE trend for the SO_2 sensor is observed. The AE signal of the SO_2 sensor contains the staircase function of the CO concentration profile, while during the O_3 calibration, the AE signal of the SO_2 sensor appears to be rather constant. Apparently, there is a disturbance of the AE-signal of the SO_2 sensor when exposed to NO_2 and CO.
- **Possibility to remove the background level from the WE signal:** It is of primary importance that $WE_{background}$ can be estimated as accurately as possible. The baseline drift is mainly due to the variation in ambient environmental conditions. For this reason, the values of WE_0 and AE_0 as used for the calibration processes are not valid for the field campaign and need to be determined for the local situation. Once this is done, the calibration performed in laboratory conditions (span calibration) appears to be usable in the field campaign. This enhances the reliability of field campaigns using low-cost gas sensors.
- **Detection and quantification limits:** The lowest concentration that the SO_2 gas sensor can distinguish from zero air is the quantity that generates a signal that is at least equal to the average value of WE_0, $\langle WE_0 \rangle$, plus 3 times the standard deviation of the signal fluctuating around $\langle WE_0 \rangle$ (i.e., limit of detection). The quantification limit is defined as $\langle WE_0 \rangle$ plus 10 times the standard deviation of the signal fluctuating around $\langle WE_0 \rangle$ [64]. This threshold can be determined with the data associated with point 12 in Figure 2c. The detection and quantification limits are 6.19 and 20.63 ppb, respectively. For measuring locations where the average SO_2 concentration is above the detection and quantification limits, the reliability of the measurements is sufficiently high. This was not the case during the field calibration in Belgium.
- **Cross-interference:** The concentrations of atmospheric pollutants CO, NO_2, and O_3 are high in Cuba. Research about the quantification of tropospheric ozone in Cuba showed that in the cold season (from February to April), the concentration of this pollutant is high [65] and exceeds the threshold of 12 ppb as established by the Cuban standard NC 39:1999 [66]. The NO_2 quantifications showed that in the cold season the concentrations are higher, although not exceeding the admissible limit according to the Cuban standard [23]. This suggests that cross-interference can be expected and is probably more important than in Belgium because the interfering gases have higher concentrations than the target gas. During the high-end calibration of low-cost O_3, NO_2, and CO sensors, the SO_2 sensor signal was recorded when exposed to the other gases This signal has been plotted as a function of the reference concentration of the pollutants (Figure 4b). The experiments show a linear response of the SO_2 sensor towards interfering gases. The SO_2 sensor has a negative linear response to O_3 and NO_2, which corresponds to the information provided by Alphasense. It also corresponds with previous research [67,68]. Some report a negative response for CO [61], while this study shows a faint but positive correlation, which is in accordance with the information supplied by the manufacturer. It is likely that the responses vary from sensor to sensor. Thus, large CO-peaks might result in false SO_2 peaks,

while NO_2 and O_3 peaks should result in SO_2 valleys. Assuming the cross sensitivities are independent, the effects of the interfering gases on the SO_2 sensor signal can be corrected by the equations shown in Figure 4b. These results demonstrate the importance of using multi-sensor systems when low-cost sensors are used, as they provide the possibility of obtaining a better estimate of the cross-sensitivity.

- **Drift of the calibration constants:** The behaviour of the sensor during field calibration reflects a common trend with fluctuations (see Figure 2e). The trend might indicate that there is some drift. Since the monitoring campaign only took 1 month, this might be too short to observe obvious drifts. However, this suggests the need for periodic calibrations to assure reliability.
- **Interference of temperature and relative humidity:** As mentioned before, the SO_2 sensor was also subjected to the calibration experiment of the AM2315 sensor (T and RH). During these conditions, the SO_2 concentration was not measured with a reference instrument; however, in Belgium, the concentration fluctuates around 1 ppb (below the detection limit of the sensor). Moreover, the indoor SO_2 concentration is usually lower than the corresponding outdoor concentration. Therefore, any variation in the SO_2 sensor signal must be caused by the changing T and RH conditions in the climate chamber. The behaviour of the AM2315 sensor during the temperature calibration (Figure 4c) and the relative humidity calibration (Figure 4d) has been plotted as a function of time. The secondary axis of both graphs shows how the WE and AE signals of the SO_2 sensor respond to these conditions. Figure 4c shows that in the range of 20 to 30 °C, the sensor signals remain stable. With decreasing relative humidity from 80% down to 40% for a temperature of 20–30 °C, both WE and AE remained stable. At higher temperatures (i.e., 50 °C), the WE signal suddenly increases while AE decreases. In the range from 40% down to 5% at a temperature of 50 °C, the WE sensor signal shows a sudden increase that might be confused with a gas-specific signal. At this moment, the AE decreases but becomes stable afterwards. After exposure to 50 °C, the WE and AE are less stable in comparison to the range from 20 to 30 °C. This means that the sensor signal appears to have remained constant when the changing conditions remained in a relative cold (<30 °C) and humid (>40%) state. However, in warmer (>30 °C) but dryer (<40%) conditions, the sensor becomes very sensitive to small changes in the environmental conditions: the lower the RH, the higher the WE signal. This suggests that the sensor shows similar behaviour in moderate and tropical climates most of the time.

Figure 5 shows the gas and particulate matter concentrations recorded by the VMM station and the signal from the working and auxiliary electrodes of the low-cost SO_2 sensor during the field calibration in Antwerp, Belgium. The highest peaks recorded by the sensor are shown in this figure. The signal generated by the auxiliary electrode shows instability over time. Its trend is related to temperature peaks and relative humidity valleys. On some occasions, the high reference concentrations of SO_2 do not coincide with the sensor peaks (see more detail in Figure 2e). This is probably due to the low detection limit of the sensor. At other moments, when the peaks of the SO_2 reference are not recorded, sensor peaks are observed. These sensor peaks are related to temperature peaks and relative humidity valleys but also to peaks of O_3 (see the red rectangles in Figure 5). Although the NO_2 concentration is higher than the SO_2, no strong relationship between the recorded peaks was observed. Only one peak of CO is simultaneously related to the peaks of NO_2 and SO_2 (see the black rectangle in Figure 5). At that moment, the VMM station also recorded a small peak of SO_2. The simultaneous peaks suggest that the pollutants are generated by the same pollution sources [68]. The occurrence of simultaneous peaks for different pollutants and the SO_2 peak caused by interference is sometimes hard to distinguish.

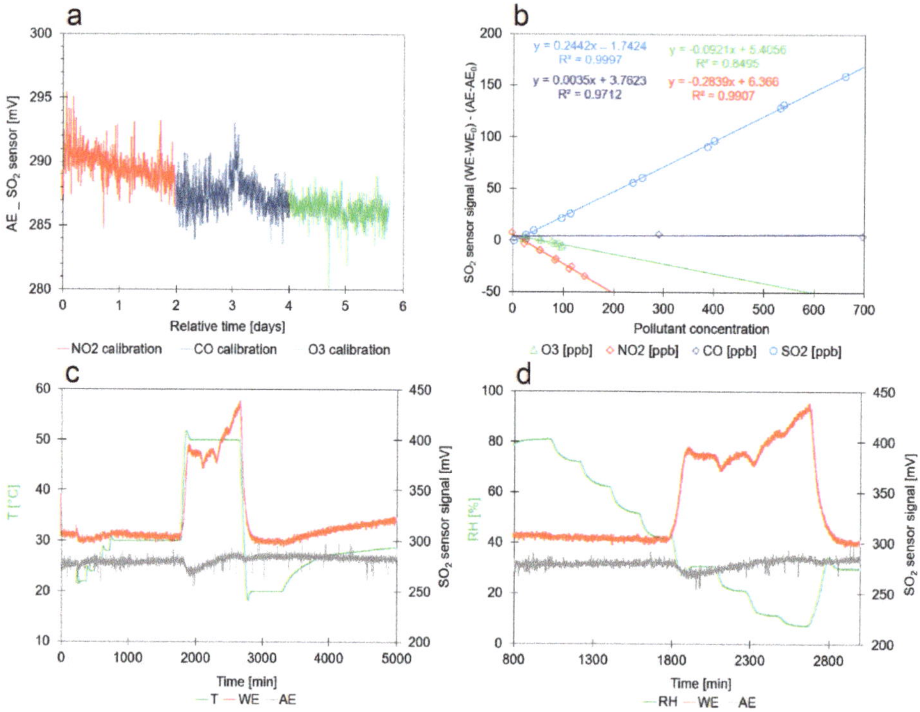

Figure 4. Impact of different pollutants and environmental conditions on the low-cost SO_2 sensor signal. (**a**) AE SO_2 sensor signal trends during the high-end calibration with NO_2, CO, and O_3; (**b**) cross-interference effect on the SO_2 sensor signal; (**c**) sensor signal trends at different temperature levels while the RH is changed in steps; (**d**) sensor signal trends at different relative humidity levels while the temperature changes are shown in (**c**).

4.2. Measuring Campaign in Cienfuegos, Cuba

When analyzing the AE values over time, this parameter shows fluctuations around an average of 284.2 mV. The minimum value for WE is 282 mV, with a corresponding AE value of 276 mV, resulting in an n_T of 1.02. The analysis of WE_0 and AE_0 during the calibration processes and the measuring campaign in Cuba also show that these values cannot be considered sensor constants because different values were obtained in all cases. Therefore, they have to be determined for every measuring campaign. As can be seen in Table 2, they change from campaign to campaign. To understand these fluctuations, it is necessary to perform more in-depth studies over a longer period of time. Some authors suggest that they are due to changes in environmental conditions such as temperature, relative humidity, pressure, and trace contaminants [69]. However, when n_T is calculated, values close to 1 are obtained in all cases.

Table 2. Overview of WE_0 and AE_0 from the different measuring campaigns. The campaigns are ranked chronologically.

	Low-Cost Calibration	High-End Calibration	Field Calibration	Measuring Campaign in Cuba
WE_0	294.6	298.6	279.0	282.0
AE_0	290.8	284.2	285.0	276.0
n_T	1.01	1.05	0.979	1.02

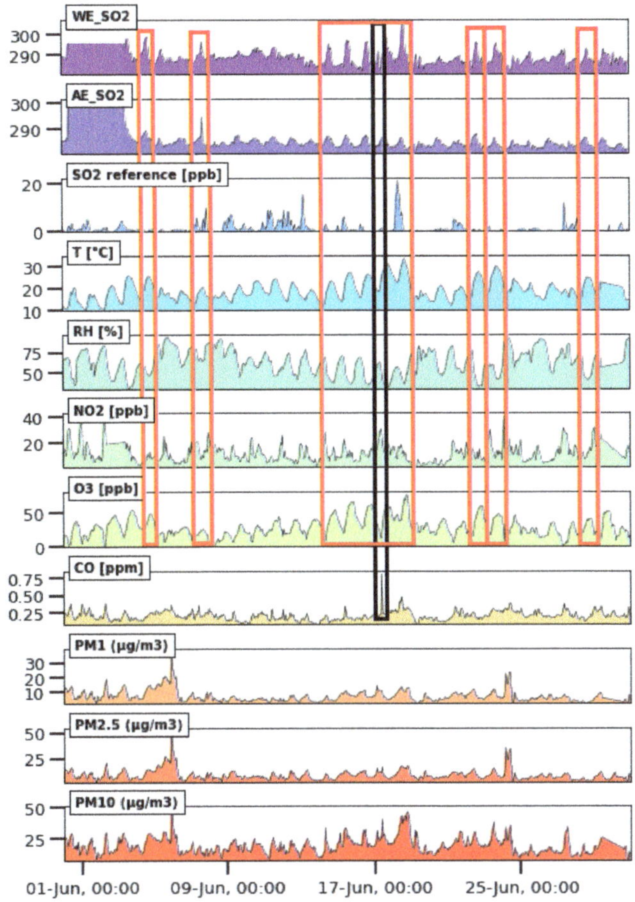

Figure 5. Concentration of gases and particulate matter recorded by the VMM station and the signal of the working and auxiliary electrodes of the low-cost SO_2 sensor during the field calibration performed in Antwerp, Belgium, from 30 May to 30 June 2022. The red rectangles indicate the simultaneous relationship of the SO_2 sensor with some parameters recorded by the station such as temperature, relative humidity, SO_2, NO_2 and O_3. The black rectangle indicates the simultaneous relationship of the SO_2 sensor with NO_2 and CO.

In the measuring campaign at Cienfuegos, the WE signal is clearly larger than the AE value. Figure 6a shows that the AE signal develops a slowly decreasing trend with superposed peaks. The minimum value for AE does not correspond with the minimum value for WE. The values for WE_0 and AE_0 are determined as the minimum value from the WE time series and the average value of the AE time series. The average value is used instead of the corresponding AE value of WE_{min}, as suggested by Equation (11), but both values are similar. In addition, the concentration-dependent signal WE_{gas} is significantly larger than zero, suggesting that the SO_2 concentration at the measuring location is above the detection limit of the sensor. When applying the low-cost and high-end calibrations, the same trend is observed (Figure 6b). The main difference is that the high-end calibration results in a shift of about 17 ppb. Despite this shift, it seems that the two calibration curves can be used for quantification if the background baseline of WE is removed in both the calibration processes and the field study. The monitoring campaign falls within the temperature range of 20–30 °C and the relative humidity range of 40–80%, so these variables do not affect the reliability of the SO_2 measurements. Figure 6 shows two periods in which the SO_2 concentration decreases appreciably. This behaviour can be related to the variation of the wind direction, which changed from east, east northeast to north, north northeast. These conditions favored the transport and dispersion of the pollutant to occur further away and in a different direction from the sampling site. In both periods, precipitation also occurred, which led to the removal of SO_2 [70]. The limited amount of information about the average SO_2 concentration in Cienfuegos, Cuba can be enriched with the average concentration as determined from the trends. Depending on the low-cost or high-end calibration, the average concentration ranges from 31.62 ppb to 48.59 ppb. These values are above the detection and quantification limits of the sensor. Therefore, the SO_2 sensor measurements recorded during the field campaign in Cienfuegos, Cuba show higher reliability than the measurements recorded in Belgium. The reliability can be further improved when the concentration range of the calibration process matches that of the measuring campaigns. The low-cost calibration process used 110 ppb as the lowest concentration.

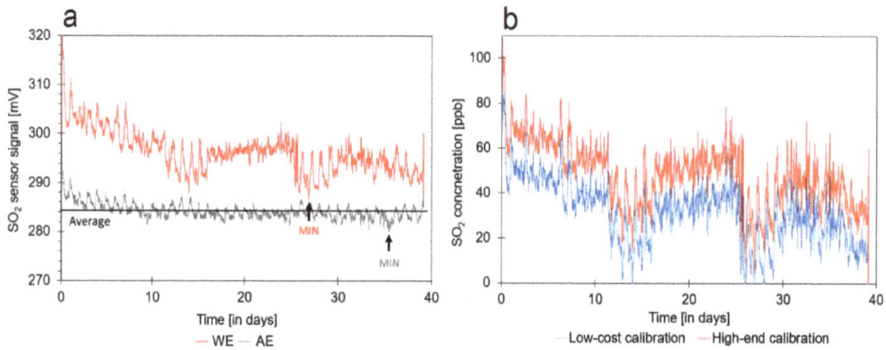

Figure 6. Measuring campaign performed in Cienfuegos. (**a**) Trends of WE and AE electrodes over time. (**b**) SO_2 concentrations have been calculated with the low-cost and high-end calibration process.

For this measuring campaign, the trends of the other gas sensors have been shown as well (see Figure 7). For this analysis, the NO_2, O_3, CO, and SO_2 concentrations were determined by using the equation obtained from the high-end calibration. It is important to mention that the O_3 signal was determined by subtracting the NO_2 signal from the O_x signal. Figure 6 shows some simultaneously occurring peaks of CO, PM, and SO_2 (see the black rectangles in Figure 7). The peaks can be related to: (1) the pollution source because the burning of, for example, diesel results in the simultaneous emission of SO_2, CO, PM, non-methane volatile organic compounds (NMVOCs), and other pollutants [71,72]; or (2) a cross-sensitivity effect due to the CO concentration in the air that is higher than that of SO_2. On the other hand, valleys in the SO_2 trend are observed when NO_2 or O_3 show peaks (see the red rectangles in Figure 7). The impact of NO_2 on the SO_2 signal is less notable since the NO_2 concentration is not considerably higher than that of SO_2. This might be explained by chemical reactions where SO_2 is being transformed into sulfates [73] or by a cross-sensitivity effect of the sensor. In addition, SO_2 peaks are observed that are not correlated with other pollutants, such as an increase in the SO_2 concentration at the sampling site or an increase in an interfering compound that is not measured (e.g., VOCs, H_2S, etc.). More background information is needed to understand how reliable the observed peaks are, but the correlation between peaks suggests the burning of diesel as a pollution source. This would have been missed if only average concentrations had been available.

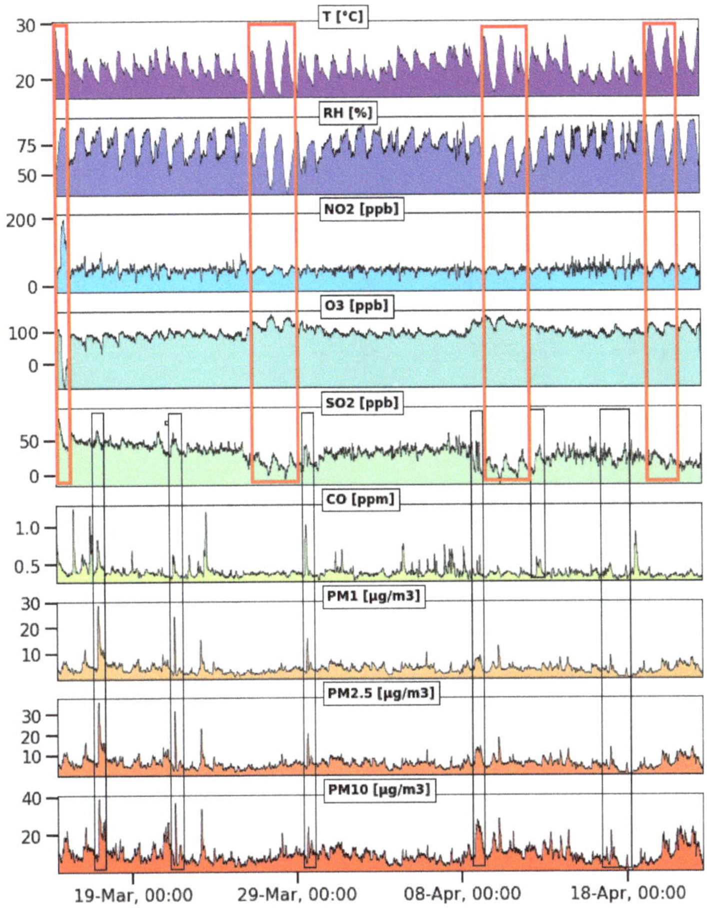

Figure 7. Results of the measuring campaign at Cienfuegos, Cuba with the low-cost monitoring system from 14 March to 22 April 2022. The red rectangles indicate the simultaneous relationship of SO_2 with some parameters such as temperature, relative humidity, NO_2 and O_3. The black rectangles indicate the simultaneous relationship of SO_2 with CO and particulate matter. All parameters have been recorded by low-cost sensors.

Figure 8 shows the correlation coefficients between the variables recorded by the low-cost monitoring system used in Cienfuegos, Cuba. The numbers vary between -1 and 1, indicating the linear relationship between two variables and, in this case, between the measured pollutants. The linear correlation of SO_2 with the rest of the variables takes values close to zero, resulting in a low correlation. The most influential pollutant was O_3, followed by CO with a correlation coefficient of -0.5 and 0.45, respectively, which might be caused by interference and be corrected with the equations in Figure 4b. This means that during the measuring campaign, the SO_2 sensor was not strongly influenced by the other pollutants and neither by temperature nor relative humidity.

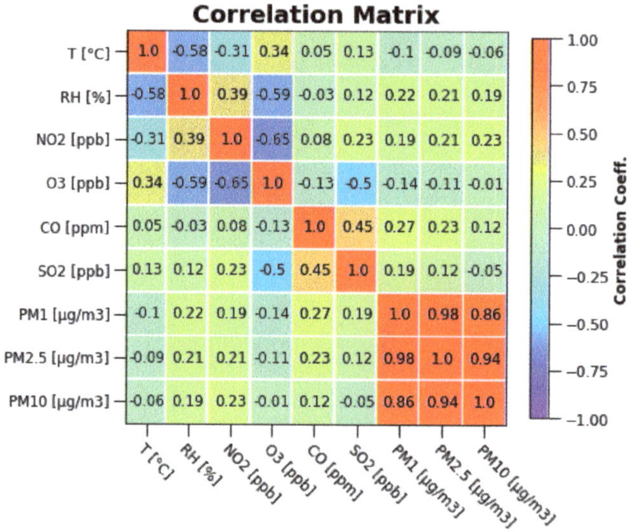

Figure 8. Correlation matrix between the different pollutants and meteorological variables recorded during the measuring campaign at Cienfuegos, Cuba.

5. Conclusions

Many publications assess the quality of low-cost sensors by comparing their measurements to those obtained from state-of-the-art reference instruments. In this study, the evaluation of low-cost sensors encompasses a broader range of indicators. While the accuracy and precision of the low-cost sensor in relation to advanced reference instruments are considered in the assessment, the focus is primarily on the extent of meaningful information that can be derived from low-cost sensors. This approach offers a significant advantage, as it enables the utilization of low-cost sensors in addressing pollution-related issues, even if they are not as accurate as state-of-the-art reference instruments.

Several indicators allowed us to analyse the reliability of the SO_2 Alphasense gas sensor. This sensor did not result in reliable results during the field calibration in Belgium because, in the city of Antwerp, the background concentrations in ambient air were too low. However, the same sensor resulted in reliable data in Cuba due to higher background concentrations. In addition, the SO_2 gas sensor could be calibrated using low-cost equipment in a tropical climate, showing a strong linear relationship with the reference concentrations. This means that low- and middle-income countries are able to calibrate the SO_2 sensor with a restricted budget. Calibration methods performed with different methods at different moments and on different continents can be compared with each other when the background baseline $WE_{background}$ can be subtracted from the signal in an accurate way. The reliability of the sensor is limited by the cross-sensitivity effect, with O_3 being the most interfering gas in the SO_2 sensor. In addition, the sensor signal appears to be more reliable when ambient conditions are below 30 °C and above 40% relative humidity. By showing the reliability of low-cost gas sensors, such as the SO_2 sensor used in this study, numerous institutions and researchers from low- and middle-income countries gain access to research infrastructure to solve local problems. This makes scientific research more inclusive.

The mathematical derivations given in this contribution show that the background level $WE_{background}$ can be calculated when WE_0 and AE_0 are known. However, the values for WE_0 and AE_0 obtained by zero calibration were strongly affected by the sampling location. Moreover, $WE_{background}$ also appeared to evolve in a fixed location over time. To improve the conversion method of sensor signals into concentration, more research is needed to understand how WE_0 and AE_0 are affected by the measuring location and the meteorological conditions and how stable they are at a fixed location. For this, more experiments in zero air are needed. The study also suggests that the span calibration is more stable in relation to location and time than the zero calibration, but this suggestion needs to be investigated in more detail by performing consecutive calibrations over time. Finally, one can wonder how in situ calibration can be done to improve sensor reliability.

Author Contributions: R.A.G.R.: Data processing and analysis of results obtained in the SO_2 sensor calibration and manuscript preparation. O.S.: Research conception and design; preparation, review, and approval of the final version of the manuscript. A.A.C.: Data processing of the results obtained in the field campaigns. E.H.R.: Design of the low-cost monitoring system. M.C.M.P.: Sampling site selection strategy and processing of the data obtained in the field campaigns. D.A.S.: Research conception and design; review and approval of the final version of the manuscript. A.M.L.: Assembly of the low-cost monitoring system; review and approval of the final version of the manuscript. W.J.: Research conception and design; review and approval of the final version of the manuscript. L.H.S.: Research conception and design; review and approval of the final version of the manuscript. All authors have read and agreed to the published version of the manuscript.

Funding: The present work has been supported by the VLIR-UOS Program as a South Initiative Project SI-2019 nr. CU2019SIN242B124 "AIR@PORT: Low-cost decision support system to evaluate the impact of ships on the air quality in the port city Cienfuegos". Additional financial support has been given by the Common Global Minds Project BE2017GMHVLHC106 of all Flemish Universities of Applied Sciences and Arts.

Institutional Review Board Statement: Not applicable.

Informed Consent Statement: Not applicable.

Data Availability Statement: All data used in this study are available upon request.

Acknowledgments: The authors thank VLIR-UOS for their financial support of the South Initiative Project SI-2019 nr. CU2019SIN242B124 "AIR@PORT: Low-cost decision support system to evaluate the impact of ships on the air quality in the port city Cienfuegos". They are also grateful for the additional support provided by the Global Minds project BE2017GMHVLHC106 "A low-cost measuring device to directly monitor exhaust gases generated by fuel engines: Design, development, validation". The researchers are grateful to the VMM for allowing them to measure near one of their monitoring stations.

Conflicts of Interest: The authors declare no conflict of interest.

References

1. Rosen, C.M. Businessmen Against Pollution in Late Nineteenth Century Chicago. *Bus. Hist. Rev.* **1995**, *69*, 351–397. [CrossRef]
2. Wisniak, J. Sodium Carbonate–From Natural Resources to Leblanc and Back. *Indian J. Chem. Technol.* **2003**, *10*, 99–112.
3. Dingle, A.E. 'The Monster Nuisance of All': Landowners, Alkali Manufacturers, and Air Pollution, 1828–64. *Econ. Hist. Rev.* **1982**, *35*, 529–548. [CrossRef]
4. Nemery, B.; Hoet, P.H.; Nemmar, A. The Meuse Valley Fog of 1930: An Air Pollution Disaster. *Lancet* **2001**, *357*, 704–708. [CrossRef]
5. Firket, J. Fog along the Meuse Valley. *Trans. Faraday Soc.* **1936**, *32*, 1192. [CrossRef]
6. Medlin, J. Toxicogenomics Research Consortium Sails into Uncharted Waters. *Environ. Health Perspect.* **2002**, *110*, A744–A746. [CrossRef] [PubMed]
7. Scott, J.A. Fog and Deaths in London, December 1952. *Public Health Rep.* **1953**, *68*, 474. [CrossRef]
8. Schwartz, J.; Marcus, A. Mortality and Air Pollution in London: A Time Series Analysis. *Am. J. Epidemiol.* **1990**, *131*, 185–194. [CrossRef]
9. Bell, M.L.; Davis, D.L. Reassessment of the Lethal London Fog of 1952: Novel Indicators of Acute and Chronic Consequences of Acute Exposure to Air Pollution. *Environ. Health Perspect.* **2001**, *109*, 389–394.
10. Stern, A.C.; Professor, E. History of Air Pollution Legislation in the United States. *J. Air Pollut. Control Assoc.* **1982**, *32*, 44–61. [CrossRef]
11. Valler, R.E.; Lawther, P.J. Further Observation of London Fog. *BMJ* **1957**, *2*, 1473–1475. [CrossRef]
12. Mylona, S. Sulphur Dioxide Emissions in Europe 1880-1991 and Their Effect on Sulphur Concentrations and Depositions. *Tellus B* **1996**, *48*, 662–689. [CrossRef]
13. Fowler, D.; Brimblecombe, P.; Burrows, J.; Heal, M.R.; Grennfelt, P.; Stevenson, D.S.; Jowett, A.; Nemitz, E.; Coyle, M.; Liu, X.; et al. A Chronology of Global Air Quality. *Phil. Trans. R. Soc. A.* **2020**, *378*, 20190314. [CrossRef] [PubMed]

14. Van Roy, W.; Van Nieuwenhove, A.; Scheldeman, K.; Van Roozendael, B.; Schallier, R.; Mellqvist, J.; Maes, F. Measurement of Sulfur-Dioxide Emissions from Ocean-Going Vessels in Belgium Using Novel Techniques. *Atmosphere* **2022**, *13*, 1756. [CrossRef]
15. Krupińska, B.; Worobiec, A.; Gatto Rotondo, G.; Novaković, V.; Kontozova, V.; Ro, C.-U.; Van Grieken, R.; De Wael, K. Assessment of the Air Quality (NO_2, SO_2, O_3 and Particulate Matter) in the Plantin-Moretus Museum/Print Room in Antwerp, Belgium, in Different Seasons of the Year. *Microchem. J.* **2012**, *102*, 49–53. [CrossRef]
16. Henschel, S.; Querol, X.; Atkinson, R.; Pandolfi, M.; Zeka, A.; Le Tertre, A.; Analitis, A.; Katsouyanni, K.; Chanel, O.; Pascal, M.; et al. Ambient Air SO_2 Patterns in 6 European Cities. *Atmos. Environ.* **2013**, *79*, 236–247. [CrossRef]
17. Magaña, M.; Sánchez, C. Efectos del petróleo crudo cubano sobre los procesos de lixiviación, sedimentación y destilación en Nicaro. *RTQ* **2015**, *35*, 54–62.
18. Veloz, L.A.; Sandoval, F.D. *Caracterización Físico-Química de Crudos Cubanos de las Estructuras en Evaluación del Bloque VIIA de la Franja Norte de Crudos Pesados*; Pueblo y Educación: La Habana, Cuba, 2011; pp. 1–9.
19. Hernández, A.R.; Quintero, J.O.L.; Clemente, A.N.; Cano, C.; Lopez, O.D. Caracterización de la fracción saturada de petróleos del yacimiento Varadero, mediante cromatografía gaseosa. *Rev. CENIC Cienc. Quím.* **2021**, *52*, 81–91.
20. EnerTechUp GmbH. Advanced Energy Technology. Available online: https://aenert.com/countries/america/energy-industry-in-cuba/general-data/ (accessed on 10 February 2023).
21. Cuesta-Santos, O.; Sosa-Pérez, C.; Iraola-Ramirez, C.; González, Y.; Nuñez-Caraballo, V.; Fonte-Hernández, A.; Imbert-Lamorú, C.; Gómez-Zamora, Y.; Portal-Castillo, D. Inventario nacional de emisiones atmosféricas de las principales fuentes fijas. *Rev. Cuba. Meteorol.* **2017**, *23*, 178–190.
22. Cruz-Montes de Oca, F.; Furet-Bridón, N.R.; Turtós-Carbonell, L.; Lorente-Vera, M. La dispersión atmosférica de contaminantes en una zona industrial de Cuba. *Rev. CENIC Cienc. Quím.* **2011**, *42*, 1–7.
23. Alejo, D.; Morales, M.C.; De la Torre, J.B.; Grau, R.; Bencs, L.; Grieken, R.V.; Espen, P.V.; Sosa, D.; Nuñez, V. Seasonal Trends of Atmospheric Nitrogen Dioxide and Sulfur Dioxide over North Santa Clara, Cuba. *Environ. Monit. Assess.* **2013**, *185*, 6023–6033. [CrossRef] [PubMed]
24. Madrazo, J.; Clappier, A.; Cuesta, O.; Belalcazar, L.C.; González, Y.; Bolufé, J.; Sosa, C.; Carrillo, E.; Manso, R.; Canciano, J.; et al. Evidence of Traffic-Generated Air Pollution in Havana. *Atmosfera* **2019**, *32*, 15–24. [CrossRef]
25. Schalm, O.; Carro, G.; Lazarov, B.; Jacobs, W.; Stranger, M. Reliability of Lower-Cost Sensors in the Analysis of Indoor Air Quality on Board Ships. *Atmosphere* **2022**, *13*, 1579. [CrossRef]
26. Narayana, M.V.; Jalihal, D.; Nagendra, S.M.S. Establishing A Sustainable Low-Cost Air Quality Monitoring Setup: A Survey of the State-of-the-Art. *Sensors* **2022**, *22*, 394. [CrossRef]
27. Hagan, D.H.; Isaacman-VanWertz, G.; Franklin, J.P.; Wallace, L.M.M.; Kocar, B.D.; Heald, C.L.; Kroll, J.H. Calibration and Assessment of Electrochemical Air Quality Sensors by Co-Location with Regulatory-Grade Instruments. *Atmos. Meas. Tech.* **2018**, *11*, 315–328. [CrossRef]
28. Drajic, D.D.; Gligoric, N.R. Reliable Low-Cost Air Quality Monitoring Using Off-The-Shelf Sensors and Statistical Calibration. *Elektron. Elektrotech.* **2020**, *26*, 32–41. [CrossRef]
29. Chojer, H.; Branco, P.T.B.S.; Martins, F.G.; Alvim-Ferraz, M.C.M.; Sousa, S.I.V. Can Data Reliability of Low-Cost Sensor Devices for Indoor Air Particulate Matter Monitoring Be Improved?—An Approach Using Machine Learning. *Atmos. Environ.* **2022**, *286*, 119251. [CrossRef]
30. Doval-Miñarro, M.; Bueso, M.C. A Comparative Study of Air Pollutant Concentrations before the COVID-19 Pandemic and in the New Normal in the Región de Murcia (Spain). *Atmosphere* **2023**, *14*, 147. [CrossRef]
31. Dimov, I.; Todorov, V.; Sabelfeld, K. A Study of Highly Efficient Stochastic Sequences for Multidimensional Sensitivity Analysis. *Monte Carlo Methods Appl.* **2022**, *8*, 1–12. [CrossRef]
32. Rhouati, A.; Berkani, M.; Vasseghian, Y.; Golzadeh, N. MXene-Based Electrochemical Sensors for Detection of Environmental Pollutants: A Comprehensive Review. *Chemosphere* **2022**, *291*, 132921. [CrossRef]
33. Farquhar, A.K.; Henshaw, G.S.; Williams, D.E. Errors in Ambient Gas Concentration Measurement Caused by Acoustic Response of Electrochemical Gas Sensors. *Sens. Actuator A Phys.* **2023**, *354*, 114254. [CrossRef]
34. Williams, D.E. Electrochemical Sensors for Environmental Gas Analysis. *Curr. Opin. Electrochem.* **2020**, *22*, 145–153. [CrossRef]
35. Silambarasan, P.; Moon, I.S. Real-Time Monitoring of Chlorobenzene Gas Using an Electrochemical Gas Sensor during Mediated Electrochemical Degradation at Room Temperature. *J. Electroanal. Chem.* **2021**, *894*, 115372. [CrossRef]
36. Todorov, V.; Ostromsky, T.; Dimov, I.; Fidanova, S. Optimized Quasi-Monte Carlo Method Based on Low Discrepancy Sequences for Sensitivity Analysis in Air Pollution Modelling. 26 September 2020. pp. 25–28. Available online: https://www.researchgate.net/publication/346122855_Optimized_Quasi-Monte_Carlo_Method_Based_on_Low_Discrepancy_Sequences_for_Sensitivity_Analysis_in_Air_Pollution_Modelling (accessed on 1 March 2023).
37. Soares, P.H.; Monteiro, J.P.; Gaioto, F.J.; Ogiboski, L.; Andrade, C.M.G. Use of Association Algorithms in Air Quality Monitoring. *Atmosphere* **2023**, *14*, 648. [CrossRef]
38. Samad, A.; Obando Nuñez, D.R.; Solis Castillo, G.C.; Laquai, B.; Vogt, U. Effect of Relative Humidity and Air Temperature on the Results Obtained from Low-Cost Gas Sensors for Ambient Air Quality Measurements. *Sensors* **2020**, *20*, 5175. [CrossRef]
39. Masic, A.; Bibic, D.; Pikula, B.; Razic, F. New Approach of Measuring Toxic Gases Concentrations: Principle of Operation. In *DAAAM Proceedings*; Katalinic, B., Ed.; DAAAM International Vienna: Vienna, Austria, 2018; Volume 1, pp. 0882–0887, ISBN 978-3-902734-20-4.

40. Zuidema, C.; Schumacher, C.S.; Austin, E.; Carvlin, G.; Larson, T.V.; Spalt, E.W.; Zusman, M.; Gassett, A.J.; Seto, E.; Kaufman, J.D.; et al. Deployment, Calibration, and Cross-Validation of Low-Cost Electrochemical Sensors for Carbon Monoxide, Nitrogen Oxides, and Ozone for an Epidemiological Study. *Sensors* **2021**, *21*, 4214. [CrossRef]
41. Tryner, J.; Phillips, M.; Quinn, C.; Neymark, G.; Wilson, A.; Jathar, S.H.; Carter, E.; Volckens, J. Design and Testing of a Low-Cost Sensor and Sampling Platform for Indoor Air Quality. *Build. Environ.* **2021**, *206*, 108398. [CrossRef]
42. Arroyo, P.; Gómez-Suárez, J.; Suárez, J.I.; Lozano, J. Low-Cost Air Quality Measurement System Based on Electrochemical and PM Sensors with Cloud Connection. *Sensors* **2021**, *21*, 6228. [CrossRef]
43. Sensor Technology House. *Alphasense Application Note AAN 803-04 Correcting for Background Currents in Four Electrode Toxic Gas Sensors*; Alphasense Limited: Heywood, UK, 2017.
44. Alphasense Application Notes. Available online: www.alphasense.com (accessed on 20 March 2023).
45. Cross, E.S.; Lewis, D.K.; Williams, L.R.; Magoon, G.R.; Kaminsky, M.L.; Worsnop, D.R.; Jayne, J.T. Use of Electrochemical Sensors for Measurement of Air Pollution: Correcting Interference Response and Validating Measurements. *Atmos. Meas. Tech.* **2017**, *10*. [CrossRef]
46. Hernandez-Rodriguez, E.; Kairuz-Cabrera, D.; Martinez, A.; Amalia, R.; Schalm, O. Low-Cost Portable System for the Estimation of Air Quality. In *Proceedings of 19th Latin American Control Congress (LACC 2022)*; Studies in Systems, Decision and Control; Springer: La Habana, Cuba, 2022; Volume 464, p. XII, ISBN 978-3-031-26361-3.
47. Martinez, A.; Hernandez-Rodríguez, E.; Hernandez, L.; González-Rivero, R.A.; Alejo-Sánchez, D.; Schalm, O. Design of a Low-Cost Portable System for the Measurement of Variables Associated with Air Quality. *IEEE Embed. Syst.* **2022**. [CrossRef]
48. Margaritis, D.; Keramydas, C.; Papachristos, I.; Lambropoulou, D. Calibration of Low-Cost Gas Sensors for Air Quality Monitoring. *Aerosol Air Qual. Res.* **2021**, *21*, 210073. [CrossRef]
49. Maag, B.; Zhou, Z.; Thiele, L. A Survey on Sensor Calibration in Air Pollution Monitoring Deployments. *IEEE Internet Things J.* **2018**, *5*, 4857–4870. [CrossRef]
50. Spinelle, L.; Aleixandre, M.; Gerboles, M. *Protocol of Evaluation and Calibration of Low-Cost Gas Sensors for the Monitoring of Air Pollution*; Publication Office of the European Union: Luxembourg, 2013.
51. Mijling, B.; Jiang, Q.; de Jonge, D.; Bocconi, S. Field Calibration of Electrochemical NO2 Sensors in a Citizen Science Context. *Atmos. Meas. Tech.* **2018**, *11*, 1297–1312. [CrossRef]
52. Topalović, D.B.; Davidović, M.D.; Jovanović, M.; Bartonova, A.; Ristovski, Z.; Jovašević-Stojanović, M. In Search of an Optimal In-Field Calibration Method of Low-Cost Gas Sensors for Ambient Air Pollutants: Comparison of Linear, Multilinear and Artificial Neural Network Approaches. *Atmos. Environ.* **2019**, *213*, 640–658. [CrossRef]
53. González Rivero, R.A.; Morera Hernández, L.E.; Schalm, O.; Hernández Rodríguez, E.; Alejo Sánchez, D.; Morales Pérez, M.C.; Nuñez Caraballo, V.; Jacobs, W.; Martinez Laguardia, A. A Low-Cost Calibration Method for Temperature, Relative Humidity, and Carbon Dioxide Sensors Used in Air Quality Monitoring Systems. *Atmosphere* **2023**, *14*, 191. [CrossRef]
54. Mattson, B. Microscale Gas Chemistry. *Educ. Quim.* **2018**, *16*, 514. [CrossRef]
55. Najdoski, M. Gas Chemistry: A Microscale Kipp Apparatus. *J. Chem. Educ.* **2011**, *16*, 295–298.
56. Najdoski, M.; Stojkovikj, S. Cost Effective Microscale Gas Generation Apparatus. *Chemistry* **2010**, *19*, 444–449.
57. Zimmerman, N.; Presto, A.A.; Kumar, S.P.N.; Gu, J.; Hauryliuk, A.; Robinson, E.S.; Robinson, A.L.; Subramanian, R. Closing the Gap on Lower Cost Air Quality Monitoring: Machine Learning Calibration Models to Improve Low-Cost Sensor Performance. *Atmos. Meas. Tech.* **2017**. [CrossRef]
58. Carro, G.; Schalm, O.; Jacobs, W.; Demeyer, S. Exploring Actionable Visualizations for Environmental Data: Air Quality Assessment of Two Belgian Locations. *Environ. Model. Softw.* **2022**, *147*, 105230. [CrossRef]
59. Lewis, A.C.; Lee, J.D.; Edwards, P.M.; Shaw, M.D.; Evans, M.J.; Moller, S.J.; Smith, K.R.; Buckley, J.W.; Ellis, M.; Gillot, S.R.; et al. Evaluating the Performance of Low Cost Chemical Sensors for Air Pollution Research. *Faraday Discuss.* **2016**, *189*, 85–103. [CrossRef] [PubMed]
60. Gamboa, V.S.; Kinast, É.J.; Pires, M. System for Performance Evaluation and Calibration of Low-Cost Gas Sensors Applied to Air Quality Monitoring. *Atmos. Pollut. Res.* **2023**, *14*, 101645. [CrossRef]
61. Pang, X.; Shaw, M.D.; Gillot, S.; Lewis, A.C. The Impacts of Water Vapour and Co-Pollutants on the Performance of Electrochemical Gas Sensors Used for Air Quality Monitoring. *Sens. Actuators B Chem.* **2018**, *266*, 674–684. [CrossRef]
62. Wei, P.; Ning, Z.; Ye, S.; Sun, L.; Yang, F.; Wong, K.; Westerdahl, D.; Louie, P. Impact Analysis of Temperature and Humidity Conditions on Electrochemical Sensor Response in Ambient Air Quality Monitoring. *Sensors* **2018**, *18*, 59. [CrossRef] [PubMed]
63. *NC 111: 2004;* Calidad del Aire—Reglas para la Vigilancia de la Calidad del Aire en Asentamientos Humanos. Cuban National Bureau of Standards: La Habana, Cuba, 2004.
64. Langford, G.O.; Winefordner, J.D. Limit of Detection. A Closer Look at the IUPAC Definition. *Anal. Chem.* **1983**, *55*, 712–724. [CrossRef]
65. Alejo, D.; Morales, M.C.; Nuñez, V.; Bencs, L.; Van Grieken, R.; Van Espen, P. Monitoring of Tropospheric Ozone in the Ambient Air with Passive Samplers. *Microchem. J.* **2011**, *99*, 383–387. [CrossRef]
66. *NC 39:1999;* Calidad Del Aire. Requisitos Higiénicos—Sanitarios. Cuban National Bureau of Standards: La Habana, Cuba, 1999.
67. Cui, H.; Zhang, L.; Li, W.; Yuan, Z.; Wu, M.; Wang, C.; Ma, J.; Li, Y. A New Calibration System for Low-Cost Sensor Network in Air Pollution Monitoring. *Atmos. Pollut. Res.* **2021**, *12*, 101049. [CrossRef]

68. Schalm, O.; Carro, G.; Jacobs, W.; Lazarov, B.; Stranger, M. The Inherent Instability of Environmental Parameters Governing Indoor Air Quality on Board Ships and the Use of Temporal Trends to Identify Pollution Sources. *Indoor Air* **2023**, *2023*, 1–19. [CrossRef]
69. Wenzel, M.J.; Mensah-Brown, A.; Josse, F.; Yaz, E.E. Online Drift Compensation for Chemical Sensors Using Estimation Theory. *IEEE Sens. J.* **2011**, *11*, 225–232. [CrossRef]
70. Instituto de Meteorología de Cuba. Available online: http://www.insmet.cu (accessed on 7 May 2023).
71. Stewart, G.J.; Acton, W.J.F.; Nelson, B.S.; Vaughan, A.R.; Hopkins, J.R.; Arya, R.; Mondal, A.; Jangirh, R.; Ahlawat, S.; Yadav, L.; et al. Emissions of Non-Methane Volatile Organic Compounds from Combustion of Domestic Fuels in Delhi, India. *Atmos. Chem. Phys.* **2021**, *21*, 2383–2406. [CrossRef]
72. İlkiliç, C.; Aydin, H. The Harmful Effects of Diesel Engine Exhaust Emissions. *Energy Sources Part A Recovery Util. Environ. Eff.* **2012**, *34*, 899–905. [CrossRef]
73. Cofer, W.R.; Schryer, D.R.; Rogowski, R.S. Oxidation of SO_2 by NO_2 and O_3 on Carbon: Implications to Tropospheric Chemistry. *Atmos. Environ.* **1984**, *18*, 243–245. [CrossRef]

atmosphere

MDPI

Article

Design Considerations for a Distributed Low-Cost Air Quality Sensing System for Urban Environments in Low-Resource Settings

Engineer Bainomugisha *, Joel Ssematimba and Deo Okure

AirQo, Department of Computer Science, Makerere University, Kampala 7062, Uganda
* Correspondence: baino@mak.ac.ug

Abstract: With rapid urbanization, hazardous environmental exposures such as air, noise, plastic, soil and water pollution have emerged as a major threat to urban health. Recent studies show that 9 out of 10 people worldwide breathe contaminated air contributing to over 7 million premature deaths annually. Internet of Things (IoT) and Artificial Intelligence (AI)-based environmental sensing and modelling systems have potential for contributing low-cost and effective solutions by providing timely data and insights to inform mitigation and management actions. While low and middle-income countries are among those most affected by environmental health risks, the appropriateness and deployment of IoT and AI systems in low-resource settings is least understood. Motivated by this knowledge gap, this paper presents a design space for a custom environmental sensing and management system designed and developed to fill the data gaps in low-resource urban settings with a particular focus on African cities. The paper presents the AirQo system, which is the first instance of the design space requirements. The AirQo system includes: (1) autonomous AirQo sensors designed and customised to be deployed in resource constrained environments (2) a distributed sensor network that includes over 120 static and mobile nodes for air quality sensing (3) AirQo network manager tool for tracking and management of installation and maintenance of nodes, (4) AirQo platform that provides calibration, data access and analytics tools to support usage among policy makers and citizens. Case studies from African cities that are using the data and insights for education, awareness and policy are presented. The paper provides a template for designing and deploying a technology-driven solution for cities in low resource settings.

Keywords: low-cost sensors; distributed urban air quality system; environmental sensing; smart city solutions; Cyber-physical systems; Internet of Things; AI for social good

1. Introduction

With rapid urbanization in Sub-Saharan Africa (SSA), hazardous environmental exposures such as air, noise and plastic pollution have emerged as a major threat to cities to become prosperous and resilient. People living in cities are in particular at a greater risk of exposure to unhealthy levels of air pollution. Recent World Health Organisation (WHO) report indicates that 99 percent of the global population breathe polluted air [1]. Furthermore, air pollution exposure is linked to over 7 million deaths per year worldwide, the majority happen in low- and middle-income countries [1].

Despite the greater risk associated with the air pollution exposures in urban spaces, there are scarcity of data on the scale and magnitude of air pollution in many cities in low- and middle-income countries. Traditional air pollution measurement approaches rely on expensive ground monitoring stations that cost thousands of dollars to set up and maintain [2–4]. In this paper, we explore the potential of emerging technologies such as low-cost air quality sensing, Internet of Things (IoT), and artificial intelligence as novel approaches to improving the monitoring and data access and their contribution to building sustainable and resilient urban spaces with a focus on low-resource settings.

Smart technology solutions have potential for accelerating actions that can be taken to tackle environmental and public health challenges in urban spaces. When harnessed appropriately smart city solutions can help track more precisely the magnitude and scale of environment exposures as well as track the progress and inform actions being taken in addressing the challenges and contribute to sustainable cities [5–7].

However, deployment and adoption of smart city solutions including Internet of Things, machine learning and artificial intelligence are less harnessed in many developing cities. Deployment and adoption of smart city solutions in many cities in low- and middle-income countries is a complex endeavour and is often impeded by a number of contextual challenges such as infrastructure bottlenecks, environmental conditions, incompatibility of the technology solutions and the city needs, cost and capacity. These factors negatively affect the deployment and implementation systems and technologies that are conceived for other contexts because the assumptions may not hold. For example, an environmental sensing system should take into account the available data communication and energy infrastructure, otherwise data completeness and device up-time can affect the reliability of the system.

Distributed low-cost air quality sensors have recently gained momentum as a complementary approach to the traditional air quality measurement instruments [8,9]. Datasets from low-cost sensor networks provide opportunity to measure at a larger spatial scale and can help inform understanding of environmental trends at a higher resolution. Traditional instruments have limited spatial coverage and heterogeneity and are inadequate for understanding air pollution exposures at individual and community level [9]. The limitations of traditional air quality monitoring have inspired the emergency of low-cost sensor approaches. Low-cost air quality sensors are leading to a paradigm shift in air quality monitoring enabling synergetic efforts between communities and the government to tackle air pollution in a city. Snyder et al. [8] predicted this paradigm, while Labzovskii et al. [10] demonstrated this shift in a practical case study of a decentralised urban air quality monitoring system. Other studies and systems that have explored for low-cost air monitoring approaches include CAIRSENSE [11], DYNAMAP [12], SentinAir [13], HazeWatch [9], MAQUMON [14], CommonSense [15], and APS [16]. The majority of these systems focus on evaluation of specific technical aspects of low-cost sensors like quality assurance, performance and feasibility.

In this paper, we present the AirQo system, an air quality smart sensing and management system that has been conceived from the ground-up with the understanding of the unique and contextual needs of African cities to generate evidence and facilitate actions for improving air quality management in urban spaces. Low-resource settings are typically characterised by many constraints ranging from limited capacity and fewer trained personnel, limited infrastructure for power, data transmission, limited funding, fewer or absence of high-end equipment, among others. The AirQo system has been deployed in major cities in East Africa and is being scaled to more cities in the African region. The digital air quality platforms are being used to facilitate citizen awareness on air quality issues, provide access to data and evidence for decision makers and used for skilling and training of students and researchers from several disciplines including Engineering, Computer Science, Public Health, among others. To our knowledge the AirQo system is the largest sensing network in production low-resource settings and thus uncovers new design and deployment requirements. Moreover, the AirQo system is linked to actions that can be taken by communities and city leaders.

2. The Design Space for an Urban Air Sensing System

This section presents the design space for an air quality monitoring and management system for cities in emerging economies and low-resource settings. Urban spaces in low-resource settings have unique characteristics and challenges related to diverse technology and power infrastructure and access, informal and upscale settlements, population density, capacity, transport systems and varying levels of pollution profiles and diverse sources.

Taken together these challenges and characteristics pose new design requirements for a smart city solution for environmental issues like air and noise pollution. The design space is informed by experiences from an iterative process of design, development and implementation of AirQo [2,17], one of Africa's largest technology-driven air quality monitoring and management system for urban spaces in low-resource settings. The design and development of the AirQo system has undergone a series of iterations and evolution following agile development methodology and practices of user-centered design [18,19]. Figure 1 presents an overview of the air quality sensing and management system. The key stakeholders and users being citizens and communities in African cities, decision makers and authorities, and researchers in local universities embedded in these cities. In Figure 1 the arrows illustrate the interaction among the different stakeholders in the ecosystem. The local University and innovators that are embedded in the city have the understanding of the technology as well as the local context and thus are well positioned to translate the needs of the city and communities into products. Citizens and communities are also critical in the product development and engineering cycle through co-creation and co-design.

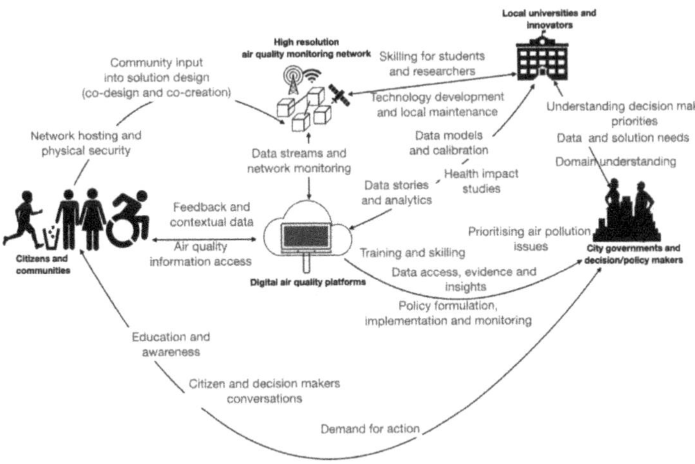

Figure 1. A conceptual design of an air quality sensing and management system urban environments.

2.1. Design Requirements for an Urban Air Quality Sensing System for Low Resource Settings

Throughout the design and development process, we identified and documented key design requirements for an effective air quality sensing and management systems aimed at having a direct impact to the United Nations Sustainable Development Goals of reducing emissions in cities as well as their impacts on human health. Table 1 provides an outline of the key design requirements (DR) for a smart city solution for air quality sensing and management in a low-resource urban setting. The design considerations are presented as a series of steps rather than the order of importance. The DRs range from air quality monitoring network design and setup, to deployment options, data transmission, network operations and adaptation to the deployment conditions, quality assurance of the data and network operations, engagement of the communities, and linking the data and digital tools to action.

Table 1. Design considerations for an urban air quality sensing system in a low-resource setting.

DR	Design Consideration
DR1	Low-cost and high resolution air quality sensing
DR2	Designing for portability and flexible deployment options
DR3	Flexible power and communication options
DR4	Resilience to local environmental conditions
DR5	Embedded quality assurance and calibration methods
DR6	Community-oriented and active engagements
DR7	Digital platforms to drive action and policy

2.1.1. DR1: Low-Cost and High Resolution Air Quality Sensing

Traditional air quality monitoring relies on reference grade monitors that have high precision and accuracy, for example, beta attenuation monitor (BAM) and tapered element oscillating microbalance (TEOM) [20]. However, these are expensive to setup [4,13,21], require specialised technical capacity to maintain, and require stable infrastructure for power connectivity and data transmission. For example, setup of a reference grade monitor in an African city context may require extension of electricity grid and data communication network infrastructure. Outdoor deployment for a reference grade monitor also requires consideration of physical security to minimise the risk of vandalism and loss. This can limit the choice of deployment sites to only those where security is guaranteed and there is access to power grid and communication infrastructure. Furthermore, sites with existing power and communication infrastructure may not necessarily be representative of the community, and so installing monitors in those locations can result in bias in the measurements. These challenges make traditional air quality sensing systems impractical for many settings in low-resource settings, particularly African urban spaces. Also because of the high cost of setup and maintenance, installation of reference grade monitors at a higher density is often not feasible even in the rich-resource settings.

The need for high resolution sensing is also motivated by the high variability of pollution profiles in urban spaces. Pollution profiles in African urban spaces can vary greatly due to the localised pollution drivers such as cooking and waste burning [2,22]. As such an air quality sensing system designed for African urban spaces should be deployed at high resolution to achieve hyper-local air quality monitoring. High resolution monitoring can also be used to promote awareness about air pollution issues and its effects [23]. For high resolution monitoring to be achieved it requires that the cost of monitoring devices and network setup and maintenance be considerably affordable. The exact resolution of an air quality sensing system should be informed by a scoping exercise to meet the monitoring goals. Sensor placement optimisation approaches should be explored to ensure that the air quality sensing network achieves the monitoring goals while at the same time optimising the cost of monitoring.

2.1.2. DR2: Designing for Portability and Flexible Deployment Options

Air quality sensing devices should be designed to meet the physical installation requirements in low-resource urban spaces. It should be possible to mount devices on the available infrastructure, including streetlight poles, schools, retail shops, trees, temporary structures, homes, or custom poles. Devices should be easy to move from location to location to meet the emerging needs, for example, moving a monitor to a new location of interest in the cases where there might be limited sensors or changing the device location because the device host has changed premises. The air sensors should also be designed to allow for mounting on mobile objects to allow for wide sensing with fewer monitors. Mobile sensing has been explored previously as a means to increase coverage of monitoring [24,25]. Particularly, in the African urban spaces there is potential to leverage mobile monitoring, for example, the use motorcycle taxis and citizens to reach wider coverage of sensing with fewer resources. The deployment needs in these contexts may vary location by location, requiring retrofitting (i.e., customisation to fit the deployment environments) using locally available materials. Whenever possible local technicians can be used to support fabrication of custom mounts for the locations in their community to ensure context-specific deployment and installation needs. This can greatly reduce the cost of shipping and transporting installation and deployment materials and also promotes local-buy-in and ownership.

2.1.3. DR3: Flexible Power and Communication Options

An environment sensing system designed for deployment in urban settings in many African countries needs to take into account the varying power and communication challenges. Access to the main electricity grid is not always guaranteed and can also be characterised by power cuts [26]. The sensing devices need to be designed to allow for flexible power options, for example, solar and main grid [27]. Devices need to have optimised battery storage and usage to allow for operation in the case of power cuts or absence of solar energy in the night hours. Incentive mechanisms need to be explored for cases where devices are powered by the main electricity grid owned by the private owners. Potential incentives include education and awareness, recognition of the community members as key stakeholders or compensation to the power costs. In the case of mobile monitoring, specialised power charging modules need to be designed. These requirements are unique to the urban spaces in low-resource settings. For example, when deploying air quality sensors on motorcycles, we find that many lack well-maintained functional batteries and thus the charging system must be designed around the alternator instead of battery or both. Motorcycle taxis often take very short trips and thus the charging system should be designed to address these unique requirements.

Autonomous data transmission is necessary for environmental sensing systems to minimise the cost and resources required for manual data collection. The Internet communication infrastructure in low-resource and emerging urban spaces is often characterised by unreliable connectivity. Slow and intermittent communications are the rule rather than an exception. Wireless connectivity is not practical and alternative communication options need to be explored. Cellular connectivity is present in many African urban spaces including rural setups and thus 2G cellular communication should be included as default option for these settings. To minimise data loss local storage and bulk data transmission options should also be key design requirements for smart environmental sensing systems in these contexts.

2.1.4. DR4: Resilience to Local Environmental Conditions

In African urban spaces, air quality sensing systems are deployed in outdoor settings characterised by high levels of pollution concentrations not common in the northern hemisphere and other settings [2,28,29]. High pollution levels have been found to affect the performance of low-cost sensors [30,31]. These environments may also be characterised by extreme tropical weather conditions and high levels of dust from unpaved roads. These conditions tend to affect the performance of air quality sensing systems and thus the design and setup needs to take into consideration these requirements. The air filters and inlets for these settings need to be designed to minimise the environmental impacts from dust, weather and insect infiltration. However, such design considerations should not affect the effectiveness of the sensors. Due to high levels of humidity specialised heating components may need to be designed. Jayaratne et al. [32] investigated the effect of humidity on the performance of low-cost air particle sensors. The operating outdoor conditions such as temperature ranges for the internal and external components should also be considered. Reference grade monitors or devices designed for other settings may need to be adapted to work in these unique conditions. Devices deployed in these settings will require regular preventive maintenance and therefore such devices need to be designed to allow for easy maintenance.

2.1.5. DR5: Embedded Quality Assurance and Calibration Methods

Environmental sensing systems deployed in outdoor urban spaces require mechanisms to ensure reliability, availability and quality of the data. An air quality sensor network cannot be entirely left to the community to maintain and do quality assurance. Therefore, a key consideration for low-cost sensor networks is calibration, in order to improve their accuracy and reliability [33–36]. In addition, the sensor network requires regular maintenance. In addition, local technical capacity needs to be built to ensure sustainability and reduced maintenance cost of sensors. This could be done through institutions such as universities, vocational schools, or private technicians within the areas of deployment. Therefore, these should be embedded in the design of the air quality sensing and management systems throughout the development and implementation processes. At the sensor development stage, in-lab analysis and co-location is necessary to ensure that all sensors behave the same way and are consistent. After production in-field comparison should also be considered to understand how the outdoor environment affects the operation of the sensors. A co-location in the outdoor environments should be considered as part of the calibration and field comparison. In the case of limited access to reference monitors, datasets from environments with similar contexts can be considered. Also mobile co-location can be considered where a device is transported to a nearby reference monitor, for example, by use of motorcycle taxis without having a fixed installation.

2.1.6. DR6: Community-Oriented and Active Engagements

Citizens and community members should have an active role in the design and implementation of a smart city sensing system for it to be sustainable and achieve uptake targets. Citizens can give input for example through co-design workshops on how air pollution affects them and desired options to consume data and hardware products. Citizens can host devices on their premises and provide device security. Citizens can host air quality sensors on their premises or their motorcycle in the case of mobile monitoring. Hosting of the sensor device should involve training and education on the purpose and goals of the air quality sensing system and how they benefit. When empowered, hosts of a sensor network can serve as champions in their communities and advocates for cleaner air. This initial engagement can also help to create demand for the resulting data and insights from the collected measurements. Where sensors are placed in schools, the air quality sensing system can be used to create awareness while data and sensors can be used in teaching mathematics and science subjects. Retention and sustained active engagement may require mechanism design approaches for example through incentives and rewards and analysis of barriers to participation [37]. In addition, communities may serve as a supplementary "sensor" of the environment, providing contextual metadata that this not trivial to collect using autonomous sensors. Members of the public can use their mobile phones to capture pictures and meta-data of air pollution events in their neighborhoods whenever high spikes are observed from the air quality sensing system. This can increase awareness of the causes of the pollution in the community. The additional datasets can be used to improve modelling and prediction of pollution events. Prediction and calibration models tend to perform poorly with the pollution spikes caused by local events, for example, spontaneous waste burning events are difficult to predict. The collection of additional metadata from community members can be used to enhance performance and accuracy of the prediction model for localised pollution events.

2.1.7. DR7: Digital Platforms to Drive Action and Policy

Environmental sensing systems collect huge amounts of datasets, which are in raw formats and are not of utility to the citizens or decision makers. Custom digital platforms need to be designed to increase timely access to air quality information and also link the data to action and facilitate data-driven decision making for leaders, communities, and citizens [38–40]. The data needs vary from a stakeholder group to another and thus segmentation of the potential users is necessary. For example, city urban planning and environmental managers may require trends and insights of variations of air quality in space and time, region-by-region comparison that can be best delivered through a web-based analytics platform. School teachers and parents require timely and near-real time information on air quality around schools or homes to understand the pollution exposures of their children. Joggers and cyclists require real-time information to find the cleanest route to be delivered to them via a mobile application on their smartwatch or smartphone. Restaurants and hotels may require API access to embedded the air quality information in their existing digital products. The access channels should also be taken into consideration to increase access and offer personalised air quality information to different user segments [41]. For example, people without smartphones could be reached through leaders in their communities or alternative communication channels such as campaigns, SMS, and radio.

3. Demonstration: The AirQo System in Selected African Cities

The above design space requirements have informed and inspired the design and implementation of the AirQo system. The AirQo system is a large-scale custom air quality sensing and management solution designed to be responsive to the needs and requirements of urban spaces in low-source settings. The AirQo system adheres to the design space requirements described above *DR1–DR7*. Initiated in 2015, the AirQo system has been implemented in major cities in Uganda, East Africa including: (1) Kampala, which is Uganda's largest and capital city with a population of over 1.6 million people and (2) Jinja, which is a city in Eastern Uganda with a population of over 0.3 million people and, (3) Fort Portal, a city with a population of over 0.06 million people and located in the South Western region of the country [42]. The AirQo system includes a network of over 120 sensors. The remainder of this section presents the key components of the AirQo system as the first instance of the design space for an air quality monitoring and management system for urban spaces in low-resource settings.

3.1. The AirQo Monitor

The AirQo monitor is a custom air quality monitoring device designed to be responsive to design space requirements: *DR1: Low-cost and high resolution air quality sensing; DR2: Portability and flexible deployment options; DR3: Flexible power and communication options; and DR4 Resilience to local environmental conditions.* One of the primary air quality indicators defined by the WHO is particulate matter due to the well studied impacts on health. AirQo devices measure both particulate matter $PM_{2.5}$ and particulate matter PM_{10} with an effective range of 0–500 $\mu g/m^3$. The device also measures other variables necessary for modelling and analysis such as longitude, latitude, internal (within the device) and external air temperature, relative humidity, and atmospheric pressure. The AirQo device is designed to support flexible power options for solar energy, mains, and capability to change from other available sources, for example, motorcycle alternator. Figure 2 shows the AirQo device. The AirQo device sensors use laser scattering technique for the measurement of particle size and concentration for $PM_{2.5}$ and PM_{10} in $\mu g/m^3$. Each device includes dual PMS5003 (https://www.plantower.com/en/products_33/74.html, accessed on 4 January 2023) sensors for measurement of $PM_{2.5}$ and PM_{10} for quality assurance and redundancy purposes. The acceptable threshold of in device dual sensor correlation is $\geq$0.99 and $R^2 \geq 0.98$. Sensors can be flagged for maintenance and/or corrective actions can be applied if correlation and R^2 values are out range of the set threshold. AirQo devices use cloud-based machine learning methods for calibration and improve the accuracy of measurements for $PM_{2.5}$ and PM_{10} to the RMSE of 7.2 $\mu g/m^3$ and 7.9 $\mu g/m^3$, respectively. The details of the evaluation of the accuracy and calibration of AirQo devices are presented in [33,43].

Dust from unpaved roads is one of the major contributors to particulate matter in many African urban spaces. However, dust deposition on sensors can lead to degradation. Resilience to degradation from dust and other outdoor conditions such as insect infestations is achieved through device design options including the choice of orientation of the sensors, air flow management, casing venting placement and alignment. The orientation of sensors and other components is optimized to minimize dust deposition on critical sensor components such as laser and photodiode. The device is designed to minimize air flow looping by separating the dual insets from dual outset with the sensing region. The casing venting is designed to be larger than sensor venting to reduce resistance in airflow. This is taken into account for both on the inset venting and outset venting. Combined sensor flow rate is 0.2 L/min. The casing venting placement is critical to further minimize dust deposition within the sensors. The design of the AirQo device also considers casing venting alignment versus the sensor venting. We also take a fine height offset from the casing venting to the sensor venting. The internal particle sensors have a mean time to failure (MTTF) of 3 years.

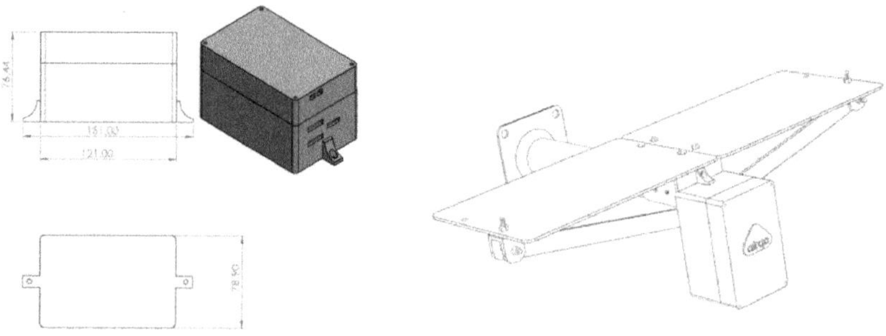

(a) (b)

Figure 2. A design of a portable low-cost air quality monitor. (**a**) AirQo monitor. (**b**) AirQo solar mount.

The 3D printed enclosures of the AirQo devices have gone through iterations of refinements, incorporating feedback from the communities, users and as well as the above portability and deployment requirements. For the mobile monitoring a custom mounting was designed with the local welding community to match the available installation options on the motorcycle. The position of the monitoring device on the motorcycle is aimed at understanding the exposure of the rider and passenger while minimising the direct emissions from the bike's tailpipe.

The data from the AirQo monitor is transmitted in near real-time to the cloud-based air quality data store. Each AirQo sensor transmits data on average every 90 s. Sensors can be optimised to transmit data in bulk where time sensitive data transmission is not a critical requirement. Bulk data transmission also improves the battery life. For each data record, the location of the pollution measurement is transmitted along with other secondary data such as battery levels used for network monitoring and maintenance. The monitor is designed to transmit data over the GSM network. The monitor relies on TCP/HTTP/HTTPS protocols to transfer and receive data accordingly. In addition, the AirQo monitor provides an option for local data storage. The AirQo monitor periodically writes data to a Micro SD card. The raw data is periodically subjected to the calibration system that is introduced in Section 3.3 and discussed in detail in [33]. The AirQo monitor is designed to provide a high resolution dataset over a 24 h period. To allow for this the monitor can be powered off mains and solar. The AirQo monitor has power input of 6V 1A. The consideration for the solar setup includes 2P 3.7V 4AH battery and 6V 1.25A panel.

3.2. *The AirQo Network*

The AirQo monitors are deployed in selected urban spaces to form a high resolution air monitoring network. The objectives of the AirQo system include: (1) collection of high resolution (spatial and temporal at average of 90 s) data on the scale and magnitude of air pollution, (2) improving accuracy of data from low-cost sensors through calibration,

(3) building a set of digital air quality tools and platforms to provide insights and evidence for decision making (4) using the data and platform to raise awareness and inform actions to mitigate air pollution. In the AirQo system, monitors are mounted on static or mobile objects including motorcycle taxis (Figure 3). Mobile monitoring is important to study pollution exposures and variations in the unique transport systems used in African urban spaces. For example, mounting of AirQo devices on motorcycle taxis (or *boda-bodas*) has allowed for understanding of exposures to emissions for the riders and passengers while at the same time providing high spatial coverage with fewer resources. To fulfil the requirement of *DR1 Low-cost and high resolution air quality sensing*, there is need to ensure that the air monitoring system is optimised for spatial resolution as well as the cost of setting up the network. From the experience of deploying AirQo sensors in cities a common question is *"How many devices are sufficient for a given city?"*. There is a risk of oversampling leading to higher cost or under-sampling leading to under-representation of the city. For the data to be useful and help drive localised actions air quality, there is need for a high resolution and a long term monitoring. This is important because it provides a representation of the state of air quality in a given geographical area of a city while adhering to the design requirement of *DR1 Low-cost and high resolution air quality sensing*.

(**a**) (**b**)

Figure 3. Flexible deployment options for high resolution air quality sensing. (**a**) Static pole deployment. (**b**) Mobile deployment.

To this end, the AirQo system includes a sensor placement model that determines the optimal number of clusters for deployment in a given geographical area [44]. The implementation of the sensor placement should be incremental in nature to allow for additional of new sensing devices depending on the need and available resources. The model uses a range of datasets and features including the geographical boundaries at the parish level (for the case of Ugandan cities), population density, household density, waste management practices in a given area, vegetation cover, km of road, power and fuel type. The output of the model is clusters of parishes with similar characteristics. Parishes in the same cluster are likely to have similar measurements and trends. The number of clusters is optimized to the available number of sensors. The sensor placement model is incorporated and deployed in the AirQo platform [44].

Figure 4 shows the output of the sensor placement model as applied to a selected urban city in Uganda. The colours represent city regions which are clustered as having similar properties (e.g., greenness and road density) while the black color dots in Figure 4a represent proposed sites for the available number of sensors. As new sensors are acquired the model can suggest new sites in regions with fewer installations as shown by the red color dots in Figure 4b.

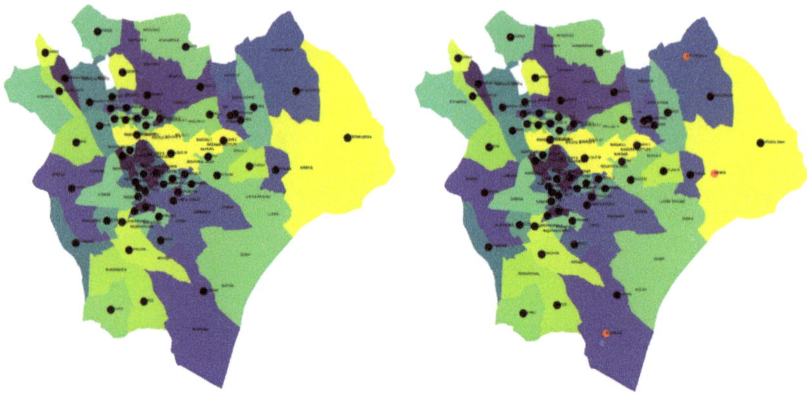

(**a**) (**b**)

Figure 4. Proposed installation sites that are recommended by the sensor placement model. (**a**) Initial sensor placement. (**b**) Refined sensor placement.

The automated placement is followed by the physical on ground validation of the suggested installation points to meet the need of practicability and feasibility. For example, the suggested location needs to meet the physical security and accessibility needs. Upon confirmation of the site, metadata about the site is collected to feed into the data models for the digital platforms. Examples of metadata include potential sources of air pollution in the area, distance from the road, installation height, contacts of the device host, and the power option selected.

3.3. Data Calibration and Quality Assurance

In order to meet the design space requirement *DR5: Embedded quality assurance and calibration methods* the AirQo systems has regular data calibration based machine learning approaches such as random forest, lasso regression model, and gradient boosting for calibration [45]. The calibration in the AirQo system is based on co-location data of low-cost monitors and reference grade monitors (BAM1022) for $PM_{2.5}$ and PM_{10}. The calibration model is developed using machine learning methods based on datasets from field co-locations. The key features considered for the model include hourly average for $PM_{2.5}$ and PM_{10}, atmospheric temperature, and relative humidity. The AirQo sensor evaluation and calibration process is presented in [33]. The AirQo calibration model has also been packaged to support third-party low-cost sensor calibration in the AirQalibrate product (https://airqalibrate.airqo.net/, accessed on 4 January 2023). Figure 5 shows the in-field comparison of data from the low-cost sensor network and the reference grade monitor for $PM_{2.5}$ deployed in an urban setting in Kampala, Uganda. Before calibration the correlation and RMSE between low-cost sensors and reference monitor are 0.9 and 18.58 $\mu g/m^3$, respectively (Figure 5a). After application of the calibration algorithm the correlation and RMSE improved to 0.96 and 7.22 $\mu g/m^3$ respectively for $PM_{2.5}$ (Figure 5b). The calibration for PM_{10} resulted in improved RMSE from 13.4 $\mu g/m^3$ to 7.9 $\mu g/m^3$ using lasso regression models (Figure 6b) [33]. These calibration tools and methods should be part of the data pipeline and in the digital platforms to facilitate continuous data calibration and quality assurance. A long term calibrated dataset from the AirQo network is described and available in [46].

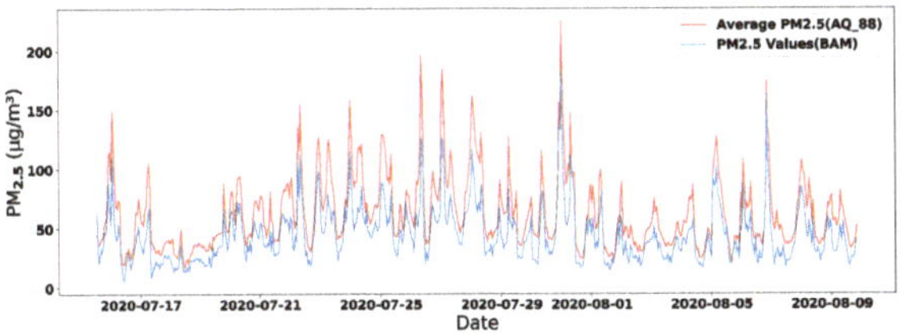

(**a**)

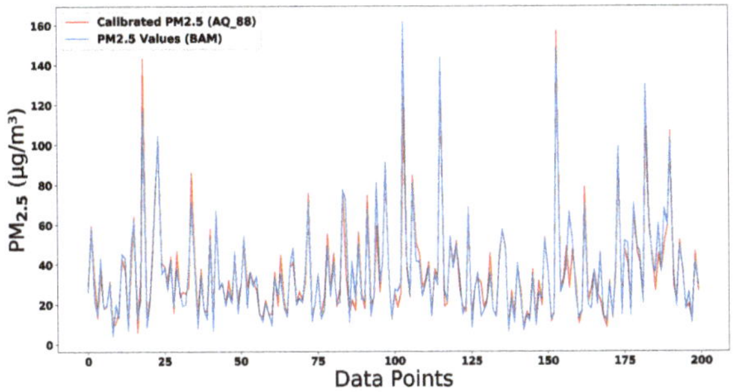

(**b**)

Figure 5. Calibration of low-cost air quality sensors using machine learning methods. (**a**) In-field comparison of AirQo low-cost sensors and reference grade before calibration. (**b**) $PM_{2.5}$ Calibrated low-cost sensor data vs reference grade monitor.

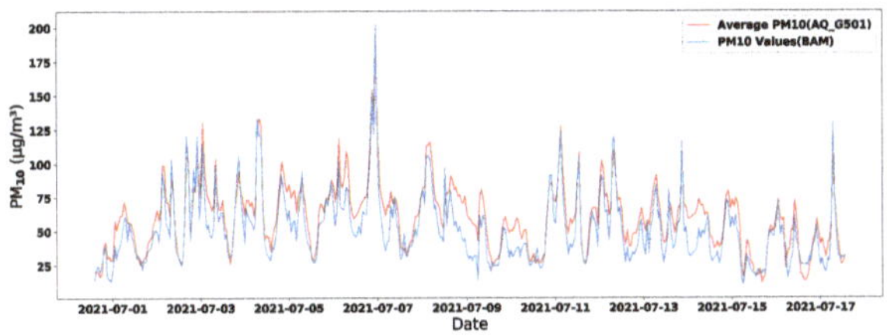

(**a**)

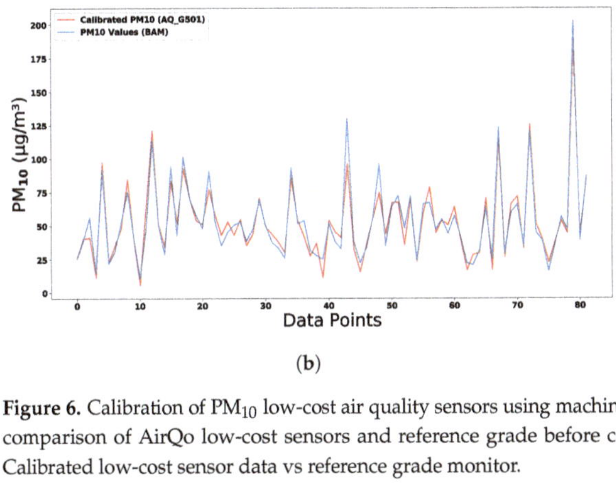

(**b**)

Figure 6. Calibration of PM_{10} low-cost air quality sensors using machine learning methods. (**a**) In-field comparison of AirQo low-cost sensors and reference grade before calibration for PM_{10}. (**b**) PM_{10} Calibrated low-cost sensor data vs reference grade monitor.

3.4. Digital Air Quality Platforms

In order to meet the design space requirement of *DR7: digital platforms to drive action and policy*, the AirQo systems has implemented a set of digital tools and platforms that fuse and extract insights from the various datasets in the data store. The AirQo digital platforms are designed to be extensible to allow for plugging-in of new datasets from other sensor networks as they become available. The design of the AirQo digital platforms follows the micro-service architecture, which allows for separation of concerns and functionalities into independent modules that exchange data through application programming interfaces (APIs). The raw data from the sensor network is sent in a cloud-based IoT data store for pre-processing that is fed into the calibration model. The resulting calibrated dataset is maintained in a separate datastore that is accessible via APIs for the different micro-services. Micro-services define specific functionality for example the AirQo platform for data access and analytics, the AirQo mobile app for presentation of air quality information in historical, location-based and near-real-time, and the network management tool for network monitoring and support. The digital tools consume the same data model but may process and render the data differently, similar to the model-view-controller model.

The following sections describe the key modules of the digital tools in the AirQo system.

3.4.1. The AirQo Platform

The AirQo platform (https://platform.airqo.net, accessed on 4 January 2023) targets decision makers (city authorities, environmental officers, politicians), researchers and advanced users. Figure 7 shows the AirQo digital analytics platform. The platform tool was developed in consultation with the city authorities from the selected cities in East Africa. The platform provides the ability to view calibrated datasets in different forms. It provides default analytics such as average pollution measurements clustered by locations within a given geographical area of a city. It also incorporates spatial and temporal insights as well as prediction in time and space for locations of interest. Users also have the ability to create custom charts for given locations, date ranges, and pollutants of interest. For advanced users and researchers the platform provides capability to download calibrated datasets in csv format for external data analysis and modelling. Through the platform, advanced users can request access to the AirQo API for data extraction using a programming language of choice.

Figure 7. Digital platforms providing trends and analysis of air quality for diverse locations in a city. The tool targets decision makers in a city to gain insights on how different parts of the city compare over a specified period of time.

3.4.2. The AirQo Mobile App

The air quality mobile app is designed and tailored for use by members of the public including parents, teachers, environmental champions, and community leaders. The mobile app (Figure 8) provides users options to view air quality insights per location of interest for the current time and forecast. The mobile app provides timely air quality information and has custom actions for users aimed at raising awareness and improving air quality. For example, depending on the advisory the user is provided with the actions they can do to minimise exposure or contribute to improving air quality in their space. The AirQo mobile app is available at Google play store and Apple App store for Android OS and iOS, respectively.

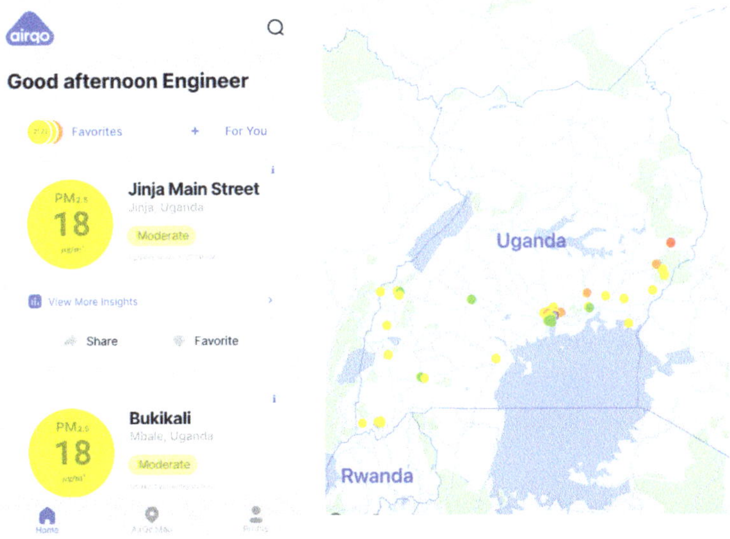

(**a**) (**b**)

Figure 8. A snapshot of the air quality data insights from the AirQo mobile app. The data is colored with advisory to inform decision making to minimise or reduce exposure to air pollution. Users have an option to favorite specific sites of interest for quick access. (**a**) AirQo mobile app targeting the public users. (**b**) Visualisation of AirQo sites across Uganda.

3.4.3. The AirQo Network Manager

As the air quality monitoring network scales, there is a need for a tool to track and manage devices, for example, sensors that are due for maintenance or not operating as expected. To address this need, the AirQo system includes a network management tool. The AirQo network manager provides capabilities to register new devices that are ready for deployment, record metadata on the site of installation, details about the sensor, host of the device, and the maintenance history. The network manager tool provides network-wide statistics and analysis that are necessary for management of a city air quality sensor network. The tool provides reports on the network up-time, individual sensor maintenance history, battery life and other parameters necessary for troubleshooting common sensor and network failures. The key target users of the tool are network technicians and people responsible for overseeing an urban air quality sensing system.

3.5. Linking Data and Technology to Action

One of the key objectives of the AirQo system is to translate the data and insights into action that can impact on improving air quality in urban spaces (*DR7: digital platforms to drive action and policy*). This is being achieved through amplifying access to the digital tools presented above as well deliberate offline actions that use data. The strategies used include citizen education, awareness, and policy engagements. Citizen education and awareness are achieved through use of data to conduct annual air quality awareness week with stakeholders from government and academia [47]. This is achieved through focus group discussions and data communication sessions that bring together participants from communities, local leaders, and industries within a specific geographical area. During the sessions key findings from the collected data and insights are shared. Further, citizen engagement is through engagement of air quality champions and community leaders who are provided access to the AirQo mobile app, educated about air pollution and use the data to drive change in their communities [48]. City technical persons are trained on the use of data and digital tools in managing the city urban environment and utilising evidence for decision making processes related to air quality management. For policy engagement, the data from the AirQo system is used to provide baseline data for formulating and monitoring of the national level or city-level air quality standards and action plans.

4. Discussion and Future Work

The AirQo system presented in this paper demonstrates the design and development of a smart city solution that is implemented in and for contexts of African urban spaces. A recent extensive scoping study [49] emphasized the need for contextualized strategies and systems to tackle air pollution in African cities. The design considerations put forward in this paper is a step towards such strategies. The technology and approaches presented have potential to be transferable to other cities in low-resource settings with similar contexts. Urban spaces in these settings share similar context with respect to the pollution profiles, transport systems, informal settlements and policy frameworks. For an urban sensing system to achieve the design impact, it should consider the seven design space requirements of: *DR1: Low-cost and high resolution air quality sensing; DR2: Designing for portability and flexible deployment options; DR3: Flexible power and communication options; DR4: Resilience to local environmental conditions; DR5: Embedded quality assurance and calibration methods; DR6: Community-oriented and active engagements; DR7: Digital platforms to drive action and policy.* Digital technology cannot achieve the desired impact in isolation and thus it is important to complement digital technology with "offline" activities that aim to widen access to data insights and information considering that many users and target stakeholders may lack access to Internet and smartphone devices in these contexts. The need for air quality sensing is present in many urban spaces and as such the sensing networks should be scaled to several urban spaces that meet the monitoring criteria, for example, dependent on the population density, pollution profile and risk. As next steps, the AirQo system is being scaled to more cities on the continent with the target more cities in Africa. Lessons gathered from scaling up of the AirQo system will further enrich and validate the design space requirements put forward in this paper. In order to impact on the air quality, approaches to enhance citizen engagement and behaviour change need to be further investigated. The datasets and tools from the AirQo systems also provide a baseline for further research on the impacts of air quality on health and economic outcomes. For example, previously the data from the AirQo system has supported studies on understanding the air pollution impacts on city competitiveness [50], macro predictions that fuse ground monitored data with other datasets [51], prediction and modelling studies [43], assessing pollution vulnerability for African cities [52], understanding impacts of air pollution on mobility [53,54], and characterisation of air pollution profiles for African cities [2]. Other areas of future work include investigation of incentive mechanisms to encourage use of data and behaviour change among citizens and decision makers. For example, the digital tools could be integrated with means to enable citizens to keep track and share individual actions done to contribute to cleaner air in their community. Reward system could be designed basing on the quality and effectiveness of individual contributions.

Author Contributions: E.B. (lead author) overall conceptualisation, writing, investigation, analysis, supervision, project administration, funding acquisition. J.S. contribution to the software and hardware development; D.O. review, project administration and network planning. All authors have read and agreed to the published version of the manuscript.

Funding: This work was supported by Google.org grant 1904-57882, EPSRC/GCRF grant EP/T00343 X/1, Belgium through the Wehubit programme implemented by Enabel Wehubit Grant Agreement BEL1707111-AP-05-20.

Institutional Review Board Statement: Not applicable.

Informed Consent Statement: Not applicable.

Data Availability Statement: The datasets used and/or analysed during the current study available at https://platform.airqo.net, accessed on 4 January 2023. Other relevant historical datasets available [46].

Acknowledgments: Authors would like to appreciate the feedback and discussions from the AirQo team and partners from government, academia, private sector and civil society. The authors appreciate the feedback received from the anonymous reviewers that helped improve the paper.

Conflicts of Interest: The authors declare no conflict of interest. The funders had no role in the design of the study; in the collection, analyses, or interpretation of data; in the writing of the manuscript; or in the decision to publish the results.

References

1. WHO. Billions of People Still Breathe Unhealthy Air: New WHO Data. 2014. Available online: https://www.who.int/news/item/04-04-2022-billions-of-people-still-breathe-unhealthy-air-new-who-data (accessed on 4 January 2023).
2. Okure, D.; Ssematimba, J.; Sserunjogi, R.; Gracia, N.L.; Soppelsa, M.E.; Bainomugisha, E. Characterization of Ambient Air Quality in Selected Urban Areas in Uganda Using Low-Cost Sensing and Measurement Technologies. *Environ. Sci. Technol.* **2022**, *56*, 3324–3339. [CrossRef]
3. Brauer, M.; Guttikunda, S.K.; Nishad, K.A.; Dey, S.; Tripathi, S.N.; Weagle, C.; Martin, R.V. Examination of monitoring approaches for ambient air pollution: A case study for India. *Atmos. Environ.* **2019**, *216*, 116940. . [CrossRef]
4. Li, H.Z.; Gu, P.; Ye, Q.; Zimmerman, N.; Robinson, E.S.; Subramanian, R.; Apte, J.S.; Robinson, A.L.; Presto, A.A. Spatially dense air pollutant sampling: Implications of spatial variability on the representativeness of stationary air pollutant monitors. *Atmos. Environ. X* **2019**, *2*, 100012. [CrossRef]
5. Basiri, M.; Azim, A.Z.; Farrokhi, M. Smart City Solution for Sustainable Urban Development. *Eur. J. Sustain. Dev.* **2017**, *6*, 71. [CrossRef]
6. Angelidou, M.; Psaltoglou, A.; Komninos, N.; Kakderi, C.; Tsarchopoulos, P.; Panori, A. Enhancing sustainable urban development through smart city applications. *J. Sci. Technol. Policy Manag.* **2017**, *9*, 146–169. [CrossRef]
7. Kandt, J.; Batty, M. Smart cities, big data and urban policy: Towards urban analytics for the long run. *Cities* **2021**, *109*, 102992. [CrossRef]
8. Snyder, E.G.; Watkins, T.H.; Solomon, P.A.; Thoma, E.D.; Williams, R.W.; Hagler, G.S.W.; Shelow, D.; Hindin, D.A.; Kilaru, V.J.; Preuss, P.W. The Changing Paradigm of Air Pollution Monitoring. *Environ. Sci. Technol.* **2013**, *47*, 11369–11377. [CrossRef] [PubMed]
9. Sivaraman, V.; Carrapetta, J.; Hu, K.; Luxan, B.G. HazeWatch: A participatory sensor system for monitoring air pollution in Sydney. In Proceedings of the 38th Annual IEEE Conference on Local Computer Networks—Workshops, Sydney, NSW, Australia, 21–24 October 2013; pp. 56–64. [CrossRef]
10. Labzovskii, L.D.; Hey, J.V.; Romanov, A.A.; Golovatina-Mora, P.; Belikov, D.; Lashkari, A.; Kenea, S.T.; Hekman, E. Who should measure air quality in modern cities? The example of decentralization of urban air quality monitoring in Krasnoyarsk (Siberia, Russia). *Environ. Sci. Policy* **2023**, *140*, 93–103. [CrossRef]
11. Jiao, W.; Hagler, G.; Williams, R.; Sharpe, R.; Brown, R.; Garver, D.; Judge, R.; Caudill, M.; Rickard, J.; Davis, M.; et al. Community Air Sensor Network (CAIRSENSE) project: Evaluation of low-cost sensor performance in a suburban environment in the southeastern United States. *Atmos. Meas. Tech.* **2016**, *9*, 5281–5292. [CrossRef]
12. Sevillano, X.; Socoró, J.C.; Alías, F.; Bellucci, P.; Peruzzi, L.; Radaelli, S.; Coppi, P.; Nencini, L.; Cerniglia, A.; Bisceglie, A.; et al. DYNAMAP—Development of low cost sensors networks for real time noise mapping. *Noise Mapp.* **2016**, *3*, 172–189. [CrossRef]
13. Suriano, D. A portable air quality monitoring unit and a modular, flexible tool for on-field evaluation and calibration of low-cost gas sensors. *HardwareX* **2021**, *9*, e00198. [CrossRef] [PubMed]
14. Völgyesi, P.; Nádas, A.; Koutsoukos, X.D.; Lédeczi, Á. Air Quality Monitoring with SensorMap. In Proceedings of the 2008 International Conference on Information Processing in Sensor Networks (IPSN 2008), St. Louis, MO, USA, 22–24 April 2008; pp. 529–530.

15. Dutta, P.; Aoki, P.M.; Kumar, N.; Mainwaring, A.; Myers, C.; Willett, W.; Woodruff, A. Common Sense: Participatory Urban Sensing Using a Network of Handheld Air Quality Monitors. In *SenSys09: Proceedings of the 7th ACM Conference on Embedded Networked Sensor Systems, Berkeley, CA, USA, 4–6 November 2009*; Association for Computing Machinery: New York, NY, USA, 2009; pp. 349–350. [CrossRef]
16. Fu, H.L.; Chen, H.C.; Lin, P. APS: Distributed air pollution sensing system on Wireless Sensor and Robot Networks. *Comput. Commun.* **2012**, *35*, 1141–1150. [CrossRef]
17. AirQo. AirQo, Africa's Leading Air Quality Monitoring and Management System. 2022. Available online: https://airqo.africa (accessed on 4 January 2023).
18. Maruping, L.M.; Venkatesh, V.; Agarwal, R. A Control Theory Perspective on Agile Methodology Use and Changing User Requirements. *Inf. Syst. Res.* **2009**, *20*, 377–399. [CrossRef]
19. Sanders, E.B.N. From user-centered to participatory design approaches. In *Design and the Social Sciences*; CRC Press: Boca Raton, FL, USA, 2002; pp. 18–25.
20. Hauck, H.; Berner, A.; Gomiscek, B.; Stopper, S.; Puxbaum, H.; Kundi, M.; Preining, O. On the equivalence of gravimetric PM data with TEOM and beta-attenuation measurements. *J. Aerosol Sci.* **2004**, *35*, 1135–1149. [CrossRef]
21. Castell, N.; Dauge, F.R.; Schneider, P.; Vogt, M.; Lerner, U.; Fishbain, B.; Broday, D.; Bartonova, A. Can commercial low-cost sensor platforms contribute to air quality monitoring and exposure estimates? *Environ. Int.* **2017**, *99*, 293–302. [CrossRef] [PubMed]
22. Petkova, E.P.; Jack, D.W.; Volavka-Close, N.H.; Kinney, P.L. Particulate matter pollution in African cities. *Air Qual. Atmos. Health* **2013**, *6*, 603–614. [CrossRef]
23. Perelló, J.; Cigarini, A.; Vicens, J.; Bonhoure, I.; Rojas-Rueda, D.; Nieuwenhuijsen, M.J.; Cirach, M.; Daher, C.; Targa, J.; Ripoll, A. Large-scale citizen science provides high-resolution nitrogen dioxide values and health impact while enhancing community knowledge and collective action. *Sci. Total. Environ.* **2021**, *789*, 147750. [CrossRef]
24. Wallace, J.; Corr, D.; Deluca, P.; Kanaroglou, P.S.; McCarry, B.E. Mobile monitoring of air pollution in cities: The case of Hamilton, Ontario, Canada. *J. Environ. Monit.* **2009**, *11 5*, 998–1003. [CrossRef]
25. Elen, B.; Peters, J.; Poppel, M.V.; Bleux, N.; Theunis, J.; Reggente, M.; Standaert, A. The Aeroflex: A Bicycle for Mobile Air Quality Measurements. *Sensors* **2013**, *13*, 221–240. [CrossRef]
26. Awokola, B.I.; Okello, G.; Mortimer, K.J.; Jewell, C.P.; Erhart, A.; Semple, S. Measuring Air Quality for Advocacy in Africa (MA3): Feasibility and Practicality of Longitudinal Ambient PM2.5 Measurement Using Low-Cost Sensors. *Int. J. Environ. Res. Public Health* **2020**, *17*, 7243. [CrossRef]
27. Zennaro, M.; Bagula, A.; Nkoloma, M. From Training to Projects: Wireless Sensor Networks in Africa. In Proceedings of the 2012 IEEE Global Humanitarian Technology Conference, Seattle, WA, USA, 21–24 October 2012; pp. 417–422. [CrossRef]
28. Amegah, A.K.; Agyei-Mensah, S. Urban air pollution in Sub-Saharan Africa: Time for action. *Environ. Pollut.* **2017**, *220*, 738–743. [CrossRef] [PubMed]
29. Kirenga, B.; Meng, Q.; van Gemert, F.; Aanyu-Tukamuhebwa, H.; Chavannes, N.; Katamba, A.; Obai, G.; Molen, T.; Schwander, S.; Mohsenin, V. The State of Ambient Air Quality in Two Ugandan Cities: A Pilot Cross-Sectional Spatial Assessment. *Int. J. Environ. Res. Public Health* **2015**, *12*, 8075–8091. [CrossRef] [PubMed]
30. Liu, H.Y.; Schneider, P.; Haugen, R.; Vogt, M. Performance Assessment of a Low-Cost PM2.5 Sensor for a near Four-Month Period in Oslo, Norway. *Atmosphere* **2019**, *10*, 41. [CrossRef]
31. Khreis, H.; Johnson, J.; Jack, K.; Dadashova, B.; Park, E.S. Evaluating the Performance of Low-Cost Air Quality Monitors in Dallas, Texas. *Int. J. Environ. Res. Public Health* **2022**, *19*, 1647. [CrossRef] [PubMed]
32. Jayaratne, R.; Liu, X.; Thai, P.; Dunbabin, M.; Morawska, L. The influence of humidity on the performance of a low-cost air particle mass sensor and the effect of atmospheric fog. *Atmos. Meas. Tech.* **2018**, *11*, 4883–4890. [CrossRef]
33. Adong, P.; Bainomugisha, E.; Okure, D.; Sserunjogi, R. Applying machine learning for large scale field calibration of low-cost PM2.5 and PM10 air pollution sensors. *Appl. AI Lett.* **2022**, *3*, e76. [CrossRef]
34. Maag, B.; Zhou, Z.; Thiele, L. A survey on sensor calibration in air pollution monitoring deployments. *IEEE Internet Things J.* **2018**, *5*, 4857–4870. [CrossRef]
35. Spinelle, L.; Gerboles, M.; Villani, M.G.; Aleixandre, M.; Bonavitacola, F. Field calibration of a cluster of low-cost available sensors for air quality monitoring. Part A: Ozone and nitrogen dioxide. *Sens. Actuators B Chem.* **2015**, *215*, 249–257. [CrossRef]
36. Giordano, M.R.; Malings, C.; Pandis, S.N.; Presto, A.A.; McNeill, V.F.; Westervelt, D.M.; Beekmann, M.; Subramanian, R. From low-cost sensors to high-quality data: A summary of challenges and best practices for effectively calibrating low-cost particulate matter mass sensors. *J. Aerosol Sci.* **2021**, *158*, 105833. [CrossRef]
37. Wehn, U.; Almomani, A. Incentives and barriers for participation in community-based environmental monitoring and information systems: A critical analysis and integration of the literature. *Environ. Sci. Policy* **2019**, *101*, 341–357. [CrossRef]
38. Silva, J.; Lucas, P.; Araújo, F.; Silva, C.; Gil, P.; Cardoso, A.; Arrais, J.; Ribeiro, B.; Coutinho, D.; Salgueiro, P.; et al. An Online Platform For Real-Time Air Quality Monitoring. In Proceedings of the 2019 5th Experiment International Conference (Exp.at'19), Funchal, Portugal, 12–14 June 2019; pp. 320–325. [CrossRef]
39. Kosmidis, E.; Syropoulou, P.; Tekes, S.; Schneider, P.; Spyromitros-Xioufis, E.; Riga, M.; Charitidis, P.; Moumtzidou, A.; Papadopoulos, S.; Vrochidis, S.; et al. hackAIR: Towards Raising Awareness about Air Quality in Europe by Developing a Collective Online Platform. *ISPRS Int. J. Geo-Inf.* **2018**, *7*, 187. [CrossRef]

40. Kumar, P.; Morawska, L.; Martani, C.; Biskos, G.; Neophytou, M.; Sabatino, S.D.; Bell, M.; Norford, L.; Britter, R. The rise of low-cost sensing for managing air pollution in cities. *Environ. Int.* **2015**, *75*, 199–205. [CrossRef] [PubMed]
41. Schulte, K. 'Real-time' air quality channels: A technology review of emerging environmental alert systems. *Big Data Soc.* **2022**, *9*, 1–25. [CrossRef]
42. Uganda Buereau of Statistics (UBOS). *Statistical Abstract, 2020*; UBOS: Kampala, Uganda, 2020.
43. Coker, E.S.; Amegah, A.K.; Mwebaze, E.; Ssematimba, J.; Bainomugisha, E. A land use regression model using machine learning and locally developed low cost particulate matter sensors in Uganda. *Environ. Res.* **2021**, *199*, 111352. [CrossRef] [PubMed]
44. AirQo. Locate: AirQo Sensor Placement. 2022. Available online: https://docs.airqo.net/airqo-handbook (accessed on 4 January 2023).
45. Bonaccorso, G. *Machine Learning Algorithms: A Reference Guide to Popular Algorithms for Data Science and Machine Learning*; Packt Publishing: Birmingham, UK, 2017.
46. Sserunjogi, R.; Ssematimba, J.; Okure, D.; Ogenrwot, D.; Adong, P.; Muyama, L.; Nsimbe, N.; Bbaale, M.; Bainomugisha, E. Seeing the air in detail: Hyperlocal air quality dataset collected from spatially distributed AirQo network. *Data Brief* **2022**, *44*, 108512. [CrossRef]
47. NEMA. Stakeholders Call for Multi-Sectoral Approach to Reduce Air Pollution. 2021. Available online: https://nema.go.ug/sites/default/files/Air%20Quality%20Awareness%20Week%20Statement.pdf (accessed on 4 January 2023).
48. AirQo. Helping Communities Combat Air Pollution through Digital Technologies. 2022. Available online: https://blog.airqo.net/helping-communities-combat-air-pollution-through-digital-technologies-6a5924a1e1e (accessed on 4 January 2023).
49. Okello, G.; Nantanda, R.; Awokola, B.; Thondoo, M.; Okure, D.; Tatah, L.; Bainomugisha, E.; Oni, T. Air quality management strategies in Africa: A scoping review of the content, context, co-benefits and unintended consequences. *Environ. Int.* **2023**, *171*, 107709. [CrossRef]
50. Bank, W. Pollution Management and the Making of Prosperous. 2020. Available online: https://documents1.worldbank.org/curated/en/716251593371242345/pdf/Final-Report.pdf (accessed on 4 January 2023).
51. Atuhaire, C.; Gidudu, A.; Bainomugisha, E.; Mazimwe, A. Determination of Satellite-Derived PM2.5 for Kampala District, Uganda. *Geomatics* **2022**, *2*, 125–143. [CrossRef]
52. Clarke, K.; Ash, K.; Coker, E.S.; Sabo-Attwood, T.; Bainomugisha, E. A Social Vulnerability Index for Air Pollution and Its Spatially Varying Relationship to PM2.5 in Uganda. *Atmosphere* **2022**, *13*, 1169. [CrossRef]
53. Galiwango, R.; Bainomugisha, E.; Kivunike, F.; Kateete, D.P.; Jjingo, D. Air pollution and mobility patterns in two Ugandan cities during COVID-19 mobility restrictions suggest the validity of air quality data as a measure for human mobility. *Environ. Sci. Pollut. Res.* **2022**, 1–16. [CrossRef]
54. Green, P.; Okure, D.; Adong, P.; Sserunjogi, R.; Bainomugisha, E. Exploring PM2.5 variations from calibrated low-cost sensor network in Greater Kampala, during COVID-19 imposed lockdown restrictions: Lessons for Policy. *Clean Air J.* **2022**, *32*, e10906. [CrossRef]

atmosphere

MDPI

Article

A Smoke Chamber Study on Some Low-Cost Sensors for Monitoring Size-Segregated Aerosol and Microclimatic Parameters

László Bencs * and Attila Nagy

Department of Applied and Nonlinear Optics, Institute for Solid State Physics and Optics, HUN-REN Wigner Research Centre for Physics, P.O. Box 49, H-1525 Budapest, Hungary; nagy.attila@wigner.hun-ren.hu
* Correspondence: bencs.laszlo@wigner.hun-ren.hu; Tel.: +36-1-392-2222 (ext. 1684)

Abstract: Low-cost sensors (LCSs) of Geekcreit $PM_1/PM_{2.5}/PM_{10}$ (based on a PMS5003 sampler) and BOHU BH-1 models A3 and B3 (based on a Pando G7 sampler) were compared for different aerosol size ranges using a research-grade instrument (Grimm 1.109) under controlled laboratory conditions. An aerosol generator was utilized to produce various sizes of monodispersed particulate matter (PM), which was introduced into a laboratory smoke chamber under resistance heating/cooling and/or varying RH conditions. In addition, the accuracy of the air temperature (T) and relative humidity (RH) sensors of the LCSs were assessed against calibrated, laboratory-grade instruments. The study LCSs showed generally accurate readings for $PM_{2.5}$, irrespectively of the slow T and/or RH changes, which provided apt conditions for accurate calibration slopes (S) and low intercepts/bias (b) of the linear fits. On the other hand, PM_1 and PM_{10} readings slightly deviated from those observed with the reference monitor, likely due to the lower detection efficacy of the LCSs towards fine and coarse PM. Varying RH influenced the S and b values, showing its impact on the detection efficacy of LCSs. Under low/medium RH, homoscedastic calibration curves of PM_x were found, whereas rather heteroscedastic calibration plots were observed at high RH. For T calibration, low RH in the smoke chamber provided more reproducible conditions in terms of lower measurement bias for LCSs as recorded against a calibrated, reference-grade thermometer.

Keywords: low-cost environmental sensor; particulate matter; calibration; optical particle counter; smoke chamber; laboratory calibration

1. Introduction

Air pollution is a research topic investigated worldwide due to the rising problems in the surrounding environment, related mostly to anthropogenic activities, manifested in human health effects [1], damage to cultural heritage [2,3], and displayed historical items [2,4], as well as the impact of anthropogenic air components on global climate [5]. For the past decade, low-cost sensors (LCSs) have been gaining a broader interest in air quality research [6–41] compared to laboratory-based instrumental methods, for instance, those applied in official air quality (AQ) stations. This is particularly due to their light weight, small size, and low electricity consumption, thus making them easier to deploy in the field [42–44]. However, LCSs can generally be operated at the expense of lower sensitivity, less accuracy, and robustness, compared to official instrumentation of air sampling and analysis [12]. Additionally, LCSs demand more frequent calibration and comparison with reference methods, depending on several factors, such as the sampling site emission and the sampled aerosol type, for instance, by means of using research-grade instruments and/or data from official AQ stations [6–12,17]. Recently, it has been shown that LCSs can be calibrated with good accuracy using commercial, DIY-kit-like electrical/chemical apparatus, which are accessible to society at large [41]. Certainly, based on these principles, it would be easier to build up nodes for AQ monitoring, even a network over areas of different anthropogenic impacts, as was already demonstrated in the literature [45].

Investigations on the laboratory testing of various types of LCSs have received considerable attention from the scientific society [6–40]. Examples of such studies towards more accurate mass calibration and determinations are reported for sensors of Plantower PMS 1003 [6,8,27,38], PMS 3003 [6,9,11], PMS A003 [7,10,18,20,36,39], PMS 5003 [8,13,15–17,21,37], PMS 6003 [13], Shinyei PPD42NS [10,14,15,34,38], Alphasense OPC-N3 [13,15,39], AirU [11,12], Nova SDS011 [14,38], Sensirion SPS30 [15,16,39], Sharp GP2Y1010AU0F [15,38], Omron B5W-LD0101 [15], Honeywell HPMA115S0 [16], BOHU BH1-A3 [17], PMS7003 [19], PurpleAir [22–26], AQMesh [28,29], SKC Split 2 [30], Microdust Pro [30], DataRam [30], Dylos [31,32], OPC-N2 [31,33,35], Dfrobot SEN0177 [32], DSM501A [34], GP2Y11AU0F [34], Innociple PSM305 [38], Nova SDL607 [38], and AirBeam2 [39]. The technical specifications of these LCSs have already been discussed in detail in comprehensive reviews, e.g., Ref. [43].

For the purposes of laboratory aerosol generation and subsequent PM mass measurements by means of LCSs, various techniques/materials have been applied, e.g., incense burning [7,25,36], oleic acid [7], NaCl [7,34,39], sucrose [34], talcum powder [7], cooking emissions [25], monodispersed polystyrene latex spheres [7], dioctyl sebacate [13,14], ammonium nitrate [9,34], aluminum oxide [9,30], industrial dust [30], dust mite [31], pollen [31], cat and dog fur [31], monodisperse silica [31], melamine resin [31], tobacco smoke [32], coal dust [37], cigarette and match lighting smoke [38], concrete dust [38], road dust [39], and poly-alpha-olefin oil [39] under controlled laboratory conditions.

In general, the concentration of PM had the most dominant effect on the sensors' responses, while the particle type and size distribution less affected the accuracy of the readings [29]. Besides these, a couple of ambient microclimatic parameters, such as air temperature (T_a) and relative humidity (RH), have been reported to influence the extent of bias for PM-mass/concentration readings, e.g., [8,17,27,34]. Further methodological improvements include the application of extended/refined calculation methods for PM mass [23], the use of deep-learning calibration techniques [40], and the application of various sets of internet of things (IoTs) accessible to society at large [41]. More details on the peculiarities of the application for LCSs can be found in extensive reviews [42–44].

In a former study [17], the accuracy of GPM and BOHU BH1-A3 sensors was assessed in the indoor and outdoor air of various urban sites. It was found that both types of sensors were fairly accurate between each other and with those aerosol monitors in the field campaigns, set in a nearby official AQ station for $PM_{2.5}$. This type of evaluation is a basic testing protocol for sensors, as suggested by US EPA [46]. Moreover, the obtained data [17] covered only a fairly short mass range of aerosols, i.e., up to about 130 $\mu g/m^3$ for PM_{10}, compared to those occurring in seriously polluted urban environments. In addition, no accuracy testing was possible for PM_1 and PM_{10} due to a lack of reference monitoring data. Laboratory measurements could be controlled in a more flexible way compared to those occurring in the field, which is in line with US EPA recommendations for advanced ("enhanced") testing of environmental PM sensors [46]. To fill in the above gap, in this study, we aimed at the laboratory evaluation and calibration of two types of low-cost sensors, i.e., PMS-5003-based Geekcreit© and Pando G7-based BOHU BH-1 (common model A3 and a newer design B3) for PM_1, $PM_{2.5}$, and PM_{10}, utilizing research-grade reference instrumentation. Additionally, we studied some of the ambient microclimatic parameters such as T_a and RH against calibrated, industrial-grade analog and digital measurement units, which is expected to provide sharper insight into the measurement accuracy of the study LCSs.

2. Materials and Methods

2.1. Instrumentation

Three types of LCSs, such as Geekcreit© model $PM_1/PM_{2.5}/PM_{10}$ (China) (further on referred to as GPM) and BOHU BH1 Models A3 and B3 (Bohu IoT Enterprise,

Shanxi, China), were compared under laboratory conditions. GPMs incorporate Plantower PMS5003 (Beijing, China) optical particulate counter (OPC), whereas each BH1 model is built with a Pando G7 OPC. Each LCS reports data in six size ranges of PM (bins), of which have lower size limits of 0.3, 0.5, 1.0, 2.5, 5.0, and 10 μm, respectively. Briefly, during operation, the OPCs utilize continuous sampling of ambient air in a low-volume measurement chamber, where light irradiation from a red laser or LED light source is applied. The multi-angle light scattering on the sampled air suspended particles is measured with a light detector. According to the recorded intensity and the angle of the scattered light, temporary varying transients are recorded and processed with a built-in 32-bit microcomputer (BH1 sensors) or with a laptop running a datalogger/control software for GPMs. The analytical results are obtained as a function of the equivalent aerodynamic diameter (EAD) of PM. The number of suspended, size-segregated PM in a unit air volume can be obtained, from which the PM mass concentration can be calculated with algorithms developed specifically for each LCS design by taking into account aerosol parameters like EAD and predefined average density of each size fraction, etc. Overview pictures of the study LCSs are plotted in Figure 1.

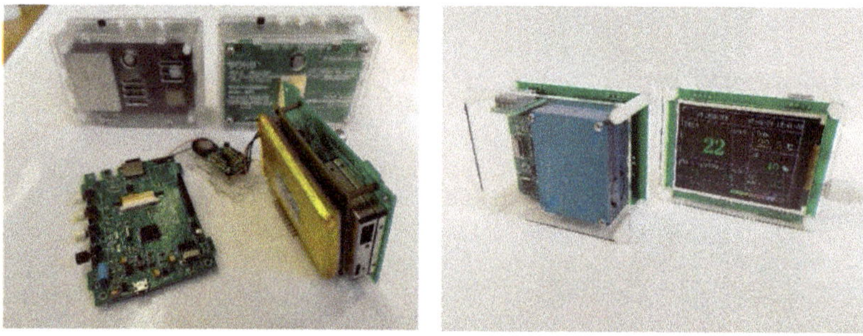

(**a**) (**b**)

Figure 1. Pictures of the study LCSs: (**a**) back panels of BH1-B3 and BH1-A3 (background), dismantled A3 sensor: front panel/micro PC board with LCD display (left), and the Pando G7 sampler/battery (grey/yellow units at the right), (**b**) front and back panel view of the GPM: PMS5003 sampler (blue module), T_a/RH sensor (grey module at the top), and interface/micro-PC board (left).

Besides suspended PM, the study LCSs could register T_a and RH of the monitored environment. For this purpose, the GPM sensor was equipped with an ASAIR AW2120 (Oneyac, Xiamen, China), while BOHU BH1 (both models A3 and B3) was fitted with an SHT20 module (Sensirion AG, Stäfa, Switzerland).

To check the accuracy of the PM_x data recorded with the LCSs and for calibration purposes, a Grimm Model 1.109 particle number counter was applied concurrently. This type of monitor is also an OPC, which collects aerosol data in 31 size ranges (bins), ranging from 0.25 μm to 32 μm, besides calculating PM_1, $PM_{2.5}$, and PM_{10} with a proprietary algorithm.

For the experiments, the LCSs were placed inside a 115 cm × 50 cm × 55 cm smoke chamber, built from 6 mm plexiglass walls, fitted with aluminum frames and rubber sealings specifically designed for aerosol investigations (Figure 2). The ambient air of the test laboratory was continuously cleaned using a Roger HEPA air purifier (Stadler Form, Zug, Switzerland) with HEPA and active carbon filters for PM and volatile organic carbon (VOC) compounds. Thus, the smoke chamber received the same clean air when it was open. In each experiment, the LCSs were operated simultaneously with synchronized clocks to monitor T_a, RH, and PM_x.

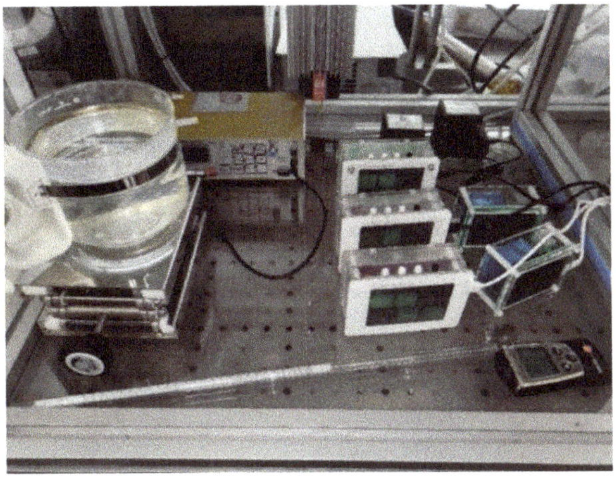

Figure 2. The smoke chamber with the installed aerosol sensors and microclimatic measurement apparatus (left side in the background: GRIMM OPC; middle: BH1-A3/B3 sensors; right side: GPM sensors and Testo).

2.2. Measurement Procedures

The T_a inside the smoke chamber was measured under controlled heating conditions by either raising the T_a with resistance heaters or preheating the cell and letting it slowly cool down. The smoke chamber is equipped with two resistance heating elements equipped with heatsinks, connected to a digital LAE 5X type LTR-5TSRD-A controller (LAE Electronic, Oderzo, Italy), which is meant to increase the internal T_a stepwise, i.e., about 0.2 °C/min. The T_a in the smoke chamber was determined via the application of the temperature sensors of the LCSs (GPM and BOHU), a Testo model 610 (Testo SE & Co. KGaA, Lenzkirch, Germany) T/RH meter, against an analog mercury-in-glass thermometer (MTA KUTESZ, Budapest, Hungary). For T_a determinations, the Testo device has a resolution of 0.1 °C and an accuracy of ±0.5 °C (measurement range: −10/50 °C), while for RH, it has a resolution of 0.1% in the range 0–100%, whereas the accuracy is ±0.5% for the range 5–95%. For T_a and RH, the GPM's sensor (AW2120) has an accuracy of ±0.5 °C and ±3%, respectively, while the SHT20 sensor has an accuracy of ±0.3 °C (range: 5–60 °C) and ±3% (range: 20–80%), respectively. The glass thermometer has a measurement range of 0–50 °C, a resolution of 0.1 °C, and an accuracy of ±0.1 °C. Each of these devices was calibrated by the manufacturer.

The indoor RH of the laboratory was generally low, ranging between 20 and 45%, due to the continuous operation of a combined air filter unit and drier. The air RH in the smoke chamber was controlled by various means. Low RH conditions (20–30%) were adjusted using the air drier, whereas high RH conditions were attained with the application of about 0.75 L of freshly boiled water, poured into a broad heat-resistant vessel (diameter: 15 cm), which was instantly inserted into the smoke chamber. Soon after, this condition provided high RH conditions, amounting to about 99%. Although RH gradually decreased with time inside the chamber, it provided near-steady-state conditions, which was advantageous for registering the gradually changing value with various LCSs and the Testo reference device (precision: ±2.5%). Medium RH conditions were registered at decreasing humidity level via performing a signal reading every 15 min.

Monodispersed polystyrene DVB microspheres (NIST Traceable, DRI-CAL, Duke Scientific Corporation, London, UK) with 500 nm or 1000 nm aerodynamic diameter (standard deviation: 10%) were applied as particle standards. The powders were solved in some mL of water and utilized as study aerosols with a GRIMM model 7.811 aerosol generator (Durag Group, Hamburg, Germany), which was operated for about 4–6 min to produce aerosol puffs in the smoke chamber. The PM concentrations obtained in this way were high enough to perform a large number of concurrent readings with the sensors. The PM experienced slow exponential decay due to a minor loss of air through small fissures of the smoke chamber. Overall, the ambient air inside the chamber provided near-steady-state conditions, which was proper for recording the studied environmental parameters. There was no significant difference observable between PM_x readings of the LCSs using the two sizes of monodispersed aerosol. The durability of the GPM and the BH1 sensors in the indoor environment, such as the smoke chamber (generally with a "clean" atmosphere) or indoor buildings [17], is very good (experienced via some years of continuous running). Nevertheless, they require cover or shelter outdoors against harsh weather conditions (wind, rain, high moisture, and other atmospheric events). Outdoors, weathering even on the sheltered/housed sensors could be observed after around one year of continuous operation, as found during field experiments in urban environments reported in Ref. [17].

2.3. Data Evaluation and Statistical Methods

The Microsoft Excel program was applied for the statistical analysis of the measurement data. First, the monitoring data were processed and filtered for outliers related to biased readings of the LCSs, e.g., the onsets of the puffs and internal sensor calibration cycles. These data were discarded from further processing. The AQI data were calculated by utilizing the stepwise function to the data sets of $PM_{2.5}$, according to US EPA recommendations [46]. Linear regression fittings to the data points were performed using the least square method. The regression equations for slopes (S) and intercepts (b) and the correlation coefficients (R values) of the linear fits were calculated to obtain insight into the possibility of using these functions as calibration curves for performing the determinations more accurately with the study LCSs. Moreover, two-paired Pearson's correlation coefficients were calculated at a 99% confidence level for each monitored variable and the AQI. The limit of detection (LOD) and the limit of quantitation (LOQ), corresponding to the average baseline concentration plus the noise/fluctuation (σ) of the baseline with 3σ and 10σ confidence, respectively, were calculated for each study LCS following the IUPAC recommendations for instrumental analytical methods [47]. For calculating the average baseline concentration, we took into account the lowest PM concentrations detected in the cleaned, aerosol-free smoke chamber by the sensors, while for σ, the fluctuation/standard deviation of the minimum readings was implied, which still gives the lowest signal increase for the study LCS.

Some additional performance indicators, such as the relative mean squared error ($RMSE$), the mean normalized error (MNE), and the mean normalized bias (MNB), were calculated to assess the deviation of concurrent readings of the study LCSs as suggested in Refs. [28,46]. These error flags were calculated for n number of data pairs according to the next equations:

$$RMSE = \sqrt{\frac{1}{n}\sum_{i=1}^{n}(y_i - x_i)^2} \quad (1)$$

$$MNE = \frac{1}{n}\sum_{i=1}^{n}\frac{|y_i - x_i|}{x_i} \times 100 \quad (2)$$

$$MNB = \frac{1}{n}\sum_{i=1}^{n}\frac{y_i - x_i}{x_i} \times 100 \quad (3)$$

where y_i and x_i are the concurrently recorded readings of the same variable from the sensors, e.g., LCS vs. the reference, respectively. In terms of sensitivity, $RMSE$ is more sensitive to outlier values, while MNE is more robust in this sense, whereas MNB shows the extent of under and overestimation. An accurate calibration procedure for the LCSs, which meets the Data Quality Objective of the European Commission (EC) [48], should give low errors in terms of $RMSE$ and MNE and should not invoke high unwanted bias, i.e., the MNB should be low as well. In this study, the US EPA recommendations for the base and enhanced testing [46] were applied for the evaluation of the sensors. Accordingly, for PM_x determinations, the following parameters were included for the linear regression fit: $S = 1.00 \pm 0.35$ (intercept of fitting between -5 and 5 $\mu g/m^3$), linearity: $R \geq 0.8366$, error: $RMSE < 7$ $\mu g/m^3$, MNE, $(MNB) < 30\%$. Similar methodological considerations were applied for evaluating the measurement data and errors for T_a and RH determinations.

3. Results

3.1. Ambient Microclimatic Measurements in the Smoke Chamber

3.1.1. Air Temperature

As the first stage of the investigation, the T_a in the smoke chamber was assessed with various thermometers by means of adjusting low, medium, or high RH. These data were collected every 5 min when heating the chamber and every 15 min when letting it cool. The low, medium, and high RH values covered the range between 25 and 40%, 40–70%, and 70–99%, respectively (Figure 3). As it appears, low RH conditions generally provide better regression fits to the concurrent measurement points, as manifested in the R values of the regression plots ranging from 0.97 to 0.99 (Figure 3a). The S of the fit equations are 1.09, 1.15, 1.1, 1.2, and 0.92 for GPM-1, GPM-2, BH1-A3, BH1-B3, and Testo, respectively. The bias for T_a, as appears in the intercepts of the fits, is slightly high for GPM-2, BH1-A3, and BH1-B3, and is 3.4, -2.1, and -2.7 °C, respectively, while it is still acceptable for Testo (1.7 °C) and GPM-1 (-0.95 °C), compared to the accuracy specified by the manufacturer.

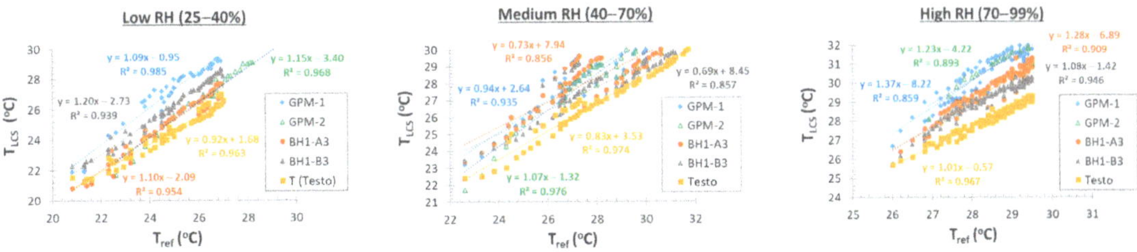

(a) (b) (c)

Figure 3. Air temperature recorded in the smoke chamber with LCSs, plotted against a reference mercury thermometer under (**a**) low, (**b**) medium, and (**c**) high RH (the color of each equation corresponds to that of the data label/marker).

Applying medium RH in the smoke chamber, the R values of the linear fits for T_a fell into the 0.925–0.988 range (Figure 3b). Interestingly, medium RH provided higher deviation in the S of the fits from the ideal unity, i.e., 0.94, 1.07, 0.73, 0.69, and 0.83, for GPM-1, GPM-2, BH1-A3, BH1-B3, and Testo, respectively. The intercepts of these fits for GPM-1 and GPM-2 (2.6 and -1.3 °C, respectively) are still acceptable, whereas, for BH1-A3, BH1-B3, and Testo, they are rather high, i.e., 7.9, 8.5, and 3.5 °C, respectively. The former two values are likely a contribution of the internal heat of the BH1 sensors from the built-in, preheated gas detector serving for VOC monitoring.

Under high RH conditions (Figure 3c), the R values of the linear fits are in the range of 0.926 and 0.983, indicating strong correlations. The S values of the linear fits are somewhat increased for GPM-1, GPM-2, and BH1-A3, i.e., 1.37, 1.23, and 1.28, respectively, whereas they approach unity well for BH1-B3 (1.07) and Testo (1.01). The intercept is negligibly small for the latter two LCSs (-1.4 and -0.57 °C), while increased values were experienced for GPM-1, GPM-2, and BH1-A3, i.e., -8.2, -4.2, and -6.9 °C, respectively.

Overall, for T calibration, it appears that the low RH provides more reproducible conditions in the present smoke chamber in terms of the lowest measurement biases for LCSs as recorded against a calibrated, reference-grade thermometer. Applying medium or high RH introduced an increased deviation of concurrently recorded readings. This could either be due to water vapor circulation or vapor loss-induced temperature gradients within the smoke chamber, e.g., a relatively sharp RH change via fissures and small openings of the chamber. This is a definite disability of the measurement procedure/system at medium and high RH, which might be overcome by reducing the volume of the smoke chamber and/or utilizing a T-insulating medium inside, as suggested in Ref. [41]. Otherwise, this source of measurement error could be overcome with the use of the outer, moisturized air in the laboratory, which could make a better balance with the internal of the smoke chamber.

3.1.2. Relative Humidity

Under low RH, a strong correlation between each LCS and the reference sensor (Testo) was found, manifested in high R values of 0.89–0.97 (Figure 4a). The S of the fits approximated unity quite well for GPM-1 (1.04) and GPM-2 (1.24), whereas lower values were observed for BH1-A3 (0.69) and BH1-B3 (0.71). The intercepts of the fits for GPM-1, GPM-2, BH1-A3, and BH1-B3 were about 4, −9, 15, and 12%, respectively.

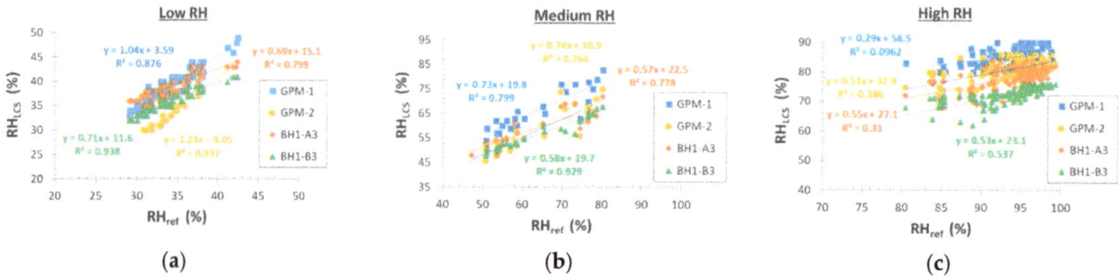

Figure 4. Relative humidity recorded in the smoke chamber with various LCSs plotted against a reference sensor (Testo) under (**a**) low, (**b**) medium, and (**c**) high RH (the color of each equation corresponds to that of the data label/marker).

When applying medium RH (Figure 4b), strong correlations were attained with $R \approx 0.87$–0.89, except BH1-B3 (R = 0.96). The S of the linear fit was ≈0.74 for GPMs, while somewhat lower values of ≈0.57 were obtained for both BH1 sensor models. Compared to low RH, the intercepts of the fits increased to 11–23%, showing lower accuracy, i.e., the worsening of the conditions for RH determinations with LCSs in the measurement system. This bias most likely arose from slow microclimatic changes in the smoke chamber and/or higher RH gradients. For high RH (>80%), the R values of the fits ranged from 0.31 to 0.73, corresponding in most cases to rather low correlations (Figure 4c). The S of the linear fits were also low, ranging from 0.29 to 0.55, while the intercepts were very high, lying between 23% and 59%. These findings point towards higher bias on RH determinations and the rather different readings of the LCSs as compared to those observed with the reference sensor. The difference in the minute monitored T_a and RH data of various LCSs are further evaluated and discussed with the sampled PM_x data as follows below.

3.2. *Monitoring Size-Segregated Aerosol in the Smoke Chamber*

3.2.1. Comparison of LCSs of the Same Design

GPM Sensors

The concentration of PM_1, $PM_{2.5}$, and PM_{10} observed in the smoke chamber under varying RH with the assistance of GPM sensors (Figure 5a–c) reached as high as 145, 300, and 400 μg/m^3, respectively, while AQI peaked at 290 (Figure 5d). It should be noted that the peak mass concentrations of the three PM_x species are different (e.g., Figure 5a,c) despite the application of a chemically stable, monodispersed polystyrene DVB for aerosol production. This is likely due to the coagulation of the evolved aerosol, which starts already in the aerosol generator and continues in the transfer tubing towards the smoke chamber. Certainly, this effect is more expressed in the smoke chamber itself, where the aerosol is of higher residence times than in the former units.

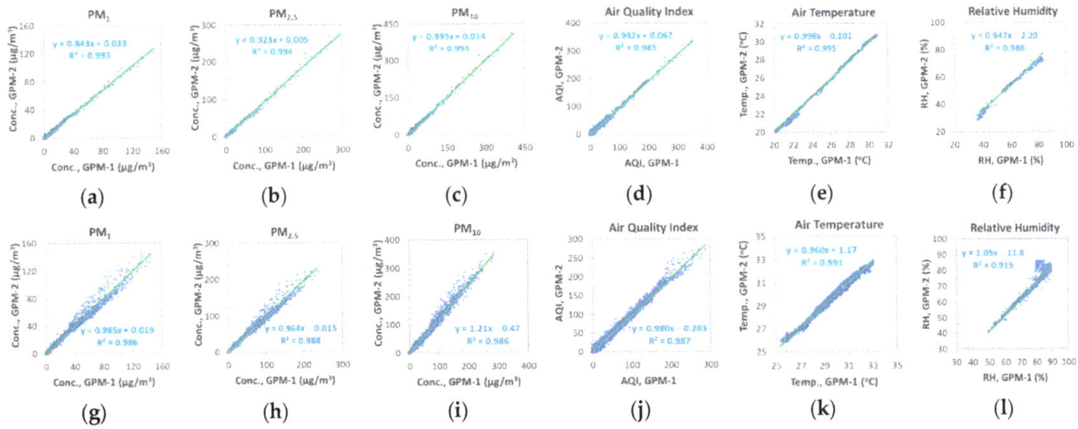

Figure 5. Correlation between monitoring data (n = 25,940) registered in the smoke chamber under low/medium RH (upper row) and medium/high RH (lower row) with two GPM sensors; PM_1 (**a**,**g**), $PM_{2.5}$ (**b**,**h**), PM_{10} (**c**,**i**), AQI (**d**,**j**), T_a (**e**,**k**), and RH (**f**,**l**).

As seen for low/medium (Figure 5a–c) and medium/high (Figure 5g–i) RH conditions, the GPM sensors show strong correlations for each PM_x range, manifested in high R values (0.993–0.997). In general, the S of the linear fits for PM_x approached unity well, ranging between 0.923 and 0.995, except for PM_1 under low RH (S = 0.84) (Figure 5a) and for PM_{10} under high RH (S = 1.21) (Figure 5i). Nevertheless, these close calibration slopes and low intercepts point towards the good intra-model accuracy of the GPMs.

As observed for PM_x, similarly sharp linear fitting was attained for AQI (S = 0.94/0.98, R = 0.993/0.995) and T_a (S = 0.94/0.98, R = 0.993/0.995) under any RHs (Figure 5d,e,j,k). On the other hand, a slightly lower correlation of the RH data was obtained under higher RH (R = 0.957) with close to an ideal slope (S = 1.05), but an increased intercept (−11.8%), compared to those observed under low/medium RH (intercept: 2.2%, R = 0.994) (Figure 5f vs. Figure 5l). Interestingly, the RH curve for GPMs starts to "saturate" and shows a higher spread for concurrent RH readings higher than 80%. This finding is not merely unexpected when one recalls the plots for high RH determinations, as reported in Section 3.1. Apparently, monitoring under high RH conditions in the smoke chamber could be characterized by higher uncertainty than those measurements performed in the low/medium RH range.

It is also noticed in Figure 5 that there is a strikingly different evolution of PM_x plots towards higher concentrations under low/medium and medium/high RHs. The former curves show homoscedasticity (Figure 5a–c), whereas the latter is rather strongly heteroscedastic (Figure 5g–i), i.e., displaying increasing variance towards higher aerosol concentrations. Similar divergence can be observed for AQI and T_a plots (Figure 5k,l).

BH1 Sensors

The maximum concentrations of PM_1, $PM_{2.5}$, and PM_{10} recorded in the smoke chamber with the assistance of BH1-A3 and B3 sensor designs under low/medium RH were as high as 183, 464, and 594 μg/m^3, respectively (Figure 6). As it appears, the BH1-A3 and B3 sensors demonstrate strong correlations for each size range of PM, shown with high R values (0.985–0.993). The S of the linear fittings for PM_1, $PM_{2.5}$, and PM_{10} was close to the unity, i.e., 1.07, 1.13, and 1.04, respectively. On the other hand, the increased intercepts of the fittings (1.6–5.3 μg/m^3) indicate an enhanced baseline bias for the model A3 sensor in each PM size range. This result is in accordance with former findings from a field monitoring study conducted with the same type of LCS [17]. The AQIs of A3 and B3 sensors correlated well, too ($R \approx 0.98$, S = 0.93), whereas the high intercept of ≈23 points to an increased positive bias of the BH1-A3 sensor, which, nevertheless, is still acceptable for AQ reporting. The concurrent values of T_a and RH show a strong correlation between the two BH1 sensor models (R = 0.979 and 0.994, respectively). Both linear fits are characterized with S of ≈1.06 and low intercepts (−1.5 °C and 0.5%, respectively). Fitting to the AQI data series of the BH1-A3 and B3 sensors also revealed that they followed a slightly S-shaped curvature, an interesting feature that has not been observed for BH1 models in a field study [17].

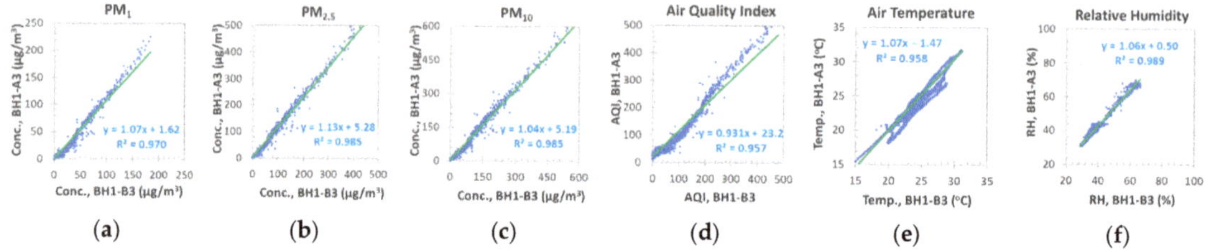

Figure 6. Correlation between monitoring data (n = 25,110) registered with BH1 sensors (models A3 vs. B3) under low/medium RH; PM_1 (**a**), $PM_{2.5}$ (**b**), PM_{10} (**c**), AQI (**d**), T_a (**e**), and RH (**f**).

Under medium/high RH (Figure 7), the BH1-B3 sensors show strong correlations for each PM_x, manifested in high R values of the regression fits (0.976–0.985). This is a somewhat lower correlation compared to those observed between sensor models of B3 and A3, which is due most likely to the different RHs applied for the tests. The S of the fitting equations for PM_1 and $PM_{2.5}$ approached unity fairly well, i.e., 0.91 and 1.24, whereas, oddly, a higher value of 1.5 was found for PM_{10}. In general, a higher spread of the data could be observed at PM_1, $PM_{2.5}$, and PM_{10} concentrations higher than about 50, 70, and 100 μg/m^3, respectively. This deviation could be due to the "saturation" under high RH conditions and/or higher inhomogeneity of PM in the smoke chamber, a possible result of internal heating, which causes an upwards convective stream of suspended particulate. In general, low intercepts have been calculated for PM_1, $PM_{2.5}$, and PM_{10}, i.e., 0.25, −0.31, and −0.42 μg/m^3, respectively, pointing to a low bias. The AQI data of the B3 sensors also correlated well (R = 0.972, S = 1.18although with a much lower intercept (−2.0) compared to that of the BH1-A3 sensor (Figure 6). The simultaneously recorded data for T_a and RH correlated strongly between the BH1-B3 sensors (R = 0.975 and 0.98, respectively), while S of 0.79 and 1.22, respectively, were obtained with slightly enhanced intercepts, i.e., 3.8 °C, and −6.8%, respectively. These biases are still acceptable for the purpose of environmental monitoring.

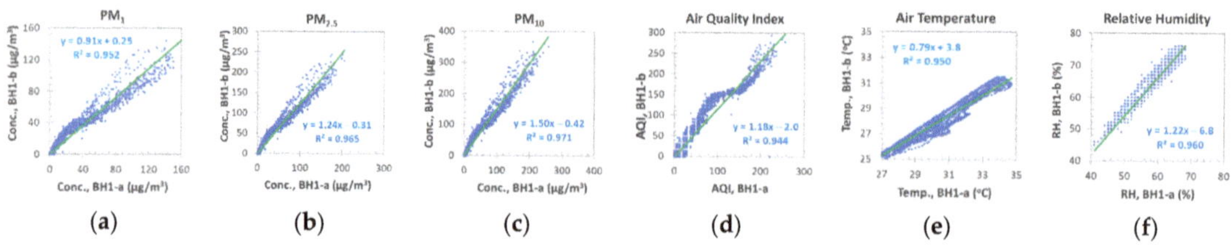

Figure 7. Correlation between monitoring data (n = 25,740) registered with BH1-B3 sensors under medium/high RH; PM_1 (**a**), $PM_{2.5}$ (**b**), PM_{10} (**c**), AQI (**d**), T_a (**e**), and RH (**f**).

3.2.2. Comparison of Sensors of Different Types

GPM versus BH1 Sensors

The fitted PM_x data recorded in the smoke chamber with the assistance of GPM and BH1 sensors under various RH conditions are plotted in Figure 8. It appears that the R values of the linear fit for each PM fraction were close to 0.993, which indicates a strong correlation; moreover, it is independent of RH (Figure 8a–c,g–i). The PM_1, $PM_{2.5}$, and PM_{10} concentrations in the smoke chamber peaked at 150, 354, and 440 μg/m^3, respectively (Figure 5a–c).

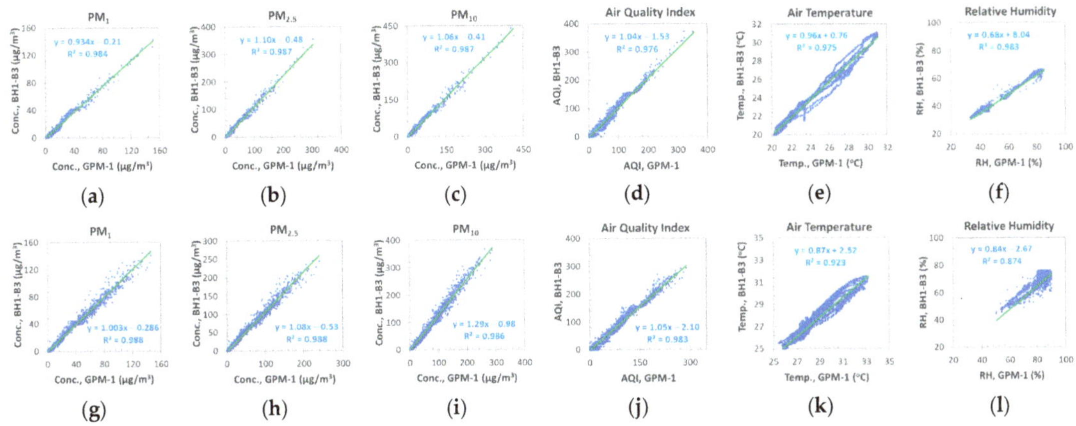

Figure 8. Correlation between the data (n = 10,240) of GPM and BH1-B3 sensors under low/medium RH (upper row) and under medium/high RH (lower row); PM_1 (**a**,**g**), $PM_{2.5}$ (**b**,**h**), PM_{10} (**c**,**i**), AQI (**d**,**j**), T_a (**e**,**k**), and RH (**f**,**l**).

The S of the linear regression fit for PM_1, $PM_{2.5}$, and PM_{10} approached the ideal quite well for low/medium RH, i.e., 0.93, 1.1, and 1.06, respectively (Figure 8a–c), whereas for medium/high RH, the S was similar or slightly higher, i.e., 1.00, 1.08, and 1.29, respectively (Figure 8g–i). The intercepts of the fits were low for low/medium RH, i.e., −0.21, −0.48, and −0.41 μg/m^3, respectively, while slightly increased bias was observed at high RH, i.e., −0.29, −0.53, and −0.98 μg/m^3, respectively. Close fitting was, in general, found for AQI data ($R \approx 0.99$, $S \approx 1.04$) with negligibly low intercepts (−1.5 and −2.1) at each RH range (Figure 8d,j). A good fit was established for the data series of T_a recorded under low/medium RH (R = 0.987, S = 0.96, intercept: 0.76 °C, see Figure 8e), while for higher RH (Figure 8k) lower correlation and higher bias was found (R = 0.96, S = 0.87, intercept: 2.5 °C). Although the RH data strongly correlated under low/medium RH (R = 0.966), the fitting deviated from the ideal (S = 0.68) with a quite high intercept (8%), pointing towards systematic errors (Figure 8f). Interestingly, a better fit was acquired for higher RH (R = 0.935, S = 0.84), with a lower intercept (−2.7%, Figure 8l) than the former (Figure 8f). Like the above observations (Section "GPM Sensors"), homoscedastic graphs were found under low/medium RH, whereas the calibration plots become heteroscedastic at medium/high RHs.

GPM versus GRIMM Monitor

The PM_x and AQI data series obtained with the GPM and GRIMM sensors (Figure 9) are characterized by strong correlations ($R \approx$ 0.984–0.988). The S of the regression fits for $PM_{2.5}$ approached the ideal accurately (1.03), whereas lower and higher values for PM_1 and PM_{10} were observed, i.e., 0.703 and 1.22, respectively. These data show that the GPM sensor has lower detection capability towards accurately predicting fine and coarse ($PM_{10-2.5}$) aerosol fractions, due likely to the much lower bin number of the PMS5003 sampler, compared to the GRIMM sensor. The intercepts of the fitting equations were negligibly low for each PM_x species and AQI, ranging from 0.2 to 0.44 μg/m^3 and 0.61, respectively. Interestingly, the results on the overestimate of PM_{10} readings are not in line with the findings in Refs. [13,14,26], reporting underestimates for coarse aerosol when applying LCSs equipped with PMS5003 sensors.

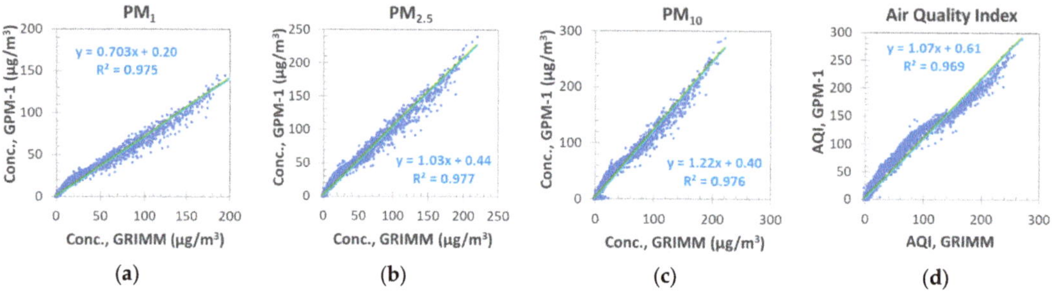

Figure 9. Correlation between measurement data (n = 25,920) recorded with the GPM sensor and the GRIMM reference monitor under medium/high RH; PM_1 (**a**), $PM_{2.5}$ (**b**), PM_{10} (**c**), and AQI (**d**).

BH1 versus GRIMM Monitor

Similar observations to those attained for GPM and GRIMM were made when comparing BH1 sensors (designs A3 and B3) with the GRIMM reference monitor under medium/high RH conditions (Figure 10). The B3 sensor approached $PM_{2.5}$ and AQI accurately ($S \approx 1.12$, $R \approx 0.98$). A very low bias was found for PM_x, ranging between -0.08 and -0.56 µg/m^3, and for AQI too (-1.52). On the other hand, slightly higher PM_{10} readings were observed with BH1-B3 (S = 1.59), compared to GPM and GRIMM, which is somewhat higher than well accepted for environmental monitoring, i.e., the S value should be a maximum of 1.35 [46]. Accordingly, the PM_{10} data recorded with the BH1-B3 requires a definite mathematical correction and calibration in the laboratory and on the field as well.

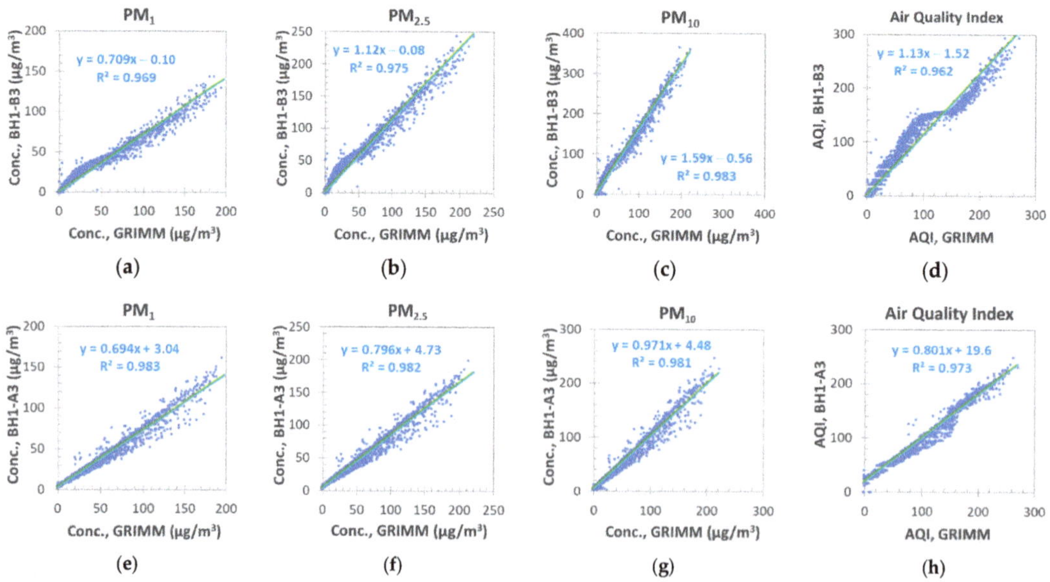

Figure 10. Correlation between measurement data (n = 25,940) registered with BOHU BH1 sensors (model B3: upper row, model A3: lower row) and the GRIMM reference monitor under medium/high RH conditions, PM_1 (**a**,**e**), $PM_{2.5}$ (**b**,**f**), PM_{10} (**c**,**g**), and AQI (**d**,**h**).

For the BH1-A3 model and the GRIMM monitor under medium/high RHs, less steep slopes of PM_1, $PM_{2.5}$, and PM_{10} and AQI were obtained with S data of 0.694, 0.796, 0.97, and 0.80, respectively, which are still acceptable for AQ reporting. On the other hand, the intercepts of the fits for PM_1, $PM_{2.5}$, and PM_{10} are fairly enhanced, i.e., 3.0, 4.7, and

4.5 μg/m^3, respectively, pointing to biased measurements with this sensor at low aerosol levels. The *LOD* values calculated for this sensor type under high RH conditions support this affirmation (see Section 3.3.2) and are in quite good agreement with the corresponding PM_x bias. Nonetheless, these calibration curves invoke errors, which is still acceptable for AQ reporting, as one recalls the US EPA recommendations [46].

3.3. *Analytical Performance of the Sensors*

3.3.1. Error of the LCSs for Ambient Microclimatic Parameters

The error indicator metrics of the GPM and BOHU LCSs, evaluated for T_a, are presented in Table 1. As can be seen, under low RH, the *RMSE* for T_a ranged from 0.7 to 1.9 °C (average: 1.2 °C, median: 1.0 °C) for GPM and BH1 sensors, while *MNE* ranged from 2.4 to 6.7% (average: 3.9%, median: 3.3%). For medium RH, the *RMSE* for T_a was in the range of 0.8 to 1.7 °C (average: 1.1 °C, median: 1.0 °C) for the two types of LCSs, whereas the *MNE* increased, lying in the range of 2.4–7.7% (average: 4.7%, median: 4.4%). Under high RH in the smoke chamber, the *RMSE* for T_a was in the range of 0.8–2.2 °C (average: 1.6 °C, median: 1.7 °C), whereas *MNE* was between 2.5 and 7.5% (average: 5.3%, median: 5.7%). These data all show that, in general, the bias of the T_a readings with the study LCSs increases slightly in the medium, especially in the high RH range, compared to low RH conditions. The *MNB* values were similar or lower compared to *MNE* for each LCS in any studied RH range.

Table 1. Error indicators for T_a under various RH conditions (*RMSE* (°C), *MNE* and *MNB* (%)).

Sensor Type/No.	Low RH			Medium RH			High RH		
	RMSE	*MNE*	*MNB*	*RMSE*	*MNE*	*MNB*	*RMSE*	*MNE*	*MNB*
GPM-1	1.9	6.7	6.7	1.7	7.7	5.4	2.2	7.5	7.5
GPM-2	0.8	2.7	2.2	0.8	2.4	2.1	2.1	7.2	7.2
BH1-A3	0.7	2.4	1.6	0.9	2.6	0.9	1.2	4.1	4.1
BH1-B3	1.1	3.8	3.5	1.2	6.2	1.5	0.8	2.5	2.5

As appears in Table 2, the LCSs (GPM and BOHU) for RH monitoring have shown larger errors than those calculated for T_a. Under low RH conditions, the *RMSE* was fairly low, i.e., it ranged from 1.2 to 5.4% (average: 3.5%, median: 3.7%), while the *MNE* was high, ranging between 3.2% and 15.5% (average: 9.8% (rel.), median: 10.4% (rel.)) for GPM and BOHU sensors. Under medium RH conditions, higher *RMSE* was experienced between 4.4% and 9.7% (average: 7.3%, median: 7.6%), while *MNE* ranged from 5.6 to 12.5% (rel.) (average: 15.6% (rel.), median: 15.1% (rel.)). For high RHs, the *RMSE* increased significantly, ranging between 9.7 and 21.6% (average: 15.2%, median: 14.6%), whereas *MNE* ranged from 9.5. to 22.8 (average: 15.6% (rel.), median: 15.1% (rel.)) for the two sensor designs. The *MNB* values were found to be similar or slightly lower compared to *MNE* data for each LCS over any studied RH range. It can also be seen that the bias in several cases is negative, corresponding to an underestimation of the true value by the use of the LCSs. Overall, the pattern is obvious for RH measurements: the higher the RH, the higher the bias of the readings obtained from LCSs. Nevertheless, both error indicators showed an aptly low level (<30%) as recommended for LCS testing by US EPA [46].

Table 2. Error metrics for various RH ranges: RMSE (% RH), MNE and MNB (rel. %).

Sensor Type/No.	Low RH			Medium RH			High RH		
	RMSE	*MNE*	*MNB*	*RMSE*	*MNE*	*MNB*	*RMSE*	*MNE*	*MNB*
GPM-1	5.1	14.5	14.5	4.4	5.6	2.8	9.7	9.5	−9.4
GPM-2	1.2	3.2	−2.5	7.5	9.0	−8.5	13.2	13.6	−13.6
BH1-A3	5.4	15.5	15.5	7.7	8.9	−8.3	16.0	16.6	−16.6
BH1-B3	2.3	6.2	5.8	9.7	12.5	−12.5	21.6	22.8	−22.8

3.3.2. Performance of LCSs for Size-Segregated Aerosol

The error indicator metrics were evaluated for various PM_x species besides the concurrent T_a and RH data, registered every 1 min with the assistance of the studied LCSs (Table 3). As can be seen, under low/medium RH, the GPM-1 and GPM-2 sensors showed quite low *RMSE* for PM_1, $PM_{2.5}$, and PM_{10}, i.e., 3.8, 6.5, and 15 μg/m^3, respectively, whereas increased bias was experienced under high RH conditions, e.g., 5.7, 8.2, and 26 μg/m^3, respectively. On the other hand, under low/medium RH, the *MNE* values of PM_1, $PM_{2.5}$, and PM_{10} were 7.0, 9.1, and 11.7%, while similar or higher *MNE*s of 8.5, 8.6, and 17.2%, respectively, were experienced for high RHs. Interestingly, for T_a and RH, the *RMSE* and the *MNE* values were of a similar extent under medium and high RH conditions for both GPMs. Under medium RH conditions, the *MNB* shows a different pattern for PM_1, $PM_{2.5}$, and PM_{10}, i.e., −4.3, −6.4, and 9.0%, respectively, whereas at high RH, its value changed differently for the three PM_x species, i.e., 4.6, 0.9, and 19.4%, respectively. These data indicate the negative bias of the LCSs for low/medium sized PM under low/medium RH, whereas the bias is positive for coarser aerosol, manifested in PM_{10} at high RHs.

Table 3. Error indicators for PM_x and microclimatic parameters between various LCSs (resolution: 1 min) calculated for aerosol levels not lower than 10 μg/m^3 under low/medium and (high) RH.

Sensor/Error Types *	Bias for Low/Medium RH (High RH)				
	PM_1	$PM_{2.5}$	PM_{10}	T_a	RH
GPM-1–GPM-2					
RMSE	3.8 (5.6)	6.4 (8.1)	15 (26)	0.2 (0.3)	7.5 (8.1)
MNE (%)	7.0 (8.6)	9.1 (8.7)	12 (17)	0.5 (0.8)	11.7 (9.1)
MNB (%)	−4.3 (4.6)	−6.4 (0.9)	9.0 (19)	−0.4 (0.1)	−10.5 (−7.7)
GPM-1–BH1-A3					
RMSE	7.1 (7.8)	11 (18)	14 (21)	1.2 (1.7)	2.3 (4.3)
MNE (%)	21 (22)	33 (41)	36 (46)	4.5 (2.8)	15 (14)
MNB (%)	−7.7 (−12.4)	−23 (−27)	−24 (−29)	−4.3 (−2.7)	−12.8 (−12)
GPM-2–BH1-A3					
RMSE	9.4 (11)	7.7 (21)	24 (44)	1.2 (1.7)	2.3 (5.0)
MNE (%)	21 (29)	25 (43)	46 (73)	4.1 (2.9)	3.0 (5.2)
MNB (%)	−3.0 (−15)	−17 (−27)	−30 (−40)	−3.9 (−2.8)	−2.6 (−4.6)
GPM-2–BH1-B3					
RMSE	14 (12)	11 (20)	23 (43)	2.2 (2.8)	13 (16)
MNE (%)	40 (46)	42 (59)	66 (94)	6.7 (8.8)	25 (26)
MNB (%)	−7.8 (20)	−21 (−32)	−35 (−44)	7.3 (−5.2)	−20 (−20)
BH1-A3–BH1-B3					
RMSE	5.4 (6.8)	6.6 (8.4)	7.3 (10)	3.1 (2.6)	11.1 (13)
MNE (%)	18 (19)	16 (17)	17 (19	10 (8.6)	22 (21)
MNB (%)	−6.9 (−6.8)	−6.2 (−6.3)	−7.6 (−7.8)	11.6 (8.1)	−18 (−16)

* The *RMSE* for PM_x, T_a, and RH is specified in μg/m^3, °C, and %, respectively, while *MNE* and *MNB* are given in relative percent (n = 920 for medium RH and n = 840 for high RH).

Under medium RHs, the GPM and BH1-A3 LCSs demonstrated somewhat higher *RMSE* as compared to that experienced between the two GPM sensors, except for PM_{10} of a similar extent. The *RMSE* for PM_1, $PM_{2.5}$, and PM_{10} was 7.1, 11.3, and 13.8 μg/m^3, respectively, while an increased bias was obtained at high RHs, i.e., 7.8, 18, and 21 μg/m^3, respectively. Similarly, the *MNE* increased compared to those found between the GPMs. Under low/medium RHs, the *MNE* for PM_1, $PM_{2.5}$, and PM_{10} was high, i.e., 21, 33, and 36%, while similar or higher values of 22, 41, and 46%, respectively, were experienced for high RH conditions. At low/medium RH, the *MNB* data showed an increasing pattern from PM_1, $PM_{2.5}$, to PM_{10} with values of −7.7, −23, and −24%, respectively, whereas at high RH, it increased to −12, −27, and −29%, respectively. For T_a and RH, the *RMSE* and the *MNE* between BH1-A3 and any of the GMPs were of similar extent, irrespective of RH in the smoke chamber, except for GPM-1 that was a five-fold higher value (14%).

The error indicators evaluated for various PM_x species, registered every 1 min with the study LCSs against the reference sensor, are listed in Table 4. As can be seen, under medium RH conditions, the GPM-1 and GPM-2 sensors showed quite similar *RMSE* for PM_1 and $PM_{2.5}$, i.e., about 20 and 10 μg/m^3, respectively, whereas somewhat increased bias was experienced for PM_{10} of GPM-2 (30 μg/m^3). A similar pattern was observed for *MNE* of PM_1 and $PM_{2.5}$, i.e., about 24% and 42%, respectively, whereas an increased value was experienced for PM_{10} of GPM-2 (64%). Under high RH conditions, the *RMSE* increased slightly for PM_1 and $PM_{2.5}$, whereas a higher increment of PM_{10} was observed. This effect is likely due to the higher rate of particle growth under increasing RHs, which brings more uncertainty in the measurement of coarse aerosol [49]. The *MNB* was quite high in the case of GPM sensors, apart from PM_1 (−2.8/−8.5%), and more or less acceptable for $PM_{2.5}$ (41/33%) but outstandingly high for PM_{10} (65/57%). These findings point to the lesser usability of the GPM sensor for coarse aerosol quantitation.

Table 4. Error flags for PM_x species of the study LCSs determined against the GRIMM reference sensor for aerosol levels not lower than 10 μg/m^3 under medium/(high) RH conditions.

Sensor/Error Type	Parameter/Bias Value *		
	PM_1	$PM_{2.5}$	PM_{10}
GPM-1			
RMSE	18 (22)	10 (11)	19 (22)
MNE (%)	24 (22)	42 (37)	64 (57)
MNB (%)	−2.8 (−8.5)	41 (33)	64 (56)
GPM-2			
RMSE	21 (21)	9.5 (10.8)	30 (40)
MNE (%)	24 (22)	36 (37)	75 (83)
MNB (%)	−6.9 (−5.0)	31 (33)	75 (83)
BOHU BH1-A3			
RMSE	14 (22)	7.5 (16)	8.0 (9.3)
MNE (%)	17 (25)	13 (19)	20 (17)
MNB (%)	−15 (−22)	5.0 (−5.2)	20 (8.0)
BOHU BH1-B3			
RMSE	12 (21)	6.8 (15)	12 (10)
MNE (%)	25 (30)	9.2 (19)	13 (16)
MNB (%)	−22 (−29)	−3.4 (−13)	8.9 (−7.0)

* The *RMSE* is expressed in μg/m^3, while *MNE* and *MNB* are in percent (n = 920 for medium RH and n = 840 for high RH).

Under medium RH, the BH1-A3 sensor featured similar or lower *RMSE* as compared to those found for GPM sensors against the reference monitor. For instance, the *RMSE* for PM_1, $PM_{2.5}$, and PM_{10} was 14, 7.4, and 7.8 μg/m^3, respectively, whereas its increase was observed at high RH, i.e., 22, 16, and 9.3 μg/m^3, respectively. The *MNE* formed a similar pattern to the *RMSE*. Under medium RH, the *MNE* of PM_1, $PM_{2.5}$, and PM_{10} for the BH1-A3 sensor was found to be 17, 13, and 20%, respectively, whereas under high RHs, it was 25, 19, and 17%, respectively. Interestingly, for BH1-A3 sensors, the *RMSE* and the *MNE* values develop a decreasing trend from PM_1 towards higher PM_x ranges, which is the opposite order experienced for the GPMs. Similar patterns of *RMSE* and *MNE* were observed for the BH1-B3 sensor (Table 4). Under low/medium RH conditions, the *MNB* displayed a varying bias for PM_1, $PM_{2.5}$, and PM_{10} with −15, 5, and 20%, respectively, whereas decreased values were found for high RH conditions.

The *LOD* and the *LOQ* data under low/medium and high RH conditions are presented in Table 5. As can be seen, at low-to-high RH, the *LOD*s of each PM_x fraction for the GPM and the BH1-B3 sensors are advantageously low (1.5 μg/m^3) compared to those calculated for the BH1-A3 (3.5–4.5 μg/m^3). This is likely due to the higher background noise of the latter design, manifested in enhanced readings (2–3 μg/m^3) in the otherwise clean air. These *LOD* data are in line with those observed for GPM and BH1-A3 in field campaigns, although the latter value was slightly higher due to the higher background PM levels [17].

Table 5. Limit of detection (*LOD*), limit of quantification (*LOQ*), and the peak concentration recorded with the study PM_x sensors under low/medium and high RH [§] conditions.

Monitor Type	Aerosol Size-Range	LOD (μg/m³)	LOQ (μg/m³)	Peak Conc. (μg/m³) *
GPM	PM_1	1.5 (1.5)	5.1 (5.1)	120 (145)
	$PM_{2.5}$	1.5 (1.5)	5.1 (5.1)	190 (240)
	PM_{10}	1.5 (1.5)	5.1 (5.1)	280 (290)
BOHU BH1-A3	PM_1	2.5 (3.5)	9 (12)	140 (160)
	$PM_{2.5}$	3.5 (4.5)	12 (15)	170 (200)
	PM_{10}	3.5 (4.5)	12 (15)	210 (250)
BOHU BH1-B3	PM_1	1.5 (1.5)	5.1 (5.1)	167 (162)
	$PM_{2.5}$	1.5 (1.5)	5.1 (5.1)	205 (198)
	PM_{10}	1.5 (1.5)	5.1 (5.1)	256 (240)
GRIMM	PM_1	0.25 (0.25)	0.85 (0.85)	165 (205)
	$PM_{2.5}$	0.25 (0.25)	0.85 (0.85)	180 (220)
	PM_{10}	0.25 (0.25)	0.85 (0.85)	185 (230)

* The maximum concentration for error calculations registered in the smoke chamber. § High RH values in parentheses.

On the other hand, for the GRIMM monitor, the *LOD* as low as 0.25 μg/m³ was experienced due to the higher sensor sensitivity towards finer PM fractions, which arises from the enhanced detection capability for lower aerosol masses, compared to GPM and BH1 sensors. In general, under low/medium RHs, lower *LOD* values of PM_x for the BH1-A3 sensors were attained, whereas the rest of the study sensors experienced the same *LOD*s at various RHs.

4. Discussion

The literature reports on several kinds of LCSs based on PMS5003, as noticed above, and a couple of AQ networks, including PurpleAir [22–26] and Clarity [13]. Moreover, this sensor has been applied in several geospatial and exposure studies, for instance, in Refs. [50,51]. A couple of studies suggest that the readings for PM_1 and $PM_{2.5}$ of this LCS are in good correlation with those measured against reference methods, but its low performance was found for coarse aerosol, e.g., $PM_{10–2.5}$ [8,14,15,50].

In a former field monitoring study [17], the GPM and the BH1-A3 sensors provided similar linearity for indoor and outdoor aerosols. But a relatively short PM mass concentration range could be covered and assessed for calibration, i.e., up to about 130 μg/m³ for PM_{10}. Moreover, the studied LCSs could be measured against a less sharp OPC sensor, which could only assess $PM_{2.5}$ levels but not give information on accuracies towards fine and coarse aerosol fractions. In this investigation, it was shown that the studied LCSs can be calibrated and tested with monodispersed spherical latex aerosol introduced into a laboratory smoke chamber for a much higher concentration range of the aerosol under controlled slow heating and/or varying RH conditions. By using an aerosol generator, as high PM_1, $PM_{2.5}$, and PM_{10} levels as about 150, 350, and 450 μg/m³, respectively, were reached in the smoke chamber. This is already about three times higher PM_x concentration than those obtained in the former field study, with similar types of LCSs [17]. In general, linear relationships were established between the concurrently registered data with the various OPCs regarding intra- and inter-model accuracy.

For this investigation, some LCSs, such as Geekcreit PM_x, built on a PMS5003 sampler, and BOHU BH1 (models A3 and B3) monitors (built with a Pando G7 OPC) were compared with a laboratory-grade sensor, GRIMM 1.109 used as a reference monitor. The results have clearly shown that the accuracy of the readings for PM_1, $PM_{2.5}$, and PM_{10} with the different and the same types of LCSs was acceptable, corresponding to those observed by means of the reference, research-grade sensor. The GPM sensors are based on PMS5003 OPC, which is characterized by low detection efficacy for coarse PM [13,14,26]. This would introduce some bias into the determination of the coarse fraction, i.e., $PM_{10-2.5}$, manifested in the underestimate of PM_{10} levels. On the other hand, in this study, an overestimate was observed for PM_{10} by the PMS-based LCSs, indicating the coagulation of particulates and a positive bias of higher RH on the PM_{10} determinations.

Controlled slow heating (0.2 °C/min) of the smoke chamber was advantageous for calibration/checking of the T_a sensors of the monitors against an industrial-grade, analog thermometer, whereas T_a data observed during the cooling stage of the smoke chamber was rather hampered by higher measurement bias.

The RH sensors of the LCSs were assessed with the application of slow evaporation of boiled/hot water in the smoke chamber. With this simple procedure, even the medium and high RH ranges could be covered for calibration. Nevertheless, the high RH ranges are less accurately approached and had the drawback of lower detection efficacy, as can be seen from the analytical performance/error indicator metrics (*RMSE*, *MNE*, *MNB*, *LOD*, and *LOQ*).

5. Conclusions

In this study, various laboratory conditions inside a smoke chamber were adjusted for the calibration of low-cost air quality devices, i.e., Geekcreit and BOHU BH1 models A3 and B3, each based on the OPC principle for PM_x components. Microclimatic variables such as air temperature and relative humidity were also monitored. It was demonstrated that the LCSs provided generally accurate readings for $PM_{2.5}$, irrespective of the air temperature and relative humidity changes in the smoke chamber. On the other hand, the PM_1 and PM_{10} readings of the LCSs (irrespective of either with PMS5003 or Pando G7 sampler) were slightly lower and higher, respectively, compared to those recorded with the research-grade air monitor. This is likely due to the lower detection efficacy of the fine and coarse fraction of ambient aerosol with the assistance of low-cost OPCs possessing a low number of bins. Moreover, at high RH, the calibration plots showed heteroscedasticity (about 10–20% increase in bias) towards higher PM_x concentrations, which was also observed for AQI, T_a, and RH calibration curves.

For temperature calibration, it appears that the low RH in the smoke chamber provided more reproducible conditions in terms of the lowest measurement bias for LCSs as recorded against a reference-grade analog thermometer. The relative humidity measurements with the LCSs performed against an industrially calibrated RH meter provided accurate readings at the low and medium RH range, but high RHs caused increased bias. Although the experimental arrangement with the smoke chamber provided relatively reliable values for PM_x and microclimatic parameters, the accuracy of the data could be improved, for instance, with the application of lower volumes of the aerosol chamber, lower heating rates, and more sharply controlled RH conditions, e.g., by means of those reported in Ref. [41].

Author Contributions: Conceptualization, L.B. and A.N.; methodology, L.B. and A.N.; validation, L.B.; formal analysis, L.B.; investigation, L.B. and A.N.; resources, L.B. and A.N.; data curation, L.B.; writing—original draft preparation, L.B.; writing—review and editing, L.B. and A.N.; visualization, L.B.; supervision, L.B. and A.N.; project administration, L.B. and A.N.; funding acquisition, L.B. and A.N. All authors have read and agreed to the published version of the manuscript.

Funding: This research was partially supported by the Hungarian National Research Development and Innovation Fund under grant No. TKP2021-NVA-04.

Institutional Review Board Statement: Not applicable.

Informed Consent Statement: Not applicable.

Data Availability Statement: The data are not publicly available due to file size limitations. The research data presented in this study are available on request from the corresponding author.

Acknowledgments: The authors thank the support provided by their research institution (Wigner RCP) for providing funds for the analytical spectroscopic and aerosol research laboratories.

Conflicts of Interest: The authors declare no conflicts of interest.

References

1. Dockery, D.W.; Speizer, F.E.; Stram, D.O.; Ware, J.H.; Spengler, J.D.; Ferris, B.G., Jr. Effects of inhalable particles on respiratory health of children. *Am. Rev. Respir. Dis.* **1989**, *139*, 587–594. [CrossRef]
2. Camuffo, D. *Microclimate for Cultural Heritage: Measurement, Risk Assessment, Conservation, Restoration, and Maintenance of Indoor and Outdoor Monuments*, 3rd ed.; Elsevier B.V.: Amsterdam, The Netherlands, 2019.
3. Spolnik, Z.; Bencs, L.; Worobiec, A.; Kontozova, V.; Van Grieken, R. Application of EDXRF and thin window EPMA for the investigation of the influence of hot air heating on the generation and deposition of particulate matter. *Microchim. Acta* **2005**, *149*, 79–85. [CrossRef]
4. Anaf, W.; Bencs, L.; Van Grieken, R.; De Wael, K.; Janssens, K. Indoor particulate matter in four Belgian heritage sites: Case studies on the deposition of dark-colored and hygroscopic particles. *Sci. Total Environ.* **2015**, *506–507*, 361–368. [CrossRef]
5. IPCC. *Climate Change 2021: The Physical Science Basis, Sixth Assessment Report*; Cambridge University Press: Cambridge, UK, 2021.
6. Kelly, K.E.; Whitaker, J.; Petty, A.; Widmer, C.; Dybwad, A.; Sleeth, D.; Martin, R.; Butterfield, A. Ambient and laboratory evaluation of a low-cost particulate matter sensor. *Environ. Pollut.* **2017**, *221*, 491–500. [CrossRef]
7. Zamora, M.L.; Xiong, F.Z.L.; Gentner, D.; Kerkez, B.; Kohrman-Glaser, J.; Koehler, K. Field and Laboratory Evaluations of the Low-Cost Plantower Particulate Matter Sensor. *Environ. Sci. Technol.* **2019**, *53*, 838–849. [CrossRef]
8. Sayahi, T.; Butterfield, A.; Kelly, K.E. Long-term field evaluation of the Plantower PMS low-cost particulate matter sensors. *Environ. Pollut.* **2019**, *245*, 932–940. [CrossRef] [PubMed]
9. Sayahi, T.; Kaufman, D.; Becnel, T.; Kaur, K.; Butterfield, A.E.; Collingwood, S.; Zhang, Y.; Gaillardon, P.E.; Kelly, K.E. Development of a calibration chamber to evaluate the performance of low-cost particulate matter sensors. *Environ. Pollut.* **2019**, *255*, 113131. [CrossRef] [PubMed]
10. Zusman, M.; Schumacher, C.S.; Gassett, A.J.; Spalt, E.W.; Austin, E.; Larson, T.V.; Carvlin, G.; Seto, E.; Kaufman, J.D.; Sheppard, L. Calibration of low-cost particulate matter sensors: Model development for a multi-city epidemiological study. *Environ. Int.* **2020**, *134*, 105329. [CrossRef]
11. Hedge, S.; Min, K.T.; Moore, J.; Lundrigan, P.; Patwari, N.; Collingwood, S.; Balch, A.; Kelly, K.E. Indoor Household Particulate Matter Measurements Using a Network of Low-cost Sensors. *Aerosol Air Qual. Res.* **2020**, *20*, 381–394.
12. Peck, A.; Handy, R.G.; Sleeth, D.K.; Schaefer, C.; Zhang, Y.; Pahler, L.F.; Ramsay, J.; Collingwook, S.C. Aerosol Measurement Degradation in Low-Cost Particle Sensors Using Laboratory Calibration and Field Validation. *Toxics* **2023**, *11*, 56. [CrossRef] [PubMed]
13. Kaur, K.; Kelly, K.E. Laboratory evaluation of the Alphasense OPC-N3, and the Plantower PMS5003 and PMS6003 sensors. *J. Aerosol Sci.* **2023**, *171*, 106181. [CrossRef]
14. Kuula, J.; Mäkelä, T.; Aurela, M.; Teinilä, K.; Varjonen, S.; González, O.; Timonen, H. Laboratory evaluation of particle-size selectivity of optical low-cost particulate matter sensors. *Atmos. Meas. Tech.* **2020**, *13*, 2413–2423. [CrossRef]
15. Vogt, M.; Schneider, P.; Castell, N.; Hamer, P. Assessment of Low-Cost Particulate Matter Sensor Systems against Optical and Gravimetric Methods in a Field Co-Location in Norway. *Atmosphere* **2021**, *12*, 961. [CrossRef]
16. Hong, G.W.; Le, T.C.; Tu, J.W.; Wang, C.E.; Chang, S.C.; Yu, Y.; Lin, G.Y.; Aggarwal, S.G.; Tsai, C.J. Long-term evaluation and calibration of three types of low-cost PM2.5 sensors at different air quality monitoring stations. *J. Aerosol Sci.* **2021**, *157*, 105829. [CrossRef]
17. Bencs, L.; Plósz, B.; Mmari, A.G.; Szoboszlai, N. Comparative Study on the Use of Some Low-Cost Optical Particulate Sensors for Rapid Assessment of Local Air Quality Changes. *Atmosphere* **2022**, *13*, 1218. [CrossRef]
18. Badura, M.; Sowka, I.; Szymanski, P.; Batog, P. Assessing the usefulness of dense sensor network for $PM_{2.5}$ monitoring on an academic campus area. *Sci. Total Environ.* **2020**, *722*, 137867. [CrossRef]
19. Jiang, Y.X.; Zhu, X.L.; Chen, C.; Ge, Y.H.; Wang, W.D.; Zhao, Z.H.; Cai, J.; Kan, H.D. On-field test and data calibration of a low-cost sensor for fine particles exposure assessment. *Ecotoxicol. Environ. Saf.* **2021**, *211*, 111958. [CrossRef]
20. Huang, C.H.; He, J.Y.; Austin, E.; Seto, E.; Novosselov, I. Assessing the value of complex refractive index and particle density for calibration of low-cost particle matter sensor for size-resolved particle count and $PM_{2.5}$ measurements. *PLoS ONE* **2021**, *16*, e0259745. [CrossRef]
21. Cowell, N.; Chapman, L.; Bloss, W.; Pope, F. Field Calibration and Evaluation of an Internet-of-Things-Based Particulate Matter Sensor. *Front. Environ. Sci.* **2022**, *9*, 798485. [CrossRef]
22. Ouimette, J.R.; Malm, W.C.; Schichtel, B.A.; Shridan, P.J.; Andrews, E.; Ogren, J.A.; Arnott, W.P. Evaluating the PurpleAir monitor as an aerosol light scattering instrument. *Atmos. Meas. Tech.* **2022**, *15*, 655–676. [CrossRef]
23. Wallace, L.; Bi, J.Z.; Ott, W.R.; Sarnat, J.; Liu, Y. Calibration of low-cost PurpleAir outdoor monitors using an improved method of calculating $PM_{2.5}$. *Atmos. Environ.* **2021**, *256*, 118432. [CrossRef]
24. Wallace, L. Intercomparison of PurpleAir Sensor Performance over Three Years Indoors and Outdoors at a Home: Bias, Precision, and Limit of Detection Using an Improved Algorithm for Calculating $PM_{2.5}$. *Sensors* **2022**, *22*, 2755. [CrossRef]

25. Park, S.-J.; Lee, S.H.; Yeo, M.S.; Rim, D.H. Field and laboratory evaluation of PurpleAir low-cost aerosol sensors in monitoring indoor airborne particles. *Build. Environ.* **2023**, *234*, 110127. [CrossRef]
26. Tryner, J.; L'Organge, C.; Mahaffy, J.; Miller-Lionberg, D.; Hofstetter, C.; Wilson, A.; Volckens, J. Laboratory evaluation of low-cost PurpleAir PM monitors and in-field correction using co-located portable filter samplers. *Atmos. Environ.* **2020**, *220*, 117067. [CrossRef]
27. Jayaratne, R.; Liu, X.T.; Thai, P.; Dunbabin, M.; Morawska, L. The influence of humidity on the performance of a low-cost air particle mass sensor and the effect of atmospheric fog. *Atmos. Meas. Tech.* **2018**, *11*, 4883–4890. [CrossRef]
28. Wahlborg, D.; Björling, M.; Mattsson, M. Evaluation of field calibration methods and performance of AQMesh, a low-cost air quality monitor. *Environ. Monit. Assess.* **2021**, *193*, 251. [CrossRef] [PubMed]
29. Castell, N.; Dauge, F.R.; Schneider, P.; Vogt, M.; Lerner, U.; Fishbain, B.; Broday, D.; Bartonova, A. Can commercial low-cost sensor platforms contribute to air quality monitoring and exposure estimates? *Environ. Int.* **2017**, *99*, 293–302. [CrossRef] [PubMed]
30. Thorpe, A.; Walsh, P.T. Comparison of Portable, Real-Time Dust Monitors Sampling Actively, with Size-Selective Adaptors, and Passively. *Ann. Occup. Hyg.* **2007**, *51*, 679–691. [PubMed]
31. Salimifard, P.; Rim, D.; Freihaut, J.D. Evaluation of low-cost optical particle counters of monitoring individual indoor aerosol sources. *Aerosol Sci. Technol.* **2020**, *54*, 217–231. [CrossRef]
32. Semple, S.; Apsley, A.; MacCalman, L. An inexpensive particle monitor for smoker behaviour modification in homes. *Tob. Control* **2013**, *11*, 295–298. [CrossRef]
33. Markowich, K.M.; Chiliński, M.T. Evaluation of two low-cost optical particle counters for the measurement of ambient aerosol scattering coefficient and Ångström exponent. *Sensors* **2020**, *20*, 2617. [CrossRef] [PubMed]
34. Wang, Y.; Li, J.Y.; Jing, H.; Zhang, Q.; Jiang, J.K.; Biswas, P. Laboratory evaluation and calibration of three low-cost particle sensors for particulate matter measurement. *Aerosol Sci. Technol.* **2015**, *49*, 1063–1077. [CrossRef]
35. Crilley, L.R.; Shaw, M.; Pound, R.; Kramer, L.J.; Price, R.; Young, S.; Lewis, A.C.; Pope, F.D. Evaluation of a low-cost optical particle counter (Alphasense OPC-N2) for ambient air monitoring. *Atmos. Meas. Tech.* **2018**, *11*, 709–720. [CrossRef]
36. Crnosija, N.; Zamora, M.L.; Rule, A.M.; Payne-Sturges, D. Laboratory Chamber Evaluation of Flow Air Quality Sensor $PM_{2.5}$ and PM_{10} Measurements. *Int. J. Environ. Res. Public Health* **2022**, *19*, 7340. [CrossRef] [PubMed]
37. Amoh, N.A.; Xu, G.; Kumar, A.R.; Wang, Y. Calibration of low-cost particulate matter sensors for coal dust monitoring. *Sci. Total Environ.* **2023**, *859*, 160336. [CrossRef]
38. Jayaratne, R.; Liu, X.T.; Ahn, K.-H.; Asumandu-Sakyi, A.; Fisher, G.; Gao, J.; Mabon, A.; Mazaheri, M.; Mullins, B.; Nyaku, M.; et al. Low-cost $PM_{2.5}$ Sensors: An Assessment of their Suitability for Various Applications. *Aerosol Air Qual. Res.* **2020**, *20*, 520–532. [CrossRef]
39. Sousan, S.; Regmi, S.; Park, Y.M. Laboratory Evaluation of Low-Cost Optical Particle Counters for Environmental and Occupational Exposures. *Sensors* **2021**, *21*, 4146. [CrossRef]
40. Si, M.X.; Xiong, Y.; Du, S.; Du, K. Evaluation and calibration of a low-cost particle sensor in ambient conditions using machine-learning methods. *Atmos. Meas. Tech.* **2020**, *13*, 1693–1707. [CrossRef]
41. Rivero, R.A.G.; Hernández, L.E.M.; Schalm, O.; Rodríguez, E.H.; Sánchez, D.A.; Pérez, M.C.M.; Caraballo, V.N.; Jacobs, W.; Laguardia, A.M. A Low-Cost Calibration Method for Temperature, Relative Humidity, and Carbon Dioxide Sensors Used in Air Quality Monitoring Systems. *Atmosphere* **2023**, *14*, 191. [CrossRef]
42. Alfano, B.; Barretta, L.; Del Giudice, A.; De Vito, S.; Di Francia, G.; Esposito, E.; Formisano, F.; Massera, E.; Miglietta, M.L.; Polichetti, T. A review of low-cost particulate matter sensors from the developers' perspectives. *Sensors* **2020**, *20*, 6819. [CrossRef]
43. Giordano, M.R.; Malings, C.; Pandis, S.N.; Presto, A.A.; McNeill, V.F.; Westervelt, D.M.; Beekmann, M.; Subramanian, R. From low-cost sensors to high-quality data: A summary of challenges and best practices for effectively calibrating low-cost particulate matter sensors. *J. Aerosol Sci.* **2021**, *158*, 105833. [CrossRef]
44. Rai, A.C.; Kumar, P.; Pilla, F.; Skouloudis, A.N.; Di Sabatino, S.; Ratti, C.; Yasar, A.; Rickerby, D. End-user perspective of low-cost sensors for outdoor air pollution monitoring. *Sci. Total Environ.* **2017**, *607*, 691–705. [CrossRef] [PubMed]
45. Yuval; Molho, H.M.; Zivan, O.; Broday, D.M.; Raz, R. Application of a sensor network of low cost optical particle counters for assessing the impact of quarry emissions on its vicinity. *Atmos. Environ.* **2019**, *211*, 29–37. [CrossRef]
46. US EPA (United States Environmental Protection Agency); Duvall, R.M.; Clements, A.L.; Hagler, G.; Kamal, A.; Kilaru, V.; Goodman, L.; Frederick, S.; Barkjohn, K.K.; VonWald, I.; et al. Performance Testing Protocols, Metrics, and Target Values for Fine Particulate Matter Air Sensors. EPA/600/R-20/280. Available online: https://www.epa.gov/ (accessed on 25 January 2024).
47. IUPAC; Inczédy, J.; Lengyel, T.; Ure, A.M.; Gelencsér, A.; Hulanicki, A. (Eds.) *Compendium of Analytical Nomenclature*, 3rd ed.; Blackwell: Oxford, UK, 1998.
48. European Commission. Directive 2008/50/EC of the European Parliament and the Council of 21 May 2008 on Ambient Air Quality and Cleaner Air for Europe. *Off. J. Eur. Union* **2008**. Available online: https://eur-lex.europa.eu/legal-content/EN/TXT/?uri=OJ:L:2008:152:TOC (accessed on 25 January 2024).
49. Won, W.S.; Oh, R.; Lee, W.; Ku, S.K.; Su, P.C.; Yoon, Y.J. Hygroscopic properties of particulate matter and effects of their interactions with weather on visibility. *Sci. Rep.* **2021**, *11*, 16401. [CrossRef]

50. Bi, J.; Wildani, A.; Chang, H.H.; Liu, Y. Incorporating low-cost sensor measurements into high-resolution $PM_{2.5}$ modeling at a large spatial scale. *Environ. Sci. Technol.* **2020**, *54*, 2152–2162. [CrossRef]
51. Kelly, K.E.; Xing, W.W.; Sayahi, T.; Mitchell, L.; Becnel, T.; Gaillardon, P.-E.; Meyer, M.; Whitaker, R.T. Community-based measurements reveal unseen differences during air pollution episodes. *Environ. Sci. Technol.* **2021**, *55*, 120–128. [CrossRef]

atmosphere

MDPI

Article

Sniffing Drones: A Promising Solution for Measuring Railroad Emissions in Urban Environments

Felipe Baglioli and Ricardo H. M. Godoi *

Department of Environmental Engineering, Federal University of Paraná (UFPR), Curitiba 81531-980, PR, Brazil; felipe.baglioli@gmail.com
* Correspondence: rhmgodoi@ufpr.br

Abstract: Locomotive emissions from railroads can particularly impact air pollution, making it crucial to understand their impacts on human health and the environment and develop strategies to reduce them. The potential of drone technology equipped with a "sniffing" system for detecting air pollution emissions is promising and can be a valuable tool for assessing dynamic emissions. This research utilized sensor-equipped drones to measure gaseous emissions from cargo and passenger trains on a railway in Curitiba, Brazil. Reference equipment evaluated the accuracy of NO_2, SO_2, and O_3 concentrations. The results showed that before the passage of trains, the average SO_2 concentration was 20 $\mu g/m^3$, with a maximum concentration of 110 $\mu g/m^3$ detected during transit. The average increase in NO_2 concentrations was from 30 $\mu g/m^3$ to 120 $\mu g/m^3$, and the average increase in O_3 concentrations was from 80 $\mu g/m^3$ to 135 $\mu g/m^3$. The vertical profiles were evaluated before and after the passage of locomotives, indicating an accumulation of pollutants above the railroad. These findings demonstrate the potential of sniffing drones to measure railroad emissions in urban environments. They also highlight the need to regulate emissions from diesel-powered locomotives to minimize atmospheric pollution and its negative impact on public health in emerging and developing countries.

Keywords: drone; air pollution; railroad emissions; electrochemical sensors; environmental monitoring

1. Introduction

The first Brazilian railroads were built in 1854 to transport coffee beans and sugar cane. Since then, the country has embraced the mode and expanded it. Most Brazilian cities have one-or-more freight or passenger trains. Diesel trains dominate these roadways, thus contributing to urban air pollution. In contrast to other vehicular emissions, air pollution from railroads is not subject to regulation by the National Environmental Council-CONAMA [1,2]. As a result, locomotives primarily use TFM S1800 diesel, which contains 1800 mg/kg of sulfur [3], contributing to emissions of sulfur oxides and other pollutants into the atmosphere. This sulfuric content indicates that exhaust gases from trains may contain high concentrations of sulfur dioxide (SO_2) and hydrogen sulfide (H_2S), but also volatile organic compounds (VOC's), methane (CH_4), nitrogen oxides (NO_X) and ozone (O_3), due to incomplete combustion [4,5].

Measuring railroad emissions in urban environments poses a significant challenge due to the complex and dynamic nature of the surrounding atmosphere. Using fossil fuels on locomotives can contribute to atmospheric pollution, which can become a public health issue. According to Abbasi's research [6], rail traffic does not affect concentrations at open ground-level stations, while underground or enclosed stations can reach ambient air several times. Non-exhaust particle emissions may be hazardous, and enclosed-rail traffic settings should be monitored for high concentrations.

Particle properties, generation methods, and exposure variables, especially rail-transport non-exhaust emissions, must be studied to create effective countermeasures. This situation is particularly concerning given the high demographic density of urban centers, which leads to increased exposure to polluted air, and a greater incidence of cardiovascular problems related to atmospheric pollution [7].

Sensor-equipped drones offer several advantages over traditional measurement techniques, including increased accuracy, mobility, and cost-effectivity. However, direct implementation of this technology presents challenges such as the representativeness of observations, continuous observation needs, and potential sensor exposure to pollution sources. Incorporating data into air quality monitoring systems requires careful technical evaluation and planning to integrate data into air quality monitoring systems to ensure reliable information for improving policy formulation and decision-making.

Emissions from diesel engines, such as locomotives [7,8], are typically linked with three gas pollutants: SO_2, NOx (consisting of NO and NO_2), and volatile organic compounds (VOCs) produced by fossil fuel combustion. In a subsequent reaction, NOx and VOCs react in the presence of sunlight to form O_3. SO_2 is the primary pollutant related to the combustion of sulfur-rich fuels [9].

For human health, exposure to this gas can cause pulmonary activity reduction, eye, nose, and throat irritation, and even premature mortality [10,11]. These problems are predominantly related to the oxidative stress that SO_2 can produce on respiratory system cells [12,13]. On the one hand, NO_2 is dangerous because of a condition named methemoglobinemia. The gas binds to hemoglobin cells in this condition and changes oxygen transport [14]. Symptoms typically related to NO_2 contamination include coughing, wheezing, and breathing difficulty [15]. In addition to these direct effects, NO_2 indirectly impacts human health, influencing tropospheric O_3 concentrations. Photochemical reactions involving both species determine if an area of the atmosphere will act as a source or a sink for O_3 [16,17]. O_3 is a strong oxidant and, similar to SO_2, has a toxic effect on human respiratory cells through oxidative stress [18]. Although O_3 is naturally found in elevated concentrations in the stratosphere, its presence at tropospheric levels qualifies it as a secondary pollutant [10,19–21].

Given the importance of comprehending these gaseous emissions in urban areas, there is a need for a sampling system to directly measure them, as it is only sometimes feasible to utilize stationary sensors atop locomotive smokestacks. In this context, sensor-equipped drones have been used, besides other applications, to evaluate methane sources and sinks, assess volcanic plumes, measure forest burn emissions, monitor urban air quality and atmospheric pollution, detect gas leaks, and monitor methane emissions from landfills [22]. Nevertheless, the applicability of such a system in evaluating railroad emissions has yet to be extensively assessed.

This study aims to evaluate the efficacy of this strategy for measuring SO_2, O_3, and NO_2 emissions using drones over diesel-powered trains in the urban area of Curitiba-Brazil, to understand the impact of railroads on air quality, as these gases are included in the air quality guidelines by the World Health Organization [23]. This investigation presents the concept of 'sniffer drones' for real-time monitoring of moving pollutant sources in urban environments and discusses the challenges associated with this technology.

2. Materials and Methods

2.1. The "Sniffing" System for Railroad Emissions

Using a drone to measure pollutant concentrations requires compact and lightweight sensors. According to Shelekhov et al. [24], measurements of turbulent fluxes and spectra in the urban atmosphere are viable and reasonably accurate using a sensor-equipped DJI Phantom 4 Standard. The authors used a smoothing procedure to study the behavior of the longitudinal and lateral turbulence spectra in the inertial and energy production ranges. The longitudinal and lateral turbulence scales were estimated by the least squares method using the von Karman model as a regression curve. It was shown that the turbulence spectra obtained with the DJI Phantom 4 Pro and AMK-03 generally coincide, with minor differences observed in the high-frequency region of the spectrum. In the power generation domain, the longitudinal and lateral turbulence scales and their ratios measured by the DJI Phantom 4 Pro and the AMK-03 agree with reasonable accuracy. In the inertial range, the behavior of the turbulence spectra shows that they obey the Kolmogorov–Obukhov "5/3" law. The airship used in this study was the DJI Phantom 3 Standard.

This model weighs 1216 g, including the battery, and can carry up to 1 kg. However, for safety reasons, it is not encouraged for the cargo weight to surpass 700 g. Additionally, flight autonomy can be decreased from 25 to less than 20 min when carrying more than 500 g [25]. Considering that the main differences between this model and the Phantom 3 are not mechanical flight technology nature [26], it was assumed that the latter model is also capable of measuring turbulent quantities.

Per our experimental design, we elected to position the sensors at the airship's center to mitigate the downwash and dilution impacts arising from the drone propellers [27,28]. In this context, three Envea Cairsens microsensors were selected for this study: one detecting SO_2, one for NO_2, and another for the O_3 measurements. These electrochemical sensors are small and have a relatively low cost, allowing the use of multiple sensors in one airship. The average weight is 42 g and includes an internal data logger that can store up to 1212 days of data with a 60 min sampling frequency. Although, considering the objectives of this study, the sampling frequency used was 1 min, meaning that up to 2 days of data could be stored in the internal data loggers [29]. For gaseous measurements, the resolution of the sensors is 1 ppb, meaning that the uncertainty related to measured concentrations is ± 0.5 ppb [29]. Considering conversion factors at 25 °C and 1 atm, this uncertainty is ± 1.31 μg/m^3 for SO_2, ± 0.98 μg/m^3 for O_3 and ± 0.94 μg/m^3 for NO_2. The assessment of the cross-sensitivity of the NO_2 and NO_2-O_3 sensors found that the sensor exhibits a response consistent with Cl_2 at levels comprising approximately 80% of the total signal output [30,31]. The study's accuracies are founded upon the performance that manufacturers have claimed. Data measured with this equipment was treated by an outlier removal process described by Junior and Dias [32].

One disadvantage of the Cairsen sensors is the lack of an internal battery, which demands an external energy feed. Also, due to the electrochemical principle used in these sensors, sharing their energy source with the drone could induce some noise in measured datasets [33]. For this reason, the drone was equipped with a small power bank connected to a USB hub. This way, the power bank had enough power to charge three sensors for up to 5 days. The energy consumption of these sensors is low, staying close to 15 mA at 5 VDC [29]. Considering the weight of the power bank, USB hub, and sensors, this configuration made the total cargo on the airship reach 218 g. This value is lower than the 500 g threshold for maintaining flight autonomy over 20 min without jeopardizing the safety of flights.

The authors conducted test flights with the assembled equipment to establish guidelines and strategies for the sampling flights. First, the flight guidelines [34] included the respect of regulations from the Brazilian Civil Aviation National Agency (ANAC) concerning flight speeds, drone operations, and distance from non-consenting people and animals. Additional constraints were incorporated into the guidelines to ensure the safety and maximize the effectiveness of the sampling flights. These constraints consider meteorological factors, such as wind speed, precipitation, visibility, and other related variables that may pose a risk to the operation.

Our study established that sampling flights were not conducted during periods of precipitation or high wind velocity due to the potential impact of these meteorological conditions on the measurements' accuracy. Our study established a protocol whereby the drone must land as soon as possible while maintaining a safe distance from non-consenting third parties if the pilot observed any meteorological conditions during a flight. This precaution was necessary due to the potential for abrupt changes in meteorological conditions during measurement, which could reduce the number of actual measurements obtained and impede continued data collection. The relatively light weight of the drone increased its susceptibility to becoming uncontrollable in high wind conditions, posing potential risks. On days deemed suitable for sampling flights, our measurement protocol dictated assessing background concentrations and emission plumes. Prior to train passage, the airship was positioned approximately 5 m above the railroad for a 15 min period, during which background concentrations were sampled. The collected data were then analyzed, with the mean values obtained for each pollutant serving as the reference point for the corresponding day. When the train was observed, take-off operations commenced. The drone was positioned at an average effective emission height of approximately 5 m above the railroad, a height chosen after considering the locomotives under study. As the train passed, the airship was slowly elevated to the upper limit of the visible emission plume before returning to its initial height of 5 m, where it remained stationary for an additional 10 min period following the train's passage. This method facilitated the examination of gas concentration effects during pre-, intra-, and post-locomotive pass-by periods, providing a comprehensive evaluation of railroad emission impacts.

To provide additional assessment, the team constructed vertical profiles of SO_2, NO_2, and O_3 above the railroad. These profiles were obtained over point 1 (details in Section 2.2) before and 5 min after passing a passenger train. For them, the drone was programmed to start sampling about 1 m above railroad level, measuring for 1 min for each meter. Therefore, the vertical resolution for these profiles was 1 m. Then, after reaching a height of 15 m, the drone was returned to the landing platform. The measured profiles were short, characterizing only a tiny part of the atmosphere above the railroad. However, due to the battery autonomy of the airship, the most extended possible profiles were 15 m in height without compromising the safety of the flights.

2.2. *Area of Study*

The authors selected the study area in Curitiba (PR), shown in Figure 1, based on three primary considerations: proximity to the laboratory at UFPR for easy equipment transportation, availability of a suitable take-off and landing platform (green area in Figure 1), and the presence of numerous points of interest. These sites include residential blocks near the railroad (yellow triangle). This religious institution aids the socially vulnerable (blue triangle), and the Marcelino Champagnat Hospital provides cardiovascular treatments (red cross). Furthermore, the authors considered a bike lane beside the railroad approximately 5 m above the ground (blue line).

The selected area had two sampling points: point 1 and point 2, which are shown as green and red triangles, respectively, in Figure 1. Point 1 was defined as the sampling point for measures over the smokestacks. As mentioned, the effective emission height was set at an average of 5 m above the railroad line. This way, that height was defined as the starting height of point 1, allowing it to be used on direct emission sampling flights.

On certain days, meteorological conditions proved unsuitable for conducting sampling flights. In such cases, point 2 was used for data collection. This designated location is above the bike path adjacent to the rail line, approximately 6 m above the level of the rail line. It represents the average train nose level for a passerby on the bike path. This selection was made to assess the impact of railroad emissions on nearby areas. Notably, the measurements at point 2 were not made with an airship, but the drone remained on the ground with the power off. Since meteorological measurements were unavailable at the site, trained personnel made local and intuitive assessments based on various considerations, such as visual observation of cloud cover and wind speed, assessment of temperature and humidity, and review of weather forecasts from nearby regions. While the data collected at point 2 provides valuable insights into the effects of railroad emissions on nearby areas, it is crucial to interpret the results with caution due to the potential influence of overestimation or underestimation of actual values.

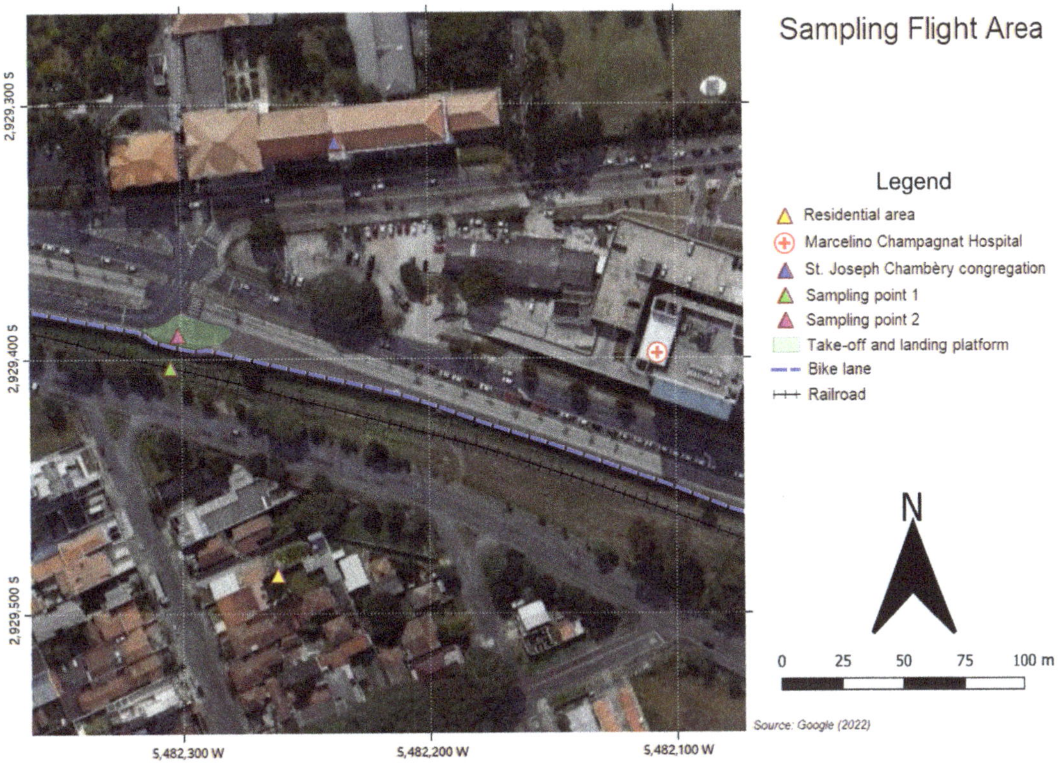

Figure 1. Area used for sampling flights. Source: [35].

During this study, the authors conducted eight sampling campaigns to measure pollutant concentrations at various points in the study area. Most sampling flights (five campaigns) were performed at point 1, with the final campaign occurring on December 25th to determine the vertical profiles of pollutant concentrations at this location. Adverse meteorological conditions resulted in two campaigns being carried out at point 2. Outliers were removed from the measured data, and the datasets were evaluated for each gaseous substance, as shown in Sections 3.1–3.3. In these sections, captions were added to the graphs to facilitate comprehension. Measurements made at sampling point one were labeled as SP 1. Meanwhile, tag SP 2 was used over measurements made on sampling point 2. Moreover, cargo trains were labeled with the letter C, and passenger trains with the letter P. Afterwards; Section 3.4 includes the results for the vertical profiles, both before and after the passage of the passenger train.

3. Results

3.1. Sulfur Dioxide

On different days, the measured background concentrations of SO_2 ranged from 10 $\mu g/m^3$ and 30 $\mu g/m^3$. After the passing of locomotives, in all the campaigns, absolute increases varying from 16 $\mu g/m^3$ and 114 $\mu g/m^3$ concentration were observed, as shown in Figure 2. The values shown in the dark blue bars represent the maximum value observed during each sampling flight.

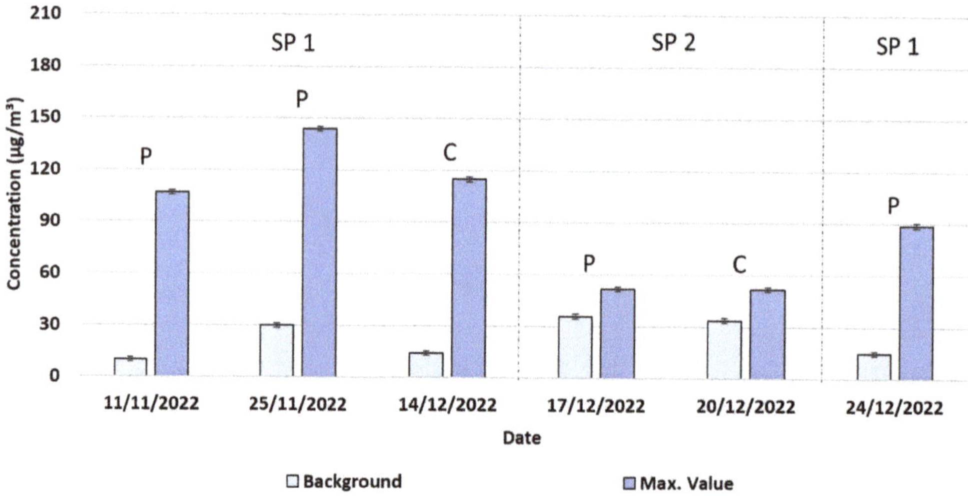

Figure 2. Background and maximum concentrations of SO_2 measured during sampling flights. Cargo trains are indicated with the letter C, while passenger trains are indicated with the letter P. SP 1 means measurements over sampling point 1, and SP 2 is related to sampling point 2.

Although the highest absolute increase, 114 µg/m^3, was observed after a passenger train on November 25, relative increases did not change between passenger trains when separately comparing the two-sampling point. In point 1, the relative increase fluctuated around 600%, while at point 2, it was much lower at around 40%. The SO_2 concentration increases measured at point 2 during train passage were considerably lower than point 1. In addition, the difference between cargo and passenger train emissions at point 2 was slight at 2 µg/m^3 and close to the sensor's uncertainty. For most campaigns, maximum concentrations of SO_2 were observed at the same time as the passage of the locomotive. Moreover, in both SP 1 and SP 2, the maximum values of SO_2 during train passage exceeded the DQA established by the WHO.

3.2. *Ozone*

As presented in Figure 3, O_3 background concentrations showed high variability in different campaigns. Average values as low as 30 µg/m^3 and as high as 60 µg/m^3 were measured during the sampling flights. Even with this high variability, the passage of trains elevated the levels of O_3 consistently in every campaign, with a relative increase over SP 1 in the order of 200%. Differently from what was noted for SO_2, the absolute increase in O_3 concentrations was higher for cargo trains compared to their passenger counterparts. While passenger trains caused a maximum absolute increase in O_3 concentration of 77 µg/m^3, cargo trains' absolute increase was over 120 µg/m^3 in both measured cases. This was especially true for the measurements at point 2, where the total increase was almost two times higher during the passage of a cargo train than the observed increase after the passenger train.

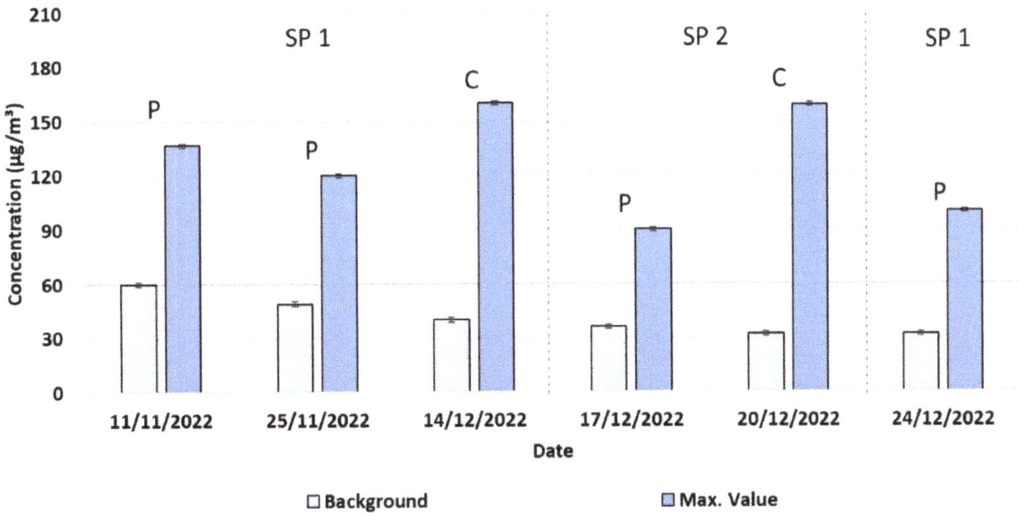

Figure 3. Background and maximum concentrations of O_3 were measured during sampling flights. Cargo trains are indicated with the letter C, while passenger trains are indicated with the letter P. SP 1 means measurements over sampling point 1, and SP 2 is related to sampling point 2.

Furthermore, maximum concentrations were regularly observed 3 to 4 min after the passage of the trains. This time lag can be related to the secondary nature of O_3 in the troposphere, as shown in Section 2, thus the highest O_3 concentrations were observed after the locomotives passed.

3.3. Nitrogen Dioxide

In contrast to what was observed for background values of SO_2 and O_3, it was found that the average concentrations obtained for NO_2 were more uniform in different campaigns. The background values of NO_2 concentration stayed between 25 μg/m^3 and 30 μg/m^3. In addition, increases in these background values were observed during train passage on all campaigns, Figure 4. An analysis of the NO_2 emissions from the cargo and passenger trains at multiple sampling points reveals notable differences in absolute increases. On SP 1, most of the measured passenger trains caused an absolute increase in the order of 85 μg/m^3. In contrast, the cargo train passage was followed by an absolute increase of 140 μg/m^3 in NO_2 concentration.

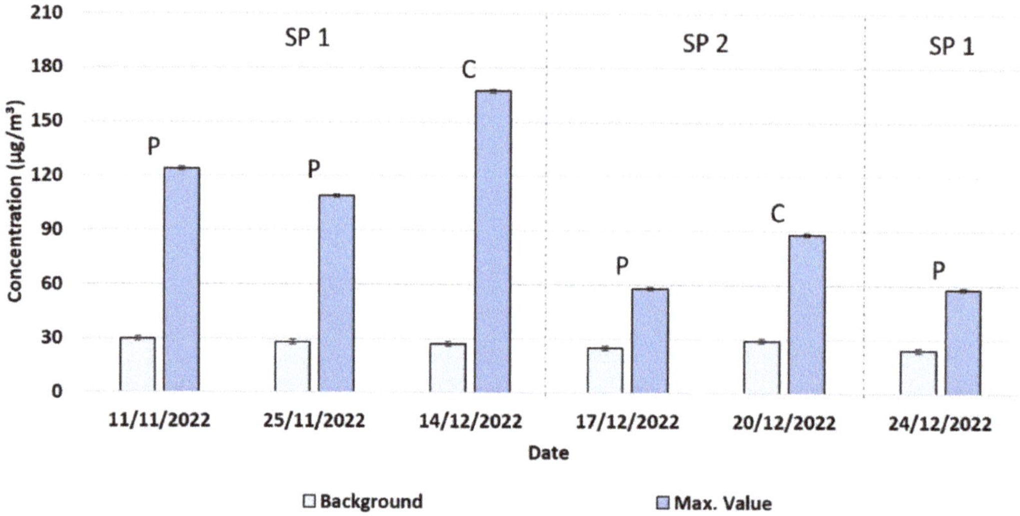

Figure 4. Background and maximum concentrations of NO_2 were measured during sampling flights. Cargo trains are indicated with the letter C, while passenger trains are indicated with the letter P. SP 1 means measurements over sampling point 1, and SP 2 is related to sampling point 2.

Specifically, cargo trains exhibit absolute increases in the concentrations of NO_2, almost twice higher than passenger trains. The maximum values measured at SP 2, 58 μg/m^3 and 88 μg/m^3, indicate that our findings show that areas close to railroad tracks are particularly vulnerable to elevated concentrations of NO_2 emissions. As well as O_3 concentrations, maximum NO_2 values were measured roughly 2 min after the passage of the trains. This time lag was consistent during every sampling campaign.

3.4. Vertical Profiles

As discussed in Section 2, the sampling flights on December 25 were used to construct vertical profiles of gaseous concentrations over point 1. These profiles were measured for SO_2, O_3, and NO_2, beginning about 1 m above railroad level and ending about 15 m above railroad level.

The first three profiles in Figure 5 represent the atmospheric concentrations of SO_2, O_3, and NO_2, in purple, without the passage of a locomotive. By examining these profiles, the authors can gain insight into the background levels of these gasses before trains pollute the environment. It is noted that concentrations are more than 20 μg/m^3 higher than the guidelines stipulated by the WHO for the former and more than 5 μg/m^3 higher for the latter. It is important to emphasize that rigorously measured concentrations cannot be directly compared to the AQGs, since these are defined as 24 h or 8 h averages for the considered gases. However, they can be treated as reference values and are therefore utilized as such in this paper.

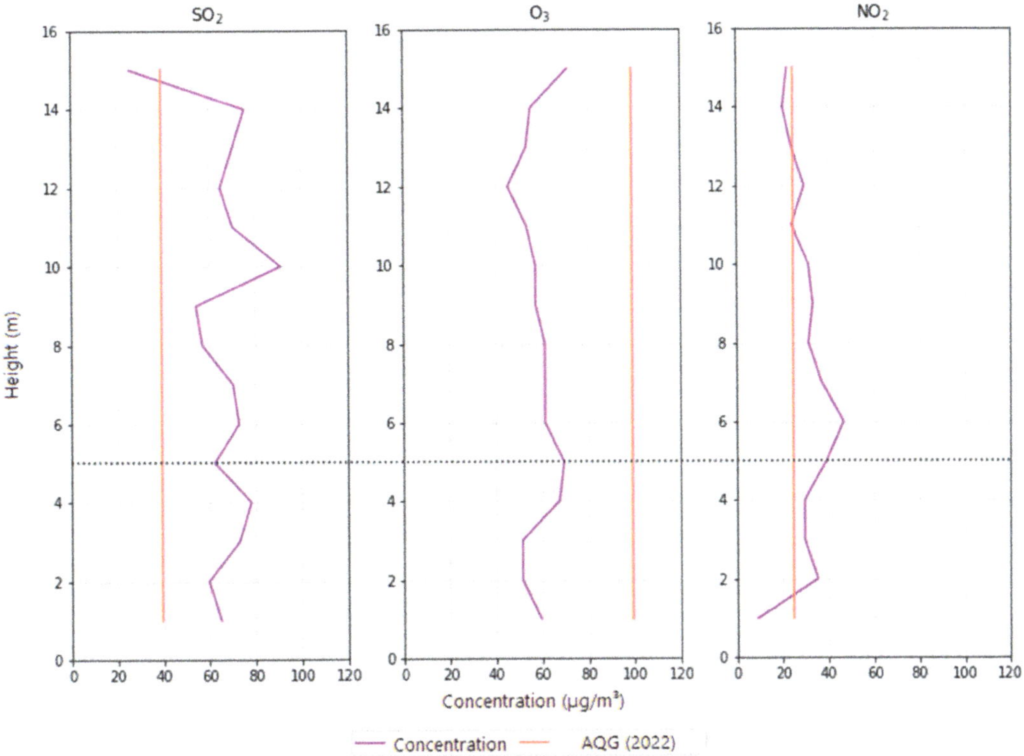

Figure 5. Vertical profiles of SO_2, O_3 and NO_2 measured over the railroad before the passage of a passenger train. In orange, the air quality guideline defined by the WHO for each pollutant.

Moreover, a minor O_3 and NO_2 accumulation with concentrations reaching up to 47 μg/m^3 can be seen near the smokestack level, 5 m above the railroad level (Figure 5 as the dotted line). This accumulation cannot be seen for SO_2. Instead, concentrations of this pollutant tend to accumulate between 10 m and 14 m above the railroad level, with values reaching 90 μg/m^3.

Shortly after 8:30 a.m. local time, a passenger train passed sampling point 1. About 5 min after this passage, the sniffing drone was deployed, measuring the vertical profile of gaseous concentrations one more time. These profiles are presented in Figure 6. In addition, Figure 6 contains the average peak concentrations of each gas measured during passenger train passages, represented by the red dots: 105 μg/m^3 for SO_2, 128 μg/m^3 for O_3, and 116 μg/m^3 for NO_2. The concentrations after the train increased for each pollutant throughout the vertical profile. The maximum increase observed for SO_2 occurred at the height of 10 m, from 90 μg/m^3 prior to the train passage to 152 μg/m^3. In addition, concentrations of O_3 and NO_2 were higher near the stack, reaching 108 μg/m^3 for the prior and 74 μg/m^3 for the latter. In contrast, SO_2 concentrations were higher at 7 and 10 m altitude.

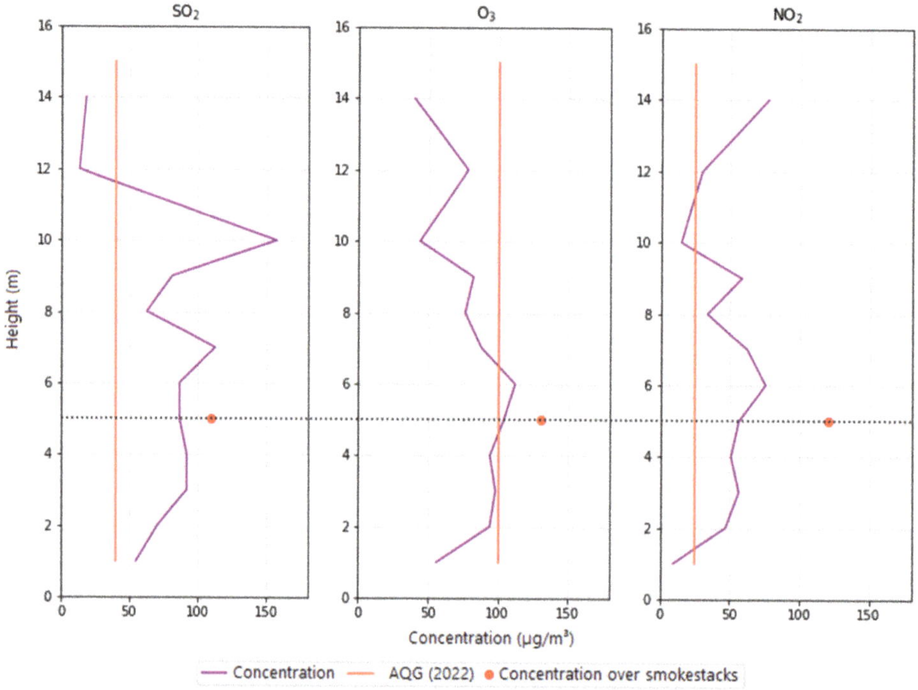

Figure 6. Vertical profiles of SO_2, O_3, and NO_2 were measured 5 min after a passenger train passed over the rail line. The red dots represent the mean concentration measured over the smokestacks of passenger trains. In orange, the WHO defines the air quality guideline for each pollutant.

In addition to the higher concentration values, the profiles in Figure 6 reveal the height of the locomotive's emission plume 5 min after its passage. According to the SO_2 profile, above 10 m, concentrations of SO_2 decrease rapidly. Similarly, the O_3 and NO_2 concentrations are proportionate up to 10 m. This behavior suggests that 5 min after a passenger train passes, the emission plume reaches a height of 10 m. Above this height, this relationship is no longer evident.

4. Discussion

The evaluation was limited to two cargo trains, while four passenger trains were assessed. The goal of comparing these types of trains was to determine which emitted higher atmospheric pollutants. This assumption was based primarily on the fact that cargo trains passing through the area of interest are typically longer and heavier than passenger trains and therefore require more fuel consumption, which leads to higher concentrations of the gases being studied. During the sampling campaigns conducted for this study, the presence of passing cargo trains in the study region presented a challenge. Unlike passenger trains, whose departure times are publicly available and easily accessible, cargo train schedules are confidential and can vary significantly on different dates. As a result, acquiring emissions data from cargo trains required more time and effort. This was true for the NO_2 concentrations and even more so for O_3 concentrations. In the latter case, absolute increases in concentration were more significant after the passage of cargo trains for measurements made at both point 1 and point 2. The total increase at point 2 due to the passage of a cargo train was almost twice as significant as the increase due to the passage of a passenger train.

Contrary to this, no substantial numerical differences in SO_2 concentrations were observed. Increases in SO_2 from passenger and cargo trains were not distinguishable. Due to the high sulfur content of locomotive fuels, it is plausible that SO_2 emissions will remain high despite a decrease in cargo loads and fuel consumption. However, it also suggests that O_3 and NO_2 emissions may be more dependent on the weight of the trainload.

Increases in SO_2 from the passenger and cargo trains were not significantly different, and SO_2 emissions may remain substantial even as cargo loads and fuel usage decrease due to the high sulfur content of locomotive fuels. Thus, O_3 and NO_2 emissions could be more sensitive to the train's cargo weight. The authors can draw analogous arguments by comparing the measurements collected at sample sites 1 and 2. Consistent with our expectations, higher concentration increases were observed more frequently at point 1, directly over the smokestacks. However, the authors also observed different behaviors on sampling point 2, with small absolute increases in gaseous concentration being the norm. We attribute this to the high temperatures at which the exhaust gases from the smokestacks are produced, causing them to rise rapidly in the atmosphere so that there is not enough time for the elevated concentrations to reach the nose of the train on the side of the railroad.

These findings suggest that the impact of railroad emissions on surrounding areas is insignificant compared to the impact on the railroad itself. The authors confirm this assumption by noting the similarity between the O_3 amounts measured at points 1 and 2. The influence of emission temperature on O_3, which acts as a secondary pollutant, is comparatively smaller than that on primary gases. Substantial O_3 concentrations can be measured even at passerby's nose levels on the railroad's side during the locomotives' passage. These results represent the first emissions data obtained from moving trains using a drone. They are the starting point for understanding the spatial distribution of railroad emissions and their potential impact on air quality in the research area.

Acquisition of vertical profiles of contaminant concentrations has provided valuable insight into the circumstances of the investigation. By comparing gas profiles before and after train passage, the authors confirmed that railroad emissions could indeed increase concentrations of SO_2, NO_2, and O_3 in the atmosphere of surrounding communities. In addition, it is important to emphasize that due to the lack of legislation, direct comparisons with the World Health Organization's air quality guidelines (WHO) are not possible in our study because the authors do not have 24 h estimates. Nevertheless, our studies have provided preliminary evidence that these parameters are exceeded. The authors observed that these gases accumulate in the lower atmosphere before a train passes, leading to concentrations that immediately exceed the AQGs in 75% for SO_2 and 25% for NO_2. It is worth noting that this does not necessarily indicate that the air quality in the research area is poor, but it does indicate a deterioration in air quality. Our findings have significant implications for policymakers and stakeholders, who must consider the impact of railroad emissions on nearby communities and the environment.

5. Conclusions

This study presents preliminary findings through eight sampling campaigns conducted in Curitiba, Brazil. The measurement technique employed in this study involves sampling both background concentrations and locomotive emission plumes of SO_2, O_3, and NO_2. As part of our study, eight sampling campaigns were conducted using a drone equipped with Envea Cairsen microsensors to accurately measure pollutant concentrations at various locations in the research area. The results demonstrate the efficacy of using compact and lightweight microsensors mounted on drones to assess the impact of train emissions on ambient air quality. Measurements taken before and after the trains passed showed increases in all three pollutants, with the most significant increase in SO_2 after passenger trains and O_3 after cargo trains. NO_2 concentrations were also higher after the cargo trains, even near the railroad. Maximum concentrations were observed 2–4 min after train passage. While drones equipped with sensors have demonstrated the ability to detect locomotive pollution, the accuracy of these measurements can be influenced by wind speed and direction; however, implementing microsensors as anemometers on drones in the future will facilitate the detection of weather changes leading to more accurate pollution sensing.

This investigation underscores the importance of monitoring air pollution from transportation sources and proposes a potential solution for achieving this aim. Leveraging drones and lightweight sensors can substantially aid in comprehending and mitigating the impact of transportation emissions on air quality and public health. The findings of this study are valuable for policymakers and transportation authorities in devising strategies to counteract the detrimental effects of railroad transportation on air quality. Further research is necessary to validate and generalize these findings to other regions.

Author Contributions: Both authors contributed equally and integrally to the development of the research and the presentation of the results. All authors have read and agreed to the published version of the manuscript.

Funding: This research received no external funding.

Data Availability Statement: No new data was created during the study, and no relevant publicly available datasets exist for this research.

Conflicts of Interest: The authors declare no conflict of interest.

References

1. Conselho Nacional do Eeio Ambiente (CONAMA). *Resolução Conama no 18, de 06/05/1986. Dispõe Sobre a Criação do Programa de Controle de Poluição do ar Por Veículos Automotores—PROCONVE*; Ministério do Meio Ambiente (MMA): Brasília, Brazil, 1986.
2. Conselho Nacional do Eeio Ambiente (CONAMA). *Resolução Conama no 315, de 29/10/2002. Dispõe Sobre a Nova Fase do Programa de Controle de Poluição do ar Por Veículos Automotores—PROCONVE*; Ministério do Meio Ambiente (MMA): Brasília, Brazil, 2002.
3. Petrobras. *Óleo Diesel: Informações Técnicas*, 1st ed.; 2021; p. 26. Available online: https://petrobras.com.br/data/files/04/93/72/4C/5A39C710E2EF93B7B8E99EA8/Manual-de-Diesel_2021.pdf (accessed on 22 August 2022).
4. Paralovo, S.L. Análise de Poluentes Gasosos e Material Particulado Fino Em Manacapuru, AM. Master's Thesis, Federal University of Paraná, Curitiba, Brazil, 2016.
5. Zanella, T.V. Navios e poluição do ar: Um estudo sobre a regulação das emissões atmosféricas por embarcações. *Rev. Esc. Guerra Nav.* **2018**, *24*, 301–328. [CrossRef]
6. Abbasi, S.; Jansson, A.; Sellgren, U.; Olofsson, U. Particle Emissions from Rail Traffic: A Literature Review. *Crit. Rev. Environ. Sci. Technol.* **2013**, *43*, 2511–2544. [CrossRef]
7. Mir, S.A.; Bhat, J.I.; Lone, F.; Rehman, M.U.; Nazir, N.; Lone, A.A.; Ali, T.; Jehangir, A. Synergistic effects of vehicular emissions (NO_2, SO_2 and SPM) on progression of *Crocus sativus* L. in saffron bowl kashmir. *Adv. Environ.* **2021**, *3*, 100033. [CrossRef]
8. D'agosto, M.d.A. *Transporte, Uso de Energia e Impactos Ambientais: Uma Abordagem Introdutória*, 1st ed.; Elsevier: Rio de Janeiro, Brazil, 2015; p. 251, ISBN 978-85-352-2821-2.
9. EPA. Sulfur Dioxide Basics. United States Environmental Protection Agency. 2022. Available online: https://www.epa.gov/so2-pollution/sulfur-dioxide-basics (accessed on 15 June 2022).
10. EPA. Health Effects of Ozone Pollution. United States Environmental Protection Agency. 2022. Available online: https://www.epa.gov/ground-level-ozone-pollution/health-effects-ozone-pollution (accessed on 22 August 2022).
11. World Bank. *Pollution Prevention and Abatement Handbook: Toward Cleaner Production*; The International Bank for Reconstruction and Development: Washington, DC, USA, 1999; p. 472. ISBN 0-8213-3638-x.
12. Di Meo, S.; Reed, T.T.; Venditti, P.; Victor, V.M. Role of ROS and RNS sources in physiological and pathological conditions. *Oxid. Med. Cell. Longev.* **2016**, *12*, 1245049. [CrossRef]
13. Reno, A.L.; Brooks, E.G.; Ameredes, B.T. Mechanisms of heightened airway sensitivity and responses to inhaled SO_2 in asthmatics. *Environ. Health Insights* **2015**, *9*, 13–25. [CrossRef] [PubMed]
14. Gaston, B.; Drazen, J.M.; Loscalzo, J.; Stamler, J.S. The biology of nitrogen oxides in the airways. *Am. J. Respir. Crit. Care Med.* **1994**, *149*, 538–551. [CrossRef]
15. EPA. Basic Information about NO_2. United States Environmental Protection Agency. 2022. Available online: https://www.epa.gov/no2-pollution/basic-information-about-no2 (accessed on 22 August 2022).
16. Seinfeld, J.H.; Pandis, A.N. *Atmospheric Chemistry and Physics: From Air Pollution to Climate Change*, 2nd ed.; John Wiley & Sons: Hoboken, NJ, USA, 2006; p. 1248. ISBN 978-0-471-72018-8.
17. Nguyen, D.-H.; Lin, C.; Vu, C.-T.; Cheruiyot, N.K.; Nguyen, M.K.; Le, T.H.; Lukkhasorn, W.; Vo, T.-D.; Bui, X.-T. Tropospheric ozone and NOx: A review of worldwide variation and meteorological influences. *Environ. Technol. Innov.* **2022**, *28*, 102809. [CrossRef]

18. Chen, T.-M.; Kuschner, W.G.; Gokhale, J.; Shofer, S. Outdoor air pollution: Ozone health effects. *Am. J. Med. Sci.* **2007**, *333*, 244–248. [CrossRef] [PubMed]
19. CCAC. Tropospheric Ozone: Climate & Clean Air Coalition. 2022. Available online: https://www.ccacoalition.org/en/slcps/tropospheric-ozone (accessed on 22 August 2022).
20. EPA. Basic Ozone Layer Science. United States Environmental Protection Agency. 2021. Available online: https://www.epa.gov/ozone-layer-protection/basic-ozone-layer-science#:~:text=the%20ozone%20layer%20lies%20approximately,earth\T1\textquoterights%20surface%2c%20in%20the%20stratosphere (accessed on 15 September 2022).
21. EPA. Ground-Level Ozone Basics. United States Environmental Protection Agency. 2022. Available online: https://www.epa.gov/ground-level-ozone-pollution/ground-level-ozone-basics (accessed on 3 November 2022).
22. Jońca, J.; Pawnuk, M.; Bezyk, Y.; Arsen, A.; Sówka, I. Drone-Assisted Monitoring of Atmospheric Pollution—A Comprehensive Review. *Sustainability* **2022**, *14*, 11516. [CrossRef]
23. WHO. Air Quality and Health. World Health Organization. 2021. Available online: https://www.who.int/teams/environment-climate-change-and-health/air-quality-and-health/health-impacts (accessed on 17 June 2022).
24. Shelekhov, A.; Afanasiev, A.; Shelekhova, E.; Kobzev, A.; Tel'minov, A.; Molchunov, A.; Poplevina, O. Low-Altitude Sensing of Urban Atmospheric Turbulence with UAV. *Drones* **2021**, *6*, 61. [CrossRef]
25. DJI. Phantom 3 Standard Specifications. 2022. Available online: https://www.dji.com/br/phantom-3-standard (accessed on 14 June 2022).
26. DJI. Phantom 3 (Pro) Vs. Phantom 4 (Pro): Review of the Image Quality Comparison. 2023. Available online: https://store.dji.com/guides/phantom-3-pro-vs-phantom-4-pro-image-quality/ (accessed on 3 March 2023).
27. Brinkman, J.L.; Davis, B.; Johnson, C.E. Post-movement stabilization time for the downwash region of a 6-rotor uav for remote gas monitoring. *Heliyon* **2020**, *6*, 9. [CrossRef]
28. Roldán, J.J.; Joossen, G.; Sanz, D.; Del Cerro, J.; Barrientos, A. Mini-UAV based sensory system for measuring environmental variables in greenhouses. *Sensors* **2015**, *15*, 3334–3350. [CrossRef] [PubMed]
29. ENVEA. Miniature Solution for Real-Time Continuous Pollution Monitoring. France. 2022. Available online: https://www.envea.global/design/medias/envea_cairsens_air-quality-odors-microsensors_en.pdf (accessed on 22 August 2022).
30. Spinelle, L.; Gerboles, M.; Aleixandre, M. *Report of the Laboratory and In-Situ Validation of Micro-Sensors for Monitoring Ambient Air Pollution-O12: CairClipO$_3$/NO$_2$ of CAIRPOL (F). EUR 26373*; JRC86479; Publications Office of the European Union: Luxembourg, 2013.
31. Spinelle, L.; Gerboles, M.; Aleixandre, M. *Report of Laboratory and In-Situ Validation of Micro-Sensor for Monitoring Ambient Air Pollution-NO9: CairClipNO$_2$ of CAIRPOL (F). EUR 26394*; JRC86499; Publications Office of the European Union: Luxembourg, 2013.
32. Junior, D.V.; Dias, N.L. Método Empírico para Determinação de Outliers em Séries de Fluxos de Dados Micrometeorológicos Pós-processados. *Ciência Natura* **2013**, *35*, 150–152.
33. Gu, Q.; Michanowicz, D.R.; Jia, C. Developing a modular unmanned aerial vehicle (UAV) platform for air pollution profiling. *Sensors* **2018**, *18*, 4363. [CrossRef]
34. Ministério da Defesa. *Aeronaves não Tripuladas para uso Recreativo—Aeromodelos*; Comando da Aeronáutica. MCA 56-2; Ministério da Defesa: Brasília, Brazil, 2020.
35. Google. Curitiba. 2021. Available online: https://www.google.com.br/maps (accessed on 29 December 2022).

atmosphere

MDPI

Article

Exploring the Spatial Variability of Air Pollution Using Mobile BC Measurements in a Citizen Science Project: A Case Study in Mechelen

Martine Van Poppel [1,*], Jan Peters [1], Stijn Vranckx [1], Jo Van Laer [1], Jelle Hofman [1], Bram Vandeninden [2,3], Charlotte Vanpoucke [2] and Wouter Lefebvre [1]

1 Flemish Institute for Technological Research (VITO), Boeretang 200, 2400 Mol, Belgium
2 Belgian Interregional Environment Agency, Gaucheretstraat 92-94, 1030 Brussels, Belgium
3 School of Public Health, Université Libre de Bruxelles, Route de Lennik 808-CP591, 1070 Brussels, Belgium
* Correspondence: martine.vanpoppel@vito.be

Abstract: Mobile monitoring is used as an additional tool to collect air quality data at a high spatial resolution and to complement data from fixed air quality stations. Citizens are interested in contributing to air quality monitoring, and while the availability of low-cost air quality sensors can create opportunities to measure the air quality at a high spatial resolution, the data are often of lower quality, and sensors that measure combustion-related aerosols (like black carbon) are not commonly available. Mobile monitoring using a mid-range instrument can fill this gap. We present the results of a mobile BC (black carbon) monitoring campaign performed by citizens in Mechelen as part of a local citizen observatory (CO), Meet Mee Mechelen, initiated as part of the European H2020 project, Ground Truth 2.0. The goal of the study was two-fold: (1) to propose and evaluate a mobile monitoring method (data collection and data processing) to construct pollution maps of BC concentrations and (2) to demonstrate how to organize community-based air quality monitoring to measure both the spatial and temporal variations in air pollution levels. Measurements were taken during peak hours in four campaigns characterized by different meteorological conditions: October–November 2017, February–March 2018, June–July 2018 and September 2018. The results show large spatial and temporal variabilities. Spatial variability is influenced by traffic volume, stop-and-go traffic and also the building environment and the distance of biking paths from road traffic. The four different campaigns show similar spatial patterns, but due to background and meteorological influences, the absolute concentrations differ between seasons. A rescaling method using data from fixed stations in the air quality monitoring network (AQMN) was presented to construct maps representative of longer periods. This paper shows that mobile measurements can be used by CO to assess the spatial variability of air quality in a city. The data can be used to evaluate mobility plans, carry out hot spot detection, evaluate the exposure of cyclists as a function of cycling infrastructure and perform model validation. However, it is important to use high-quality instruments and apply the correct measurement methodology (number of repetitions, season) to obtain meaningful data.

Keywords: air pollution; mobile measurements; citizen science; black carbon; spatial variability

1. Introduction

Citizens are becoming more and more aware of air quality and the impact of air quality on health. The increasing availability of low-cost air quality sensors is giving citizens the opportunity to start measuring their own environment. However, the quality of the sensors used and the setup of the monitoring campaign need special attention.

Citizen science projects face challenges in relation to data credibility (e.g., data fragmentation, data inaccuracy, lack of experimental design) and the utility of citizen science data [1]. If large datasets are produced and are not scientifically sound, these may become a headache for those who are responsible for managing air quality [2]. In addition, the results of citizen science projects need to be explained in a comprehensible way to keep stakeholders engaged. Therefore, citizens, researchers and policymakers need to work together to make a citizen observatory successful. Low-cost portable devices are available and have been increasingly used to monitor air quality. While these sensors can have potential when used in fixed networks to assess the spatial variability for some pollutants when appropriate calibration models are used, some issues remain if they are used for mobile applications. In this study, we used a validated monitoring approach (www.airqmap.com) that resulted in scientifically sound data with respect to the monitoring device and the method with which the measurements were made [3].

Traditionally, air quality is measured at fixed air quality monitoring stations (AQMSs) with a low spatial resolution. The main advantages of fixed reference stations are their measurement accuracy and long-term time series. However, they do not necessarily represent citizens' exposure to air pollution. To obtain a better estimate of the spatial variability of air pollution, more locations need to be sampled within a city or region. Two different measurement approaches can be used: a dense network using low-cost monitoring instruments (sensors or passive samplers) or, mobile measurements using high-grade or medium-grade monitors [4–7] or low-cost sensors [8–11]. While a dense network of stationary locations has a trade-off between the cost of instruments and the quality of the data, mobile measurement approaches can use higher-quality instruments at higher costs because one instrument can cover a large area. Therefore, mobile measurements are complementary to fixed stations with high-end instruments in terms of spatial coverage and are preferred to dense networks when low-cost sensors are not available for the pollutant or are of questionable quality.

Mobile measurements are often a snapshot of the situation, e.g., in a specific season or at a specific time of the day. Given the spatio-temporal variability of air quality, the resulting data from mobile measurements need to be sampled and analyzed in a different way compared to fixed measurements [12]. In this study, we performed measurements on different days and in different seasons to reduce the statistical noise related to the differences in meteorological conditions, resulting in a representative map of the spatial variability of pollution.

This study focuses on black carbon (BC), a compound of fine particles that are emitted from combustion sources [13]. BC is a good indicator of combustion-related air pollution, and exposure to BC is recognized by the World Health Organization to be associated with cardiovascular and cardiopulmonary mortality [14]. A recent study has provided important evidence of the associations between long-term exposures to low concentrations of, for example, BC and various health endpoints [15]. In addition, studies of short-term health effects suggest that BC is a better indicator of harmful particulate substances from combustion sources (especially traffic) than undifferentiated particulate matter (PM) mass. Data on the speciation of PM, and more specifically, BC, are scarce because BC concentrations are not regulated by the EU Air Quality Directive, and instrumentation is relatively complicated and expensive. Recent research has developed empirical models based on, for example, machine learning ([16,17] and references therein) or land use regression [18], where BC is estimated based on other pollutants or interpolated at other locations to cover a broader area. While these modeled data can address the need for better exposure estimates, they still require validation that is based on measured data with a high spatial resolution. In Flanders, BC was measured at 25 stations in the AQMN at the time of the study, and no station was situated in the city of Mechelen.

In previous research, we showed the potential of mobile measurements to assess the personal exposure of cyclists [19] or other transport modes [4,20] and to assess the spatial variability of urban air quality of different pollutants (e.g., UFP, BC and heavy metals in [5]; UFP and PM_{10} in [21]; UFP and BC in [6]; BC in [7]). We examined the number of repetitions needed and the impact of background concentrations [3,22]. Other papers also showed the potential of mobile measurements [23–26] for personal exposure in studying the spatial variation of air pollution. In a number of recent studies, mobile BC measurements have been made to assess the exposure of bicycles and other commuters [5,27–33] in relation to the traffic situation [34] and the effect of bicycle infrastructure on the exposure levels of bicycle commuters [35,36]. A similar exercise was conducted for $PM_{2.5}$ by distributing 500 low-cost sensors to voluntary participants in Utrecht, the Netherlands [11].

In this paper, we performed a study on the spatial variability of BC during peak hours in different seasons in the city of Mechelen. The goal of the study was two-fold: (1) to evaluate the mobile monitoring method (data collection and data processing) to construct pollution maps of BC concentrations and to propose a scientific sound method, and (2) to demonstrate how to organize community-based air quality monitoring to measure both the spatial and temporal variations in air pollution levels.

The measurements were performed by citizens of Mechelen who took part in a local citizen observatory (CO), "Meet Mee Mechelen"; the measurement requirements, setup and execution were performed by the citizens in close consultation with the research partner VITO and the city council. Using this unique dataset, we compared campaigns that were performed under different meteorological conditions. A rescaling methodology based on fixed monitoring stations of the Flemish Environment Agency was applied and evaluated. Spatial patterns were analyzed and interpreted in terms of the current mobility plans, in particular with respect to cyclist exposure. The results of a mobile monitoring campaign were compared to state-of-the-art modeling results. Based on simultaneous mobile measurements, the currently used approach of measuring BC using medium-cost instruments was compared to measurements taken using a low-cost sensor for PM that is often used in citizen science projects.

2. Material and Methods

2.1. A Citizen Observatory for Air Quality

As part of the Ground Truth 2.0 project (http://gt20.eu/, accessed on 16 June 2024), citizen observatories (COs) have been initiated in six European and African countries involving citizens, policymakers and scientists, following a co-design process developed as part of the project. The main findings on best practices for the setup of a CO are discussed in two other publications [37,38]. In Mechelen, Belgium, a group of enthusiastic volunteers, local policymakers and scientists initiated Meet Mee Mechelen (translated as "Join our measurements Mechelen"), a CO with a focus on improving the local air quality to enhance health, quality of life and social cohesion.

The participants of the CO defined the following aim of the measurement campaign to measure local air pollution levels in Mechelen (1) to identify the impact of road traffic and obtain a set of benchmark measurements (prior to proposing or implementing specific actions) and (2) to identify the spatial and temporal changes in air quality levels. Researchers helped to translate the citizens' needs into a scientifically sound measurement setup. They also trained the coordinators of the effective measurement campaign and instructed participants on how to operate and wear the instruments, as well as how to perform the measurement campaign with a special focus on driving speed and direction, number of repetitions, and synchronization of the measurements on the different routes.

2.1.1. Instrumentation

Citizens of Mechelen carried out the measurements using airQmap (www.airqmap.be, accessed on 16 June 2024), a user-friendly monitoring tool developed by VITO to map BC at street level; airQmap is a mobile platform that measures BC via repeated measurements by bicycle and is based on a scientifically justified method [3]. It comprises a measurement unit consisting of a microAeth® (AE51, AethLabs, San Francisco, CA, USA), a GPS and an automated data processing infrastructure to construct the aggregated BC maps. The microAeth® measures the attenuation of light (880 nm LED) through a Teflon-coated borosilicate glass fiber filter on which light-absorbing particles are sampled continuously. The difference in attenuation (per time base) is converted to a BC concentration using a specific absorption coefficient sigma = 12.5 m^2 g^{-1} (at 880 nm) and the actual sample flow rate. The sample flow rate is set at 150 mL min^{-1}, and the measurements are made at a temporal resolution of 1 s. To reduce the signal noise at the one-second time resolution (especially at low concentrations), the ONA (optimized noise-reduction averaging) algorithm was used with an attenuation threshold of 0.05 [39]. Prior to each monitoring campaign, the flow of the instruments is calibrated, and the different instruments used are intercompared by co-location (for at least one day) to check that the inter-instrument variability is less than 10%. A periodic comparison with the AE33 (Magee) is performed.

To assess the potential of using low-cost sensors and different metrics, the AirBeam was aligned with the aethalometer and used simultaneously on some trips. Useful for citizen science, the AirBeam is a PM sensor developed by HabitatMap, a US non-profit environmental health justice organization whose goal is to raise awareness of the impact the environment has on human health. HabitatMap builds tools to support grassroots environmental organizing, including AirCasting. AirCasting is an open-source, end-to-end solution for collecting, displaying and sharing health and environmental data using your smartphone. The platform consists of wearable sensors that detect changes in your environment and physiology, including a palm-sized air quality monitor called the AirBeam, the AirCasting Android app, the AirCasting website and wearable LED accessories. The AirBeam and AirCasting infrastructures are used worldwide. Citizen scientists from around the world take measurements and contribute to the air quality crowd map. The AirBeam sensor was assessed by the AQ-SPEC (the Air Quality Performance Evaluation Center), and while it showed an overestimation of the PM values, it also showed a good correlation with more expensive FEM instruments. In a recent study [40], AirBeam showed acceptable quantitative data for different sources and the authors concluded that AirBeam was of sufficient accuracy and reliability to be able to detect sources and assess increased exposure to $PM_{2.5}$ in homes. Another study showed the fit-for-purpose of the AirBeam for mobile monitoring applications in urban environments, based on extensive lab- and field benchmarking [41].

2.1.2. Description of the Measurement Campaign

Measurements were performed in different seasons, resulting in four measurement campaigns: Campaign 1 (October/November 2017), Campaign 2 (February/March 2018), Campaign 3 (June/July 2018) and Campaign 4 (September 2018). The meteorological conditions (air temperature, precipitation, windspeed and wind direction) during the four campaigns were compared to normal values from a 30-year reference period based on data from the Royal Meteorological Institute of Belgium (RMI). Air temperatures were (very) high for the time of the year during Campaigns 1 and 3 (about 2–3 °C higher) and low during Campaign 2 (about 2 °C lower). Precipitation was exceptionally low during large parts of Campaigns 1, 2 and 3. The wind conditions were normal for the times of the year, except for the dominance of a northerly wind in Campaign 3 when southerly winds are usually dominant. We do not expect that the differences in meteorological conditions that were observed during the individual campaigns compared to long-term averages for these times of the year will affect the representativity of the results because the measurements were performed over different seasons.

In each of the campaigns, 4 routes representing 4 different areas of the city were measured. The total combined distance of the four routes was roughly 40 km. The different routes are shown in Figure 1. In this way, each of the routes was measured within one hour and the measurements could be performed in a restricted time frame within traffic rush hours. The resulting concentration maps are considered a representative hourly average concentration over the peak hour measured. The route selection was made by local citizens and policymakers based on the guidance of air quality experts. This guidance included having a good mix of background locations and busy streets and variations in traffic intensities and street configurations.

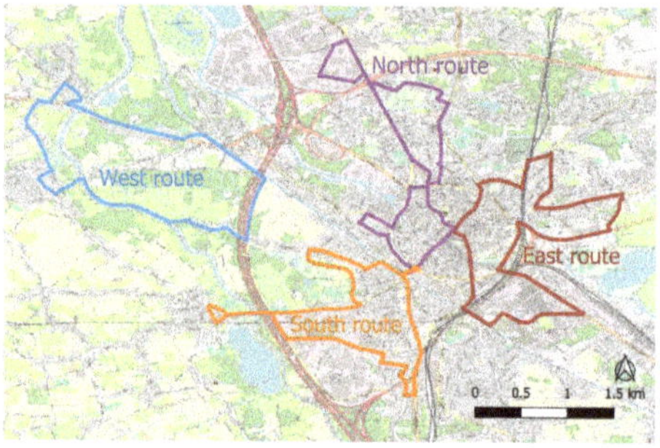

Figure 1. Street map of Mechelen with the four different routes.

During the first campaign, the South route was not measured simultaneously with the other three routes; the South route was measured from 23 October 2017 to 5 November 2017, prior to the other routes (North, East and West routes), which were measured from 6 November 2017 to 19 November 2017. During Campaign 3 (25 June–8 July 2018), only the North and East routes were measured because not enough volunteers were available.

Measurements were performed during the morning and evening rush hours. Repeated measurements are needed, as shown in previous publications [3,21]. Data experiments with a large dataset of 256 repeated runs showed that when 13 repetitions are performed, the BC concentrations can be assessed with a deviation of less than 50% at 90% of the 20 m road segments [3]. While the referenced study did not focus on rush hours, it does give some idea of the number of repetitions needed. In this study, we used approximately 25 repetitions (runs) for each route and campaign. Details on the number of measurements (in the morning and the evening) for the four routes over the different monitoring campaigns are given in Table 1. The aim of this study was to assess the air quality during peak traffic hours, so participants were instructed to carry out the mobile measurements between 8:00–9:00 and 17:30–18:30 local time. In practice, most measurements taken during the morning were started between 7:45 and 8:15, with some outliers. However, the time window for the evening measurements was slightly broader, between 16:45 and 17:45, as it was decided that it suited the volunteers better to start a bit later after a few days into the first campaign. Figure 2 shows the frequency distribution of the starting times for the measurements performed over the four routes and the four campaigns. Overall, the monitoring efforts were adequately aligned between the different teams of volunteers and over the four campaigns.

Table 1. Overview of the number of runs (morning or evening) for the different campaigns and the different routes.

	Campaign 1 Morning/Evening	Campaign 2 Morning/Evening	Campaign 3 Morning/Evening	Campaign 4 Morning/Evening
North	11/11	11/12	12/14	12/11
East	11/13	13/10	12/11	12/12
South	13/14 [(a)]	10/9	-	11/10
West	11/13	11/11	-	12/12
All	12/13	13/12	14/14	12/12

[(a)] Measurements at this route were not performed simultaneously with the other routes during this campaign and are not included in the 'Total' number of runs in the campaign.

On four bicycle trips (runs), simultaneous measurements were taken with the aethalometer and the AirBeam for comparison (5 and 6 March 2018, morning and evening rush hours on both days, West route). The sensors were installed at identical positions.

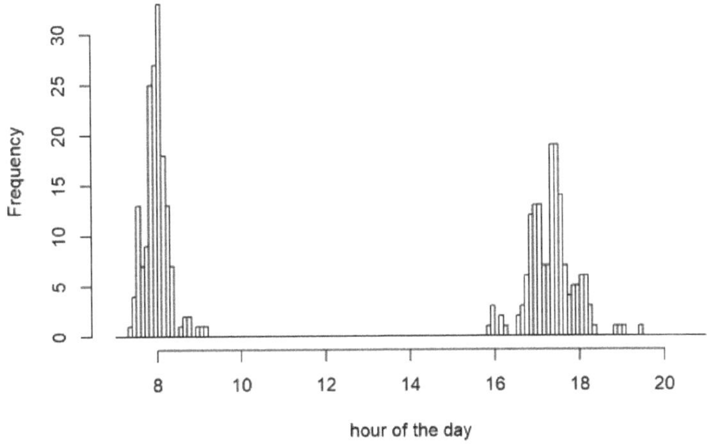

Figure 2. Frequency distribution of the starting time of all measurements (4 campaigns and 4 subroutes).

2.1.3. Collected Dataset

Black carbon concentration measurements were collected by bicycle along four routes in four different campaigns over two weeks. A raw dataset of 1,110,771 geo-tagged measurements was collected. In the process of data validation and aggregation, the number of data points was drastically reduced to approximately 2300 data points per campaign. In the aggregation, geo-tagged measurements are attributed to fixed points that are 20 m apart from each other, and a trimmed mean of the collection of measurements per fixed point is calculated. The resulting dataset is overlaid onto a street map. For more details on the data processing (validation and aggregation), we refer to a previous paper [3]. For GPS data, we used the projection on predefined street segments of the route based on the shortest distance to correct for inaccurate GPS data. For the BC concentrations, we applied the ONA correction (see also Section 2.2) and used the trimmed mean per spatially aggregated point. We did apply a filter-loading correction but prevented high filter loading by frequently replacing the filter tickets (every six runs).

We did not apply any data validation on the $PM_{2.5}$ data of the AirBeam that was collected on four of the routes.

2.1.4. Data Processing

The dataset of the four campaigns was used to assess two rescaling approaches: (i) to construct a 'yearly average' map representative of the peak rush hours by using a weighted average to aggregate the results, and (ii) to evaluate a rescaling strategy in order to better visualize the results of the individual campaigns.

Weighted averaging was performed to aggregate the results of the four different campaigns into one map of an estimated yearly average BC concentration during peak traffic hours. The virtual urban traffic station (see Section 2.2) was used as a reference since data from the mobile measurements were collected at street level, although some locations along the routes are more typical of an urban background. The weighting factors are proportional to the BC concentration during the campaigns in relation to the year-round BC concentration and are calculated as follows (pseudo-code):

Input:

- BC data from the four mobile campaigns $\mathbf{X} = \{X_1, X_2, X_3, X_4\}$;
- BC data for a virtual traffic station from the reference dataset of the Flemish Environment Agency, **Y**, for the period October 2017 until October 2018 (including the period from start $\mathbf{X_1}$ until end $\mathbf{X_4}$).

Output: Map of the estimated yearly average BC concentration during peak traffic hours from the mobile campaigns.
Method:

- Subset **Y** ($\mathbf{Y_s}$) for the morning and evening rush hour measurements, in line with the hours of the day of the mobile monitoring (**X**, see Figure 2);
- Calculate the weekly averages from $\mathbf{Y_s}$ ($\hat{y}_{w,i}$ for i = 1,. . .,52);
- Calculate the averages from X_i ($\hat{x}_i$ for i = 1, . . ., 4);
- Calculate a frequency histogram of $\hat{y}_{w,i}$ in 0.5 μg m^{-3} bins;
- Obtain the frequency of occurrence of the bins that coincide with $\hat{x}_i$ (f_i for i = 1, . . ., 4, $F = \sum f_i$);
- Obtain a weighting (w_i for i = 1, . . ., 4) per campaign as f_i/F;
- Apply these weighting factors to estimate the yearly average BC concentration ($\sum w_i X_i$).

Subsequently, we tried a rescaling strategy to assess whether the measurements performed under different background concentrations or meteorological conditions would yield similar patterns when the concentrations at fixed monitoring stations are taken into account (see Section 2.2). For the period of each of the four mobile campaigns, the average BC concentration at the urban traffic AQMS was calculated ($\hat{y}_{UT,i}$ for i = 1, . . ., 4). The yearly average BC concentration was calculated for the AQMS ($\hat{y}_{UT,year}$). The ratio of the yearly average to the campaign averages $\left(\hat{y}_{UT,year}/\hat{y}_{UT,i}\right)$ was used to rescale (multiply) the legend categories (lower and upper bounds) of the BC map of each of the four campaigns. Applying this method will not result in an additional bias of the resulting BC concentrations. Any relative bias between the AE51 and BC monitor of the AQMN is canceled out in the ratio $\left(\hat{y}_{UT,year}/\hat{y}_{UT,i}\right)$ and as a result, the rescaled values will have the same relative bias as the original data. Only when there is a large offset of the AE51 compared to the BC monitor of the AQMN can the rescaling introduce an additional bias and result in an over- or underestimation of the rescaled values. Previous studies showed that the AE51 corresponds well with the BC monitors of the AQMS, with acceptable slopes (0.83–1.13), an MAE below 0.3 μg/m^3 after slope correction [3] and a relative bias of 1–4% [42]. No indications for a clear offset between the AE51 and the monitors of the AQMN were observed.

2.2. *Air Quality Monitoring Stations and Data*

The black carbon concentrations on the 'virtual' rural (R), 'virtual' urban background (UB) and 'virtual' urban traffic (UT) stations of the official monitoring network in Belgium (Figure 3) for the four different sampling seasons are summarized in Table 2. The term 'virtual' refers to the average concentrations of a selection of typical stations of this type; the stations included for calculating the average virtual station are N029 (Houtem) and N016 (Dessel) for the rural station, R803 (Park Spoor Noord), R801 (Borgerhout-achtergrond), R701 (Gent-Baudelo) and R817 (Antwerpen-Groenenborgerlaan) for the urban background and R802 (Borgerhout-straatkant), R805 (Antwerpen-Belgiëlei) and R702 (Gent) for the urban traffic station, respectively. The 'virtual' stations give an indication of the concentrations at a typical location without relying on just one station and these 'virtual' stations are also used in the Flemish Environment Agency's annual reports [43]. The R817 station is categorized as a suburban station but is included here to calculate the 'virtual' urban background value.

Table 2. Average BC concentrations at the 'virtual' rural background, urban background and urban traffic monitoring stations in Belgium during the four different campaigns.

Concentration (µg m^{-3})	Rural (R)	Urban Background (UB)	Urban Traffic (UT)
Campaign 1 [(a)]	1.24 (1.17)	2.41 (1.86)	3.31 (2.69)
Campaign 2	1.30	1.83	2.41
Campaign 3	0.55	0.90	1.38
Campaign 4	0.80	1.82	2.48
Annual average 2017	0.9	1.5	1.9

[(a)] Numbers between brackets are average concentrations during the measurements at the South route, which was performed just before the other three subroutes.

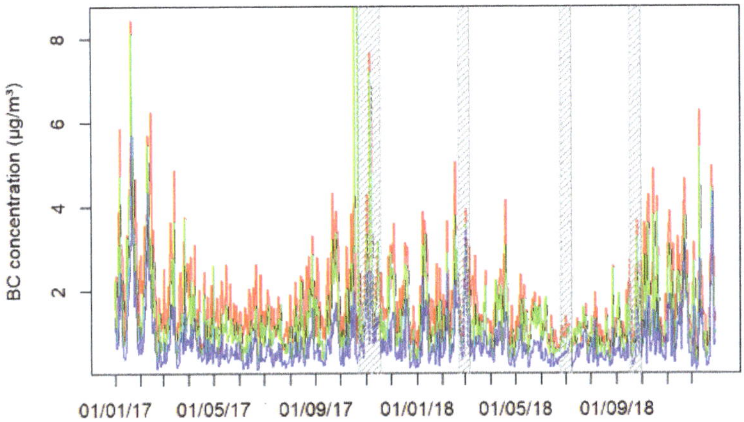

Figure 3. Daily average BC concentrations at the 'virtual' rural background (blue), urban background (green) and urban traffic (red) monitoring stations in Belgium for the years 2017 and 2018. The four campaigns of mobile BC monitoring are indicated in gray.

2.3. Model

An integrated model chain, ATMO-Street, has been set up to assess the air quality at the local scale (street scale), including both regional variability as well as the local variation in sources of pollution. High-spatial-resolution air quality information is available for the Flemish region in Belgium from the ATMO-Street model chain at an hourly resolution. ATMO-Street starts with the regional background concentrations obtained by intelligent interpolation of the hourly observations from the monitoring network using land-cover data as a spatial driver. These background concentrations are combined, taking into account double counting, with the output of a dispersion model running on a dense, irregular receptor grid for the emissions from point sources and line sources, including traffic emissions, with a resolution of up to 15 m close to the emission sources. This output is coupled to a street canyon module to capture the street canyon increments. The model results at the receptor locations are interpolated to raster maps. This integrated model chain has been validated against many validation datasets (e.g., [44,45], validation reports from http://www.irceline.be/nl/documentatie/modellen/validatie, accessed on 16 June 2024). The citizen science data presented in this publication offer an additional validation target.

3. Results

3.1. Black Carbon Concentration Maps

The maps of the four individual campaigns are shown in Figure 4. Each individual map represents the average BC concentration during the morning and evening rush hours over a two-week period. Concentrations are shown for every 20 m road segment. Note that in the third monitoring campaign (June/July), only two subroutes were included, as fewer volunteers could join during the summer holiday period. Table 3 summarizes the statistics of the four campaigns.

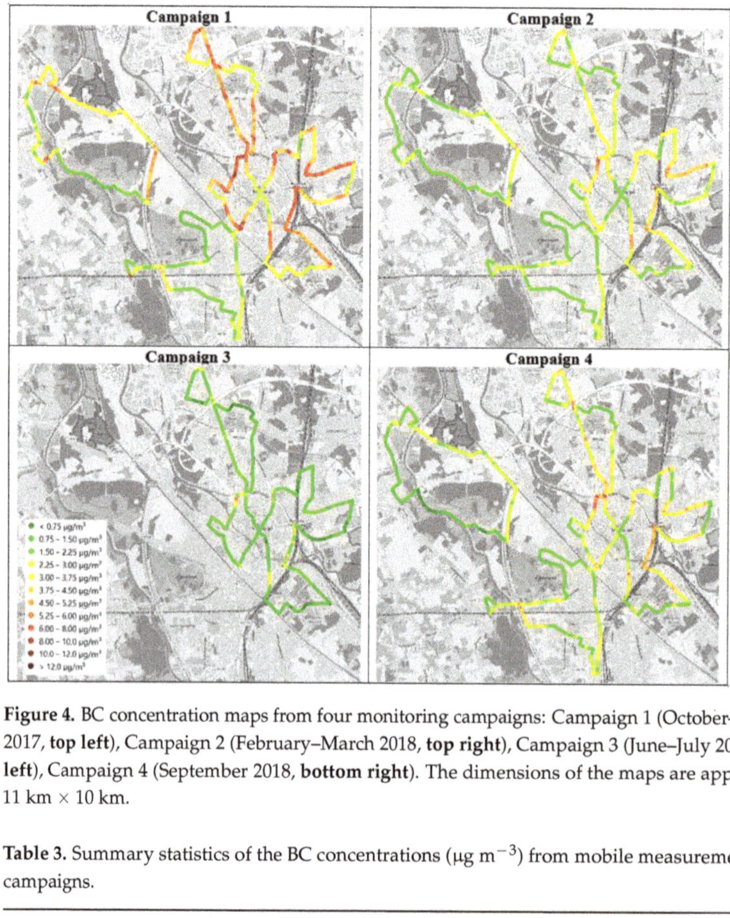

Figure 4. BC concentration maps from four monitoring campaigns: Campaign 1 (October–November 2017, **top left**), Campaign 2 (February–March 2018, **top right**), Campaign 3 (June–July 2018, **bottom left**), Campaign 4 (September 2018, **bottom right**). The dimensions of the maps are approximately 11 km × 10 km.

Table 3. Summary statistics of the BC concentrations (μg m^{-3}) from mobile measurements in four campaigns.

	Min.	First Qu.	Median	Mean	Third Qu.	Max.
Campaign 1	1.0	2.4	3.3	3.7	4.7	12.4
Campaign 2	1.3	2.0	2.3	2.5	2.9	10.5
Campaign 3	0.5	0.9	1.2	1.4	1.7	6.2
Campaign 4	0.5	1.8	2.5	2.5	3.1	8.3

In the first monitoring campaign, one of the subroutes (South route) was not measured at the same time as the other three. Nevertheless, a comparison of the results of the measurement campaigns in the four different seasons showed that BC concentrations have strong spatial and temporal variabilities. The highest concentrations were measured in October–November 2017, and the lowest concentrations in June–July 2018. The difference in concentrations between the different measurement campaigns can be explained by the different background concentrations, different dispersion conditions and different emissions (additional sources in winter and less traffic in the summer months). Given the seasonal differences, the spatial patterns are similar between the different campaigns. However, when background concentrations and meteorological conditions result in low concentrations (Campaign 3), the spatial variability is less apparent with the same fixed color scale. Table 2 shows that the average BC concentration at the 'virtual' traffic station for Campaign 3 is lower than the average BC concentration at the 'virtual' urban background station for Campaigns 1, 2 and 4. Low concentrations during Campaign 3 are also due to a lower traffic volume in the holiday season. Therefore, the conditions during Campaign 3 are less suitable for assessing hot spots due to traffic.

3.2. Black Carbon Concentrations at AQMS

At the AQMSs, the highest concentrations were measured during the first measurement campaign (October–November 2017), and the lowest concentrations were measured in the third campaign (June–July 2018), which was during the summer holiday season (Table 2). The 'virtual' rural station showed similar concentrations (0.06 μg m^{-3} higher) during the second campaign (end of the winter season) compared to the first campaign. During the second and fourth campaigns, the concentrations at the 'virtual' urban background and the 'virtual' urban traffic stations were very similar, while at the 'virtual' rural station, the concentrations during Campaign 2 were higher. Concentrations at the UB and UT stations were around 25% lower during Campaign 2 and Campaign 4 compared to Campaign 1 or were reduced by 0.58 μg m^{-3} and 0.59 μg m^{-3} at UB, and 0.91 μg m^{-3} and 0.83 μg m^{-3} at UT, respectively. At the R station, concentrations were 5% and 35% higher for Campaign 2 and Campaign 4, respectively, when compared to Campaign 1, or they increased by 0.06 μg m^{-3} and reduced by 0.43 μg m^{-3}, respectively. We observed that the higher concentration at the 'virtual' rural station during Campaign 2 was due to slightly higher concentrations at one of the two stations (N029) that were used to calculate the rural value; the higher value can be explained by an easterly (NE) wind direction during Campaign 2 compared to a westerly (SW–NW and SW, respectively) one during Campaign 1 and Campaign 4. In the third campaign, concentrations were more than 50% lower (56%, 63% and 58% at R, UB and UT, respectively) or reduced by 0.69 μg m^{-3} at R, 1.51 μg m^{-3} at UB and 1.93 μg m^{-3} at UT. Table 2 also shows the yearly average concentration for 2017 for a comparison. However, we can expect the yearly average during peak hours to be even higher. The averages of the four campaigns measured during peak hours are, respectively, 8%, 16% and 25% higher than the yearly average.

3.3. Aggregated and Rescaled Maps

The maps (Figure 4) show that the timing of monitoring can have a large impact on the result and that the monitoring effort can be optimized. On the one hand, the resulting map from Campaign 3 (June/July), for example, shows almost no spatial variability and is not as useful in identifying hot spots in the city. On the other hand, periods of low concentration and low spatial variability do occur and need to be accounted for when the monitoring campaign is aiming to acquire an estimate of the average concentrations over longer periods, e.g., a yearly average. The variability of the concentrations over time can also be seen from the results of the fixed monitoring stations. Table 2 shows large differences in the concentrations at the fixed monitoring stations over the four campaigns, and the trends observed at the UB and UT stations correspond to the trends of the average concentrations of the mobile measurements (Table 3): lower concentrations during Campaign 3, the highest concentrations during Campaign 1 and intermediate concentrations during Campaigns 2 and 4. The data from the fixed monitoring stations were used in the rescaling strategy presented here.

Two data processing methods have been applied (see Section 2.1.4) to create a yearly average map and a method to enhance the contrast for a more informative visualization of the individual campaign maps. The map of yearly average concentrations (Figure 5) aggregates the maps of the four campaigns into one single map, taking the differences in background concentrations during the campaigns into account.

Figure 5. Map of the estimated yearly average BC concentration during peak hours calculated as a weighted average over the different campaigns.

For the rescaling of the individual maps, ratios of the yearly average to campaign averages for urban traffic were applied (Table 4), as explained in Section 2.1.4. The results of the rescaled maps for the different seasons (see Figure 6) show a better correlation between the different campaigns. Also, the rescaled maps provide better visualization of the spatial variability with concentrations of the entire color scale, and improvements can be observed for periods 1 and 3, especially.

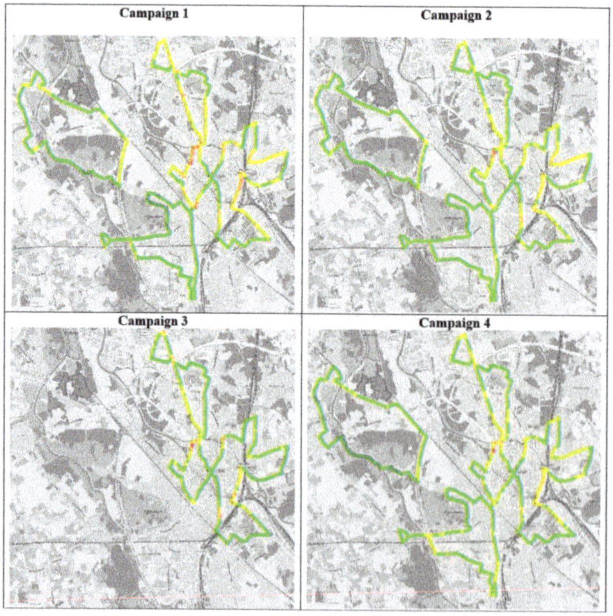

Figure 6. BC concentration maps from four rescaled monitoring campaigns (same scale as Figure 4): Campaign 1 (October–November 2017, **top left**), Campaign 2 (February–March 2018, **top right**), Campaign 3

(June–July 2018, **bottom left**) and Campaign 4 (September 2018, **bottom right**).

Table 4. Ratio of the yearly average BC concentration to 'campaign average' BC concentrations during measurement hours at 'virtual' rural background (R), urban background (UB) and urban traffic (UT) monitoring stations in Flanders during the four different campaigns.

Ratio (Yearly av./Campaign av.)	R	UB	UT
Campaign 1	0.73	0.62	0.57
Campaign 2	0.69	0.82	0.79
Campaign 3	1.64	1.66	1.38
Campaign 4	1.12	0.82	0.77

3.4. Comparison with Different Metric and Measurement Technology: BC by AE51 versus PM2.5 by AirBeam

The boxplots of the four bicycle trips (runs) where simultaneous measurements were taken with the aethalometer and the AirBeam are presented in Figure 7. The boxplots clearly show lower BC concentrations than PM concentrations, with BC included in the PM. The concentration trends between the different trips are similar for both pollutants, i.e., the median (and average) concentrations are ranked identically over the four trips for both (trip 3 > trip 1 > trip 4 > trip 2). A higher concentration range during the morning rush hour trips was observed for both pollutants. However, the BC measurements have a more skewed distribution than the PM measurements, with higher upper extremes compared to the majority of the data, whereas the PM concentrations show a more symmetrical distribution. This is explained by the higher contribution of local traffic to the BC measurements in contrast to the PM measurements, where traffic is also a contribution, but the background variations are more pronounced. This is also observed in the autocorrelation analysis, where the BC measurements show a much more rapid drop in the autocorrelation than the PM measurements. Again, this is explained by the more localized variations in BC. Overall, the correlation between the BC and PM measurements is about 0.3 (with a lower value of 0.15 for one of the trips), showing no clear linear trend between both.

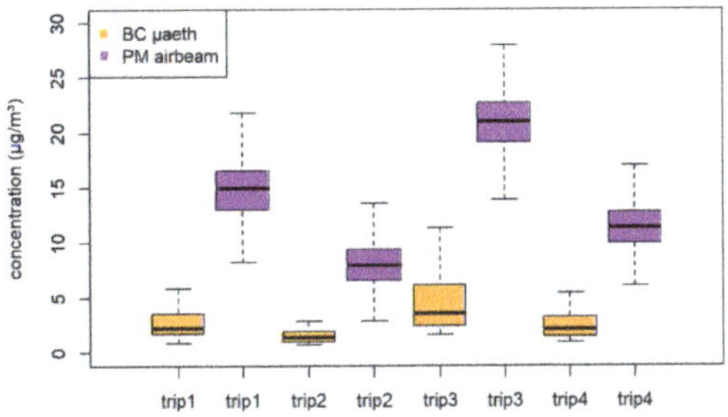

Figure 7. Boxplots for the four BC and PM runs measured on the West route during Campaign 2. The black line in the boxes represents the median, the vertical size of the box is the interquartile range (IQR). Whiskers extend to minimum and maximum values not exceeding 1.5 times the IQR from the median.

To compare the spatio-temporal data collected by the two methods, the BC and PM measurements (of the four campaigns) were plotted using the same color scale based on percentiles (with five categories as follows: P0–20, P20–40, P40–60, P60–80 and P80–100). These maps show street sections where colors coincide but other sections where the colors indicate low BC and high PM values or vice versa (maps not shown in this paper).

3.5. Comparison with Model Results

The ATMO-Street model is applied to model the BC concentrations in Flanders in high resolution, including street canyon increments, at an hourly resolution. The model chain uses a variety of input data, including the BC observations at monitor stations, meteorological data and BC emissions from line sources such as traffic. The model validation was performed for the first campaign in 2017. The model results for the North, East and West routes are presented in Figure 8. The South route is also modeled but not shown in this figure. For details on the four routes, please refer to Figure 1. The modeled results showed similar patterns for the measured data (Figure 4).

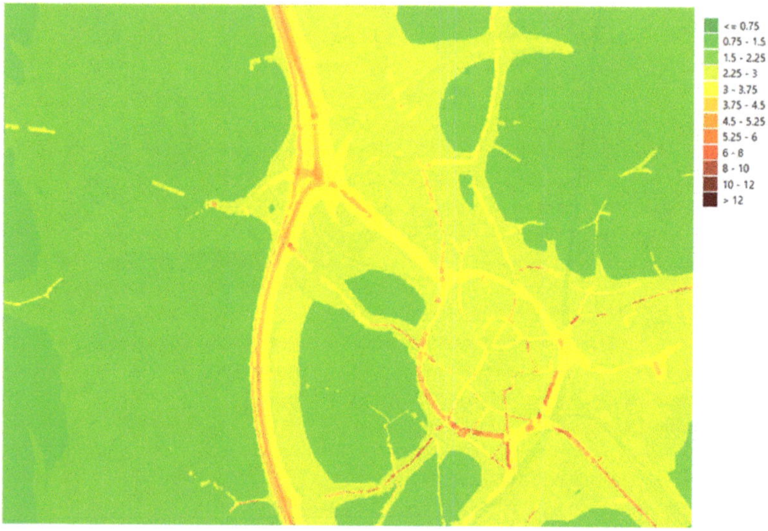

Figure 8. Modeled results for Campaign 1 (North, East and West routes).

The model clearly has a fair correlation with the measurements at background locations, where cyclists are not directly exposed to the exhaust from local traffic. The model appears to underestimate the peak concentrations observed at locations near dense areas. This is also shown by the scatter plots of all data and selected urban background data (Figure 9).

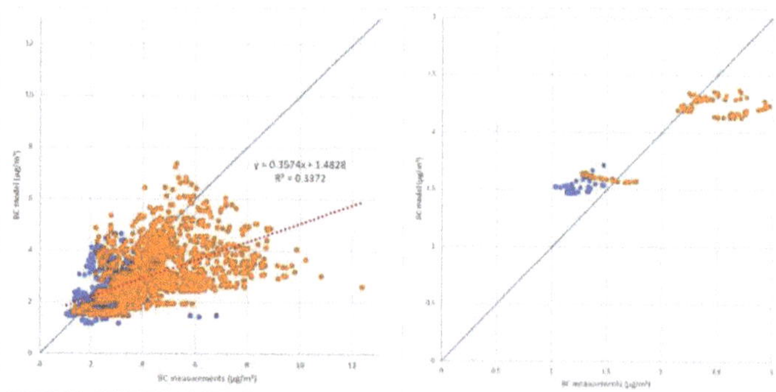

Figure 9. Scatter plots of the BC ATMO-Street model concentrations against the BC measurements for all locations (**left**) and urban background locations only (**right**) in Mechelen, Belgium, in October–November 2017. The southern trajectory is plotted in blue, as it had been measured two weeks earlier. The blue line represent the 1:1 line. Units: μg m^{-3}.

The scatter plots compare the average concentration levels from the two-week campaigns during the morning and evening rush hours for 20 m road segments with modeled

BC concentrations for the same location and averaged for exactly the same hours. Data for the South route are plotted in blue, and the three other trajectories are in orange, as these were measured during a different two-week period. It must be mentioned that all mobile measurements are considered as an hourly average during peak hours over the measured days and are compared with the corresponding modeled hourly averages over the measurement days, whereas the measurements were only the result of a few seconds of sampling at each spot of every run.

The scatter plot shows that the ATMO-Street model concentrations are significantly lower than the measured values, except for the lower values, with a normalized mean bias of −24%, a relative root-mean-square error of 46% and a coefficient of determination of 0.34. Significant differences and scatter between the model and measurement results can be observed, as can be expected given the different time resolutions. A subset of background locations was selected and is presented in Figure 9. This is a subset of measurements in eight city parks, traffic-free zones and dead-end streets. The scatter plot shows a clear correlation between the model and the observations for this set of locations, with a normalized mean bias of 2% and a relative root-mean-square error of 16%.

4. Discussion

4.1. Comparison of Different Campaigns

Strong seasonal differences in the BC concentrations can be observed from a comparison of the maps over the four campaigns and from the summary statistics (Table 3). The seasonal variation, with the highest concentrations in Campaign 1 (end of the autumn season), the lowest concentrations during the summer season (Campaign 3), and intermediate concentrations in winter and at the end of the summer, are reflected in the maps and correspond to the measurements at the fixed monitoring sites of the air quality network. The seasonal differences in BC concentrations with higher concentrations on some days in winter and autumn are typical in the research area. As a consequence, maps were generated based on mobile monitoring to represent the situation when the data collection occurred. These maps cannot be generalized and interpreted using another time base, for example, interpreted as if it were a map of the yearly averaged concentration.

A similar conclusion can be drawn for the experienced variability within one day. As BC concentrations fluctuate during the day in accordance with the source, dilution and dispersal dynamics, the maps represent the situation at the hours of the day when the measurements were made. In this particular study, the maps represent average BC concentrations during morning and evening peak traffic hours. Maps of the nighttime concentrations, for example, would look completely different (lower concentrations, much less spatial variation).

For mobile monitoring of air pollution, it is advisable to restrict the monitoring to a predefined period of the year and predefined hours of the day. As such, the collected data are much less sparse and become easier to understand. As the greatest advantage of mobile monitoring in urban areas is being able to highlight streets and zones of high concentrations (hot spots) and low concentrations (background area), the most suitable hours of the day to perform the monitoring are the hours when contributions from local sources are the largest. For BC, this coincides with the rush hours when traffic volumes are highest. The amount of time spent on transport is one of the main determinants of personal exposure, and the traffic micro-environment can contribute largely to overall exposure. [20] Mobile monitoring during rush hours can, therefore, provide valuable information on the spatial variability of BC in relation to exposure.

Similarly, when selecting the best season to perform monitoring, the season with the highest local contribution is the most suitable. Holiday periods with significantly lower traffic volumes appear less useful in this sense. Following these suggestions, the outcome of the monitoring (e.g., BC concentration maps) is biased toward episodes of higher concentration and cannot be interpreted as maps of yearly averaged concentration, for example.

4.2. Rescaling

In this study, four monitoring campaigns were conducted, spread over different seasons. As discussed before, the aggregated map of Campaign 3 (June/July) shows almost no spatial variability, whereas the aggregated map of Campaign 1 shows so many hot spots that the spatial differences (especially in the North and East routes) are difficult to see. Could the results of these campaigns be aggregated to obtain a map of yearly average BC concentrations? Or is it possible to use rescaling so that campaigns performed over different seasons are comparable? Two approaches were evaluated: one in which the legend of the map from one season was rescaled based on the 'virtual' traffic station of the air quality monitoring network, and another where the 'annual' average of the map was constructed based on data from the fixed monitoring network.

The results from the rescaled maps for the different seasons show better correspondence between different campaigns. The rescaled maps also provide a better visualization of the spatial variability with concentrations of the entire color scale, and improvements can be observed for periods 1 and 3, especially. The rescaled map for Campaign 3 shows a better spatial variability along the ring road with hot spots in the NE, somewhat lower concentrations in the SE and lower concentrations in the W. Also, differences between two access roads, H and G (see also Figure 5), can be better distinguished, and the impact of stop-and-go traffic at location F is clearer. In addition, the impact of the car-free city center is more visible in the rescaled maps. The rescaled results of the first campaign also show a better spatial resolution, meaning that different parts of the ring road can be distinguished.

While the spatial patterns of different campaigns are very similar, the absolute values are still slightly different. This could partly be due to varying emissions, e.g., the period of Campaign 3 is typically characterized by less traffic due to the end of secondary school and the start of the holiday season. In the colder season, we can expect additional emissions from household heating.

An aggregated map based on measurements in different seasons results in a more representative picture. To assess the air quality in a city, it is recommended that measurements be taken in different seasons. However, this is not always possible due to constraints, such as the budget or availability of volunteers. In addition, meteorological conditions can result in extremely high or low concentrations and are not always known in the planning stages of the monitoring campaign. In addition, to assess hot spots in a city, one campaign may be sufficient as long as the situation when the measurements are taken is representative. For example, holiday seasons with reduced traffic should be avoided if the aim is to assess the impact of traffic, and when comparing different maps, these should always be interpreted on the basis of the ambient concentrations at fixed stations in relation to a reference, e.g., the yearly average.

4.3. Interpretation of Results with Respect to Mobility Planning

The map (Figure 10) shows a large difference in BC concentrations at street level. The results show the clear impact of traffic on the air quality. Relatively low concentrations are measured in the pedestrian zone in the city center (A), in residential areas (B) or on quiet roads (C) with only limited local traffic. Dedicated biking paths (separated from road traffic), in particular, show low concentrations (D), as observed in earlier studies [5]. In contrast, high concentrations are measured near the ring road (E) and locations with more stop-and-go traffic (F). However, the building environment and the distance of the biking path from the traffic also have a significant impact on the measured concentration and exposure. Small streets with high buildings ('street canyons') can show higher concentrations (G) compared to more open environments near busy traffic roads (H). In addition, the distance of the biking path from the traffic is also a significant factor. If the biking lane is separated from the traffic (e.g., by a river, such as in I), the cyclist's exposure to BC is much lower than compared to a biking lane next to a busy road (J). When no biking lane is available, exposure is higher compared to a biking lane separated from road traffic. Below, the BC concentrations are interpreted with respect to the type of road categories in relation to the traffic circulation plan of the city.

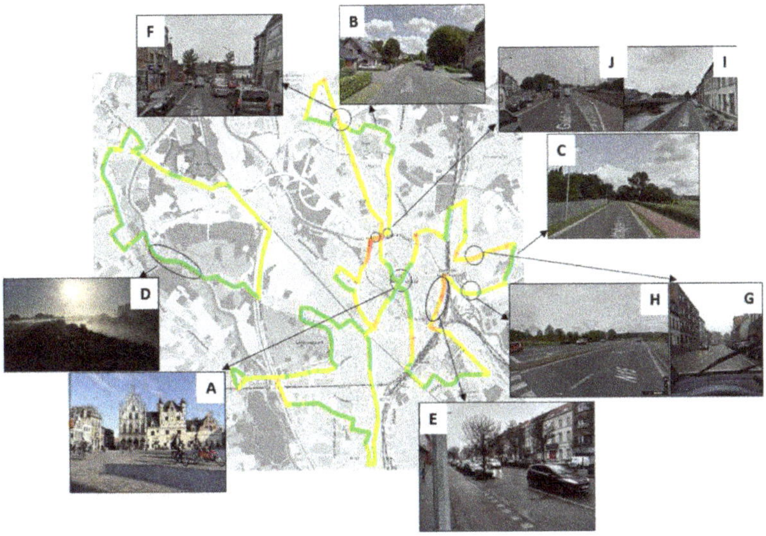

Figure 10. BC concentration map from monitoring Campaign 4 with some pictures of typical situations relating to traffic and the environment (same scale as Figure 5).

The traffic circulation in the city has changed over recent years. Access to the city is gained via a number of entrance roads that end on the circular ring road. Nowadays, the city center is a low-traffic area as cars are not allowed in these streets between 11 am and 6 pm. There are also traffic-free streets in this area (e.g., the main pedestrian shopping street). Public transport buses are allowed to enter certain low-traffic and traffic-free streets. There are four circulation routes for traffic entering the city center. The entrances to and exits from the circulation routes are located on the circular ring route. Areas other than the low-traffic center are accessible from the circulation routes or directly from the ring road. This is a predominantly residential area. The majority of streets are one-way. There are also background areas near the city, e.g., a park in the southeast part of the city. Finally, the bicycle infrastructure includes bicycle streets, i.e., streets with a speed limit of 30 km h^{-1} where cars are not allowed to overtake cyclists. One of the three bicycle streets also functions as part of a traffic circulation loop.

Based on the traffic circulation plan, a number of categories were defined, including an access road, ring road, circulation street, low-traffic street, pedestrian street, bicycle street, residential area and background area. The streets and roads where the BC measurements took place were classified according to these categories. However, some streets could not be attributed to one of the categories and/or were not withheld for this analysis, for example, streets in neighboring villages. The average BC concentration was calculated for each of the categories based on all the data from the validated measurements of the four campaigns (Figure 11, showing the average values and SD); the aggregated dataset was used as described above.

The BC concentration is highest on the ring road (4.3 $\mu g\ m^{-3}$), where two-by-two-lane traffic intensity is high and (mainly residential) buildings border the ring road, which likely causes the street canyon effect to some extent. The access roads to the ring road have markedly lower concentrations (3.3 $\mu g\ m^{-3}$) on average (although the error bars do overlap). Here, the bicycle lane is often further away from the road, and some roads are much more open than the ring road with higher dispersion rates of BC pollution. The bicycle streets have remarkably high BC concentrations. Two reasons could probably explain this. Firstly, the fact that one of the bicycle streets is also used as an entrance street to the city center in a circulation loop; hence traffic intensity is quite high here. Secondly, there is a mix of bicycles and cars on this street, and cyclists often end up right behind a car near the exhaust. Of course, this issue would require additional research, but these data do suggest that bicycle streets may lead to increased exposure to traffic pollution when they are laid out in streets with high traffic intensity. The data do not allow an estimate to be made of what the traffic intensity threshold would be, and this is probably also very street-specific due to other influencing factors. Yet, it would appear to be beneficial to separate the entrance streets to the city center for bicycles and traffic.

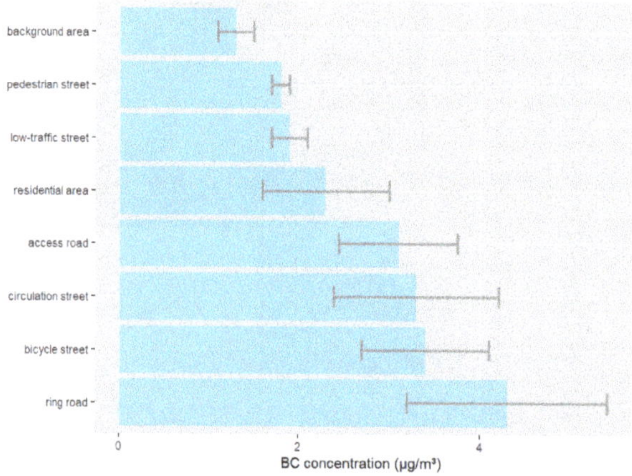

Figure 11. Average BC concentrations from bicycle measurements in different categories of roads and streets.

The BC concentration in the residential area, where most streets are one-way, is 2.3 µg m^{-3} on average and is significantly lower than on the circulation streets. However, the BC concentration on low-traffic and traffic-free streets in the city is lower (1.8 µg m^{-3}). City areas with decreased traffic intensities clearly benefit from these measures in terms of decreased air pollution and decreased exposure of residents and pedestrians. The background areas, such as a green park in this case, have the lowest BC concentration (1.5 µg m^{-3}) and may represent ideal locations for leisure activities.

4.4. Comparison with Model Results

The results of the mobile measurements can be used for model validation. While model validation was not the main aim of this paper, we have shown that a mobile campaign, as presented here, can be useful for model validation, and we carried out an initial exercise using the results of Campaign 1. Validating the spatial resolution of model results is difficult due to the lack of measurements at a high spatial resolution. Often, diffusive samplers are used to validate model results (e.g., [43]). Mobile measurements of good data quality can be used to fill this gap. The results of Campaign 1 were used to evaluate the model results for the same time period. The ATMO-Street model was applied to model the BC concentrations in Flanders in high resolution, including street canyon increments, at an hourly resolution. As the model results for 2018 were not yet available, validation could only be performed for the first campaign in 2017 at the time of writing. The comparison shows that the model results are clearly correlated with the measurements at background locations without the direct impact of local traffic. The model appears to underestimate the peak concentrations observed at locations near dense traffic.

Several aspects of the modeling approach explain the differences observed in the measurements. The majority of the measurements are taken when cyclists are cycling along traffic lanes very close to the traffic or in between traffic. The bi-Gaussian modeling of local emissions applies an irregular receptor grid. This is a line-source-following grid, with the highest resolution close to emission sources and fewer receptor points at greater distances from the emission sources. Receptor points are located at least 15 m from the sources. Only at a distance from the source can the concentrations be well represented by a Gaussian profile. Furthermore, Gaussian models are not well-suited to explain concentration levels in the turbulent environment between road traffic. The concentration levels at the receptor points are converted to concentration maps using a TIN interpolation but will thus underestimate the concentrations on and next to the roads. Street canyon locations are modeled separately and yield concentration levels on the building façade. The measurements in canyons are much closer to the traffic in many locations. Additionally, the quality of the model output also depends on the quality of the model input. The traffic emissions are calculated based on traffic volumes and the traffic fleet. Some streets still have significant traffic volumes and are not captured. Furthermore, the traffic model yielding the traffic volumes is not fully up to date with the current traffic conditions. The city center of Mechelen has been changed to a zone with limited traffic by closing it to through-traffic and making several streets one-way. These recent changes are not captured in the traffic data. Some outliers are found at locations that are not treated as street canyons but where significant street canyon contributions can be expected. As these limitations affect the validation for locations with significant traffic volumes, an analysis for the set of locations at a significant distance from road traffic has been made.

However, mobile measurements also have drawbacks, and extreme conditions during the mobile measurements could also be the reason for a difference between the measurements and results. Therefore, it is better to compare model results and measurements of a high spatial resolution on a more aggregated level on different days. Despite the repeated measurements and the use of a trimmed mean, on some occasions, a remaining outlier can result in higher values, especially at road crossings when the number of data points used for the averages is limited. This was the case for one of the three outliers (BC measurements $> 10\ \mu g\ m^{-3}$ in Figure 11), whereas the two other outliers were representative values of very high average concentrations (the BC concentrations were $>10\ \mu g\ m^{-3}$ in 6–7 of the measurement runs).

The comparison with model results presented here should only be seen as a first attempt to examine the added value of mobile measurements in model validation. Further work on model validation with mobile measurements is planned when the data from other campaigns are available and will be a topic for future publication.

4.5. Mobile Measurements Using airQmap in Citizen Science Projects

The results show that mobile measurements can be used in citizen science projects. The participants of the CO can define the aim of the measurement campaign and the area that needs to be investigated. Next, researchers can assist in fine-tuning the research question even further into a scientifically validated measurement approach. After a short training session, citizens can perform the measurements. By using a scientifically validated measurement approach and instruments that provide high-quality data, results can be used by policymakers; in this study, the results were used to evaluate the current mobility plans. The resulting high-resolution maps can also be used for hot spot detection, personal exposure assessments, evaluations of other policy measures and validation of pollutant dispersion models. The resulting high-resolution maps can also justify the need for additional permanent stations within the AQMN.

Mobile monitoring, with the help of citizens, can complement fixed regulatory stations and can also complement low-cost fixed sensor networks. Low-cost sensor networks require maintenance and calibration efforts that may result in higher overall costs or lower data quality if this is not done properly. Because mobile monitoring only requires one (or a few) instrument(s) to measure a large study area, medium-grade instruments can be used, which can measure emerging pollutants and generate data of higher quality than low-cost sensors. Data quality is also relevant in CS projects because it is very important that participants know the value of the dataset in order to keep them engaged.

In this paper, we also compared the current tool, airQmap, which measures BC, with another low-cost tool, AirBeam, which measures PM. The main conclusion from this assessment is that BC and PM maps do not result in exact agreement.

In citizen science projects, the AirBeam cannot be used as a low-cost alternative for the micro-aethalometer, as they are simply two different tools. In a citizen science context, the benefit of the methodology for BC measurements is that BC measurements will provide a better visualization of the contribution that traffic makes to air pollution. The downside is the price of the sensor as well as the lack of interfacing and automated data visualization and sharing. The AirBeam is a tool that is embedded in a data management system, which makes it more useful for citizens. However, one should bear in mind that the PM maps that are generated only partly reflect the local traffic contribution and that background variations will have a greater impact.

5. Conclusions

This study used a mobile measurement approach to assess local differences in air quality in the city of Mechelen. Using a user-friendly measurement platform, airQmap, engaged people from the CO Meet Mee Mechelen were able to set the agenda, perform the measurements and discuss the results.

The results show that repeated measurements are needed and would preferably be performed in different seasons to obtain a representative result. If it is not possible to perform measurements in different seasons, it is recommended that the data be compared with fixed stations and rescaling applied in order to obtain a good indication of the hot spots and green zones.

The results of this CO are used to assess the current mobility plans in relation to the exposure of cyclists. The analysis shows that BC concentrations at the street level are highly impacted by traffic intensity and stop-and-go traffic, as well as proximity to traffic and the building environment (open versus street canyon).

The dataset from the first measurement campaign was used to validate a dispersion model and demonstrated that, in general, the model showed a good spatial representation. However, in some cases, the model underestimated the concentrations because of street canyons, slight misclassifications in location or missing traffic data. In addition, we should mention that the models can assess pollutant concentrations at street level but are not built to assess the concentrations between traffic flows due to possible higher turbulence, which cannot assessed by the model. In some cases, deviations between the measurements and the model are due to an overestimation of the concentration by the measurements due to the high impact of extremely high concentrations during one of the campaigns, despite using the trimmed mean.

Measurements were performed with a monitoring platform, airQmap, which is based on mid-grade instruments that are not low-cost. During a selection of the measurements, a low-cost PM sensor was compared with the current approach, and this comparison demonstrated that the BC measurements showed a larger variability in concentrations, which can be explained by the higher contribution of local traffic to the BC measurements in contrast to the PM measurements. The results of this study show that cyclists' (and pedestrians') exposure can be reduced by keeping traffic out of the city center, providing green cycling highways where there is no motorized traffic, trying to reduce the traffic in street canyons, increasing the distance between the street and the biking lanes and avoiding a mix of vehicles and cyclists in busy streets.

This paper shows that mobile measurements can be used by COs to assess the spatial variability of air quality in a city. The data can be used to assess mobility plans, evaluate the exposure of cyclists with respect to cycling infrastructure and validate models. However, it is important to use high-quality instruments and apply the correct measurement methodology (number of repetitions, season) in order to obtain meaningful data.

Author Contributions: Conceptualization, M.V.P., J.P. and S.V.; methodology, M.V.P., J.P. and S.V.; software, J.P. and W.L.; validation, S.V., J.P. and W.L.; formal analysis, J.P. and S.V.; investigation, J.V.L. and S.V.; resources, C.V. and B.V.; writing—original draft preparation, M.V.P., J.P. and S.V.; writing—review and editing, M.V.P., J.P., S.V. and J.H.; visualization, J.P., S.V. and W.L.; supervision, M.V.P.; funding acquisition, J.P. and S.V. All authors have read and agreed to the published version of the manuscript.

Funding: This research was partly funded by the European Union's Horizon 2020 project Ground Truth 2.0, grant agreement No. 689744, and partly funded by the FLAMENCO project, a research project financially supported by Agentschap Innoveren & Ondernemen (VLAIO), contract number IWT-SBO 150044. This study is also supported by the European Union's Horizon 2020 Green Deal project RI-URBANS project (Research Infrastructures Services Reinforcing Air Quality Monitoring Capacities in European Urban & Industrial Areas), grant agreement No. 101036245, and the methodology is further validated in the RI-URBANS project.

Institutional Review Board Statement: Not applicable.

Informed Consent Statement: Not applicable.

Data Availability Statement: The data that support the findings of this study are available from the corresponding author, M.V.P., upon reasonable request.

Acknowledgments: We are grateful to the volunteers of the CO Meet Mee Mechelen for their assistance in the collection of the mobile data and their interesting discussions. We would also like to thank all co-workers in the project for their assistance in the measurement campaign and the city of Mechelen for facilitating the project from the policy side.

Conflicts of Interest: The authors declare no conflicts of interest.

References

1. Conrad, C.C.; Hilchey, K.G. A review of citizen science and community-based environmental monitoring: Issues and opportunities. *Environ. Monit. Assess.* **2011**, *176*, 273–291. [CrossRef] [PubMed]
2. Lewis, A.; Edwards, P. Validate air pollution sensors. *Nat. N. Comment* **2016**, *535*, 29–31. [CrossRef] [PubMed]
3. Van den Bossche, J.; Peters, J.; Verwaeren, J.; Botteldooren, D.; Theunis, J.; De Baets, B. Mobile monitoring for mapping spatial variation in urban air quality: Development and validation of a methodology based on an extensive dataset. *Atmos. Environ.* **2015**, *105*, 148–161. [CrossRef]
4. Dons, E.; Laeremans, M.; Orjuela, J.P.; Avila-Palencia, I.; Carrasco-Turigas, G.; Cole-Hunter, T.; Anaya-Boig, E.; Standaert, A.; De Boever, P.; Nawrot, T.; et al. Wearable Sensors for Personal Monitoring and Estimation of Inhaled Traffic-Related Air Pollution: Evaluation of Methods. *Environ. Sci. Technol.* **2017**, *51*, 1859–1867. [CrossRef] [PubMed]
5. Hofman, J.; Samson, R.; Joosen, S.; Blust, R.; Lenaerts, S. Cyclist exposure to black carbon, ultrafine particles and heavy metals: An experimental study along two commuting routes near Antwerp, Belgium. *Environ. Res.* **2018**, *164*, 530–538. [CrossRef] [PubMed]
6. Peters, J.; Van den Bossche, J.; Reggente, M.; Van Poppel, M.; De Baets, B.; Theunis, J. Cyclist exposure to UFP and BC on urban routes in Antwerp, Belgium. *Atmos. Environ.* **2014**, *92*, 31–43. [CrossRef]
7. Van Poppel, M.; Peters, J.; Levei, E.A.; Mărmureanu, L.; Moldovan, A.; Hoaghia, M.-A.; Varaticeanu, C.; Van Laer, J. Mobile measurements of black carbon: Comparison of normal traffic with reduced traffic conditions during COVID-19 lock-down. *Atmos. Environ.* **2023**, *297*, 119594. [CrossRef] [PubMed]
8. Apte, J.S.; Messier, K.P.; Gani, S.; Brauer, M.; Kirchstetter, T.W.; Lunden, M.M.; Marshall, J.D.; Portier, C.J.; Vermeulen, R.C.H.; Hamburg, S.P. High-Resolution Air Pollution Mapping with Google Street View Cars: Exploiting Big Data. *Environ. Sci. Technol.* **2017**, *51*, 6999–7008. [CrossRef] [PubMed]
9. Hofman, J.; Panzica La Manna, V.; Ibarrola-Ulzurrun, E.; Peters, J.; Escribano, M.; Van Poppel, M. Opportunistic Mobile Air Quality Mapping using Sensors on Postal Service Vehicles: From Point Clouds to Actionable Insights. *Front. Environ. Sci.* **2023**, *2*, 1232867. [CrossRef]
10. Messier, K.P.; Chambliss, S.E.; Gani, S.; Alvarez, R.; Brauer, M.; Choi, J.J.; Hamburg, S.P.; Kerckhoffs, J.; LaFranchi, B.; Lunden, M.M.; et al. Mapping Air Pollution with Google Street View Cars: Efficient Approaches with Mobile Monitoring and Land Use Regression. *Environ. Sci. Technol.* **2018**, *52*, 12563–12572. [CrossRef]
11. Wesseling, J.; Hendricx, W.; de Ruiter, H.; van Ratingen, S.; Drukker, D.; Huitema, M.; Schouwenaar, C.; Janssen, G.; van Aken, S.; Smeenk, J.W.; et al. Assessment of PM2.5 Exposure during Cycle Trips in The Netherlands Using Low-Cost Sensors. *Int. J. Environ. Res. Public Health* **2021**, *18*, 6007. [CrossRef]
12. Blanco, M.N.; Bi, J.; Austin, E.; Larson, T.V.; Marshall, J.D.; Sheppard, L. Impact of mobile monitoring network design on air pollution exposure assessment models. *Environ. Sci Technol.* **2023**, *57*, 440–450. [CrossRef]
13. Report to Congress on Black Carbon. Available online: https://19january2017snapshot.epa.gov/www3/airquality/blackcarbon/2012report/fullreport.pdf (accessed on 16 April 2024).

14. Janssen, N.A.; Gerlofs-Nijland, M.E.; Lanki, T.; Salonen, R.O.; Cassee, F.; Hoek, G.; Fischer, P.; Brunekreef, B.; Krzyzanowski, M. *Health Effects of Black Carbon*; Technical Report; World Health Organization: Geneva, Switzerland, 2012.
15. Brunekreef, B.; Strak, M.; Chen, J.; Andersen, Z.J.; Atkinson, R.; Bauwelinck, M. Mortality and Morbidity Effects of Long-Term Exposure to Low-Level PM2.5, BC, NO_2, and O_3, an Analysis of European Cohorts in the ELAPSE Project. *Res. Rep. Health Eff. Inst.* **2021**, *2021*, 208.
16. Rovira, J.; Paredes-Ahumada, J.A.; Barceló-Ordinas, J.M.; García-Vidal, J.; Reche, C.; Sola, Y.; Viana, M. Non-linear models for black carbon exposure modelling using air pollution datasets. *Environ. Res.* **2022**, *212*, 113269. [CrossRef]
17. Fung, P.L.; Savadkoohi, M.; Zaidan, M.A.; Niemi, J.V.; Timonen, H.; Pandolfi, M.; Alastuey, A.; Querol, X.; Hussein, T.; Petäjä, T. Constructing transferable and interpretable machine learning models for black carbon concentrations. *Environ. Int.* **2024**, *184*, 108449. [CrossRef] [PubMed]
18. Kerckhoffs, J.; Hoek, G.; Vlaanderen, J.; van Nunen, E.; Messier, K.; Brunekreef, B.; Gulliver, J.; Vermeulen, R. Robustness of intra urban land-use regression models for ultrafine particles and black carbon based on mobile monitoring. *Environ. Res.* **2017**, *159*, 500–508. [CrossRef] [PubMed]
19. Berghmans, P.; Bleux, N.; Int Panis, L.; Mishra, V.K.; Torfs, R.; Van Poppel, M. Exposure assessment of a cyclist to PM10 and ultrafine particles. *Sci. Total Environ.* **2009**, *407*, 1286–1298. [CrossRef]
20. Dons, E.; Int Panis, L.; Van Poppel, M.; Theunis, J.; Wets, G. Personal exposure to Black Carbon in transport microenvironments. *Atmos. Environ.* **2012**, *55*, 392–398. [CrossRef]
21. Peters, J.; Theunis, J.; Van Poppel, M.; Berghmans, P. Monitoring PM10 and ultrafine particles in urban environments using mobile measurements. *Aerosol Air Qual. Res.* **2013**, *13*, 509–522. [CrossRef]
22. Van Poppel, M.; Peters, J.; Bleux, N. Methodology for setup and data processing of mobile air quality measurements to assess the spatial variability of concentrations in urban environments. *Environ. Pollut.* **2013**, *183*, 224–233. [CrossRef]
23. Blanco, M.N.; Bi, J.; Austin, E.; Larson, T.V.; Marshall, J.D.; Sheppard, L. Characterization of Annual Average Traffic-Related Air Pollution Concentrations in the Greater Seattle Area from a Year-Long Mobile Monitoring Campaign. *Environ. Sci. Technol.* **2022**, *56*, 11460–11472. [CrossRef] [PubMed]
24. Dekoninck, L.; Botteldooren, D.; Int Panis, L.; Hankey, S.; Jain, G.S.K.; Marshall, J. Applicability of a noise-based model to estimate in-traffic exposure to black carbon and particle number concentrations in different cultures. *Environ. Int.* **2015**, *74*, 89–98. [CrossRef] [PubMed]
25. Kerkhoffs, J.; Hoek, G.; Gehring, U.; Vermeulen, R. Modelling nationwide spatial variation of ultrafine particles based on mobile monitoring. *Environ. Int.* **2021**, *154*, 106569. [CrossRef] [PubMed]
26. Languille, B.; Gros, V.; Nicolas, B.; Honoré, C.; Kaufmann, A.; Zeitouni, K. Personal Exposure to Black Carbon, Particulate Matter and Nitrogen Dioxide in the Paris Region Measured by Portable Sensors Worn by Volunteers. *Toxics* **2022**, *10*, 33. [CrossRef]
27. Apparicio, P.; Carrier, M.; Gelb, J.; Seguin, A.M.; Kingham, S. Cyclists' exposure to air pollution and road traffic noise in central city neighbourhoods of Montreal. *J. Transp. Geogr.* **2016**, *57*, 63–69. [CrossRef]
28. Jereb, B.; Batkovic, T.; Herman, L.; Sipek, G.; Kovse, S.; Gregoric, A.; Mocnik, G. Exposure to Black Carbon during Bicycle Commuting-Alternative Route Selection. *Atmosphere* **2018**, *9*, 21. [CrossRef]
29. Hankey, S.; Marshall, J.D. On-bicycle exposure to particulate air pollution: Particle number, black carbon, PM2.5, and particle size. *Atmos. Environ.* **2015**, *122*, 65–73. [CrossRef]
30. Kaminska, J.A.; Turek, T.; Van Poppel, M.; Peters, J.; Hofman, J.; Kazak, J.K. Whether cycling around the city is in fact healthy in the light of air quality–Results of black carbon. *J. Environ. Manag.* **2023**, *337*, 117694. [CrossRef] [PubMed]
31. Okonen, E.O.; Yli-Tuomi, T.; Turunen, A.W.; Taimisto, P.; Pennanen, A.; Vouitsis, I.; Samaras, Z.; Voogt, M.; Keuken, M.; Lanki, T. Particulates and noise exposure during bicycle, bus and car commuting: A study in three European cities. *Environ. Res.* **2017**, *154*, 181–189. [CrossRef]
32. Targino, A.C.; Rodrigues, M.V.C.; Krecl, P.; Cipoli, Y.A.; Ribeiro, J.P.M. Commuter exposure to black carbon particles on diesel buses, on bicycles and on foot: A case study in a Brazilian city. *Environ. Sci. Pollut. Res.* **2016**, *25*, 1132–1146. [CrossRef]
33. Targino, A.C.; Krecl, P.; Danziger, J.E.; Segura, J.F.; Gibson, M.D. Spatial variability of on-bicycle black carbon concentrations in the megacity of Sao Paulo: A pilot study. *Environ. Pollut.* **2018**, *242*, 539–543. [CrossRef]
34. Targino, A.C.; Gibson, M.D.; Krecl, P.; Rodrigues, M.V.C.; dos Santos, M.M.; Correa, M.D. Hotspots of black carbon and PM2.5 in an urban area and relationships to traffic characteristics. *Environ. Pollut.* **2016**, *218*, 475–486. [CrossRef]
35. Betancourt, R.M.; Galvis, B.; Balachandran, S.; Ramos-Bonilla, J.P.; Sarmiento, O.L.; Gallo-Murcia, S.M.; Contreras, Y. Exposure to fine particulate, black carbon, and particle number concentration in transportation microenvironments. *Atmos. Environ.* **2017**, *157*, 135–145. [CrossRef]
36. Lonati, G.; Ozgen, S.; Ripamonti, G.; Signorini, S. Variability of Black Carbon and Ultrafine Particle Concentration on Urban Bike Routes in a Mid-Sized City in the Po Valley (Northern Italy). *Atmosphere* **2017**, *8*, 40. [CrossRef]
37. Wehn, U.; Almomani, A. Incentives and barriers for participation in community-based environmental monitoring and information systems: A critical analysis and integration of the literature. *Environ. Sci. Policy* **2019**, *101*, 341–357. [CrossRef]
38. Gharesifard, M.; Wehn, U.; van der Zaag, P. Context matters: A baseline analysis of contextual realities for two community-based monitoring initiatives of water and environment in Europe and Africa. *J. Hydrol.* **2019**, *579*, 124144. [CrossRef]
39. Hagler, G.S.; Yelverton, T.L.; Vedantham, R.; Hansen, A.D.; Turner, J.R. Postprocessing method to reduce noise while preserving high time resolution in Aethalometer real-time black carbon data. *Aerosol Air Qual. Res.* **2011**, *11*, 539–546. [CrossRef]

40. Singer, B.C.; Delp, W.W. Response to consumer and research grade indoor air quality monitors to residential sources of fine particles. *Indoor Air* **2018**, *28*, 624–639. [CrossRef]
41. Hofman, J.; Lazarov, B.; Stroobants, C.; Elst, E.; Smets, I.; Van Poppel, M. Portable Sensors for Dynamic Exposure Assessments in Urban Environments. *State Sci.* **2024**, preprint. [CrossRef]
42. Van den Bossche, J.; De Baets, B.; Botteldooren, D.; Theunis, J. A spatio-temporal land use regression model to assess street-level exposure to black carbon. *Environ. Model. Softw.* **2020**, *133*, 104837. [CrossRef]
43. Jaarrapport Lucht. Emissies 2000–2016 en Luchtkwaliteit 2017. Available online: https://www.vmm.be/bestanden/VMM-2017-LKT_TW.pdf (accessed on 16 April 2024).
44. Lefebvre, W.; Van Poppel, M.; Maiheu, B.; Janssen, S.; Dons, E. Evaluation of the RIO-IFDM-street canyon model chain. *Atmos. Environ.* **2013**, *77*, 325–337. [CrossRef]
45. Hooyberghs, H.; Lefebvre, W.; Vranckx, S.; Maiheu, B.; Trimpeneers, E.; Vanpoucke, C.; Janssen, S.; Meysman, F.; Fierens, F.; De Craemer, S. Validation and optimization of the ATMO-Street air quality model chain by means of a large-scale citizen-science dataset. *Atmos. Environ. Int. J.* **2022**, *272*, 118946. [CrossRef]

atmosphere

MDPI

Article

City-Scale Air Quality Network of Low-Cost Sensors

Adnan Masic

Faculty of Mechanical Engineering, University of Sarajevo, Vilsonovo 9, 71000 Sarajevo, Bosnia and Herzegovina; masic@mef.unsa.ba

Abstract: In this paper, we present a city-scale (possibly global) air pollution network made of low-cost sensors for particulate matter concentration in the air. The components of the presented system are based on our research and experience from previous studies focused on air quality instruments and sensors for addressing urban air pollution. Sensor nodes are produced locally and distributed over Sarajevo, Bosnia and Herzegovina. The system collects, calibrates, and displays particulate matter concentrations in real-time. In comparison to the available measurements from governmental institutions, our system demonstrated good agreement of measured parameters, with several advantages including, higher resolution in space and time, lower costs, both horizontal and vertical measurements. One of the capabilities of the system is the real-time air pollution city map with animations. By installing multiple sensor nodes over a slope, we receive vertical measurements of temperature, humidity, and particulate matter concentration in real time, which gives a valuable insight into the dynamics of temperature inversion episodes and air pollution below the inversion layer.

Keywords: particulate matter; low-cost sensors; air pollution; temperature inversion

1. Introduction

The city of Sarajevo, capital of Bosnia and Herzegovina, is heavily affected with air pollution with one primary component of pollution: particulate matter (PM). All other components of air pollution, for example, heavy metals [1], are mostly within the safe exposure level. According to the report of the World Bank [2], it is estimated that 3300 premature deaths per year can be related to the air pollution in Bosnia and Herzegovina, while the economic costs are between 1 and 1.8 billion US dollars.

In order to illustrate typical exposure levels of the concentration of particulate matter smaller than 2.5 μm in diameter ($PM_{2.5}$), we present Table 1 with statistics for the last 5 years (this was measured using the system presented in this article). One of the reasons for a such strong air pollution is the topology of the terrain. With an average altitude above sea level (ASL) of 550 m, Sarajevo is situated in a valley surrounded by mountains, and it is affected by temperature inversions during the winter time. Our previous research on using drones for investigation of temperature inversion [3] shows that the temperature gradient in the inversion layer is very strong and the height at which temperature inversion appears is low, often just 100 m to 150 m above the ground. This prevents natural convection and pollutants emitted by households, traffic and other sources are accumulated below the inversion layer. Figure 1 shows the drone image of central part of Sarajevo, taken during a research mission.

Our system occasionally detected daily average values of $PM_{2.5}$ concentrations greater than 500 μg/m^3, which is alarming according to the local regulation. In order to address this issue, we need measurements not only of air pollutants concentrations but also those related to the temperature inversion phenomenon (vertical profiles of temperature across the inversion layer). Then, we can quantitatively describe the influence of temperature inversion on the air pollution in the city.

Table 1. Yearly statistics for concentration of $PM_{2.5}$ (in $\mu g/m^3$) at the location of Faculty of Mechanical Engineering in Sarajevo.

Year	Average	Median	5th Percentile	95th Percentile
2019	33.9	22.0	8.1	97.6
2020	43.4	17.5	6.3	209.9
2021	29.3	18.9	6.9	90.2
2022	35.1	20.3	8.6	121.8
2023	35.6	25.5	8.9	128.8

Governmental institutions have a sparse network of stations for monitoring several components of air pollution. The spatial and temporal resolution of such measurements is low, and there is no information about vertical distribution of pollutants and temperature. Our system which is named MAQS (Mobile Air Quality System) is designed to provide high resolution measurements of air pollution components and meteorological parameters, with both horizontal and vertical variations of sensor locations.

Figure 1. Central part of Sarajevo and surrounding mountains. Image taken from drone on 31 January 2022. Faculty of Mechanical Engineering is indicated with red circle.

2. About MAQS

MAQS is a system for collecting, storing and displaying measurements of air pollution and meteorological parameters in real time. It is based on internet cloud, as shown in Figure 2. The software components of MAQS are designed using open source projects. The code for server, which is responsible for direct communication with sensors, is developed in-house and written from scratch in C++ programming language with significant use of the Boost library [4]. It is highly optimized code designed to be flexible and scalable. Our approach is based on multi-threaded asynchronous server code with parallelized database connections. The database load balancing is implemented using the round-robin (RR) algorithm. In that case, we can have a large number of concurrent connections (active sensors) using very moderate computer resources. In order to test the scalability of MAQS server code, a simulator was created to test the performance of the MAQS server under the concurrent activity of many thousands of sensors, and the results were outstanding: with 10,000 simulated sensors, our server had CPU load of around 23% using only two cores (out of 16) on Intel Xeon E5-2620 v4 processor and produced 1.43 Mbit/s of network traffic on average. The MAQS server is so efficient that it can run on a Raspberry Pi 3 single board computer (with very limited hardware resources). In reality, the MAQS server uses the

Ubuntu Linux operating system installed on a mid-range desktop PC (Intel Core i5-4460 with 6 GB of DDR3 RAM). The measurements are stored in a database. We use open source relational database PostgreSQL [5]. Clients with appropriate permissions can access the database using SQL queries.

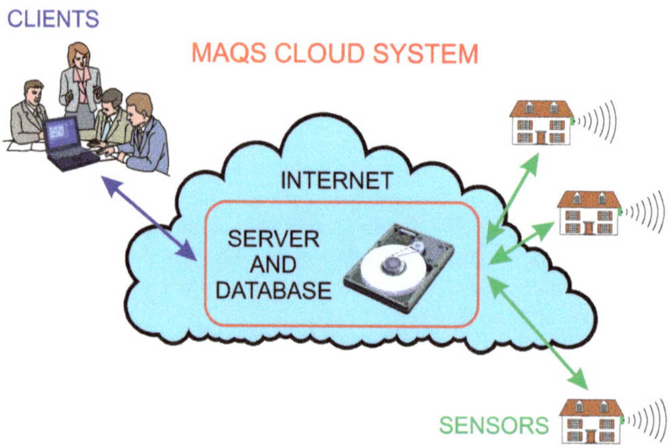

Figure 2. Components of MAQS: sensors, server with database, and clients connected via internet cloud.

MAQS sensors represent the most important component of the system. We have developed two generations of sensors:

- Generation 1—each sensor from this generation is powered by an Arduino [6] microcontroller (Mega or Due) with external ESP8266 Wi-Fi network chip.
- Generation 2—this generation omits Arduino and uses ESP32 for both network and processing tasks [7].

Any combination of sensors from first and second generation is supported by our server (both generations are still in use, but Generation 1 is being phased out gradually). Communication between sensors and the server is implemented using our own protocol and TCP/IP connection. This choice is crucial to achieve the reliability and robustness of the system. In reality, Wi-Fi connections are unstable (particularly at some of our locations of measurements) and we cannot afford to lose data because of that. TCP has mechanism of retransmission: if data packet is damaged or lost, it will be resent.

The MAQS protocol defines messages with a header (of fixed size) and payload of variable size, which contains the measurements. It is protected with a cyclic redundancy check (CRC-16) error-detecting code. There are four types of control messages: new measurement, transfer the oldest measurement, delete the oldest measurement, and keep alive message. Measurements are stored locally in RAM or flash memory of the sensor until they are transferred to the server. If the connection is interrupted for too long, once the local buffer is full, the oldest measurement is overwritten.

At the beginning, there was a batch of 10 sensors, while at the time the manuscript is written, there are 17 outdoor sensors. We have a list of desired locations with priorities given to the settlements with a significant population and absence of any nearby monitoring device or to the locations for observing important natural phenomena (temperature inversion and microclimate characteristics). Then, we have the list of available locations, where we can install the sensors. Using these two lists, we install the sensors starting from the most desired and available locations. Most MAQS sensors are installed at public institution like schools, faculties, and cable car, but also at private objects like houses and small family companies.

2.1. Particulate Matter Sensor

The choice of particulate matter sensor for the MAQS project was the most important decision. We previously evaluated PM sensors for mobile measurements [8] and found suitable sensor for this project: PMS5003 (Plantower, Nanchang, China). Apart from being the low-cost sensor, PMS5003 demonstrated surprisingly good correlations with reference measurements of $PM_{2.5}$ under conditions of strong urban pollution in Bosnia and Herzegovina, which is illustrated in Figure 3. In order to verify performance of PMS5003 sensor, we organized campaign from 2 December 2019 until 2 March 2020 with reference (gravimetric) measurements of $PM_{2.5}$ concentrations. Sampling was performed with a Gemini sampler (Dadolab, Milan, Italy) with an air flow rate of 2.3 m^3/h. The sampling started at 12:00 (local time, CET zone) each day and ended at 11:58 the next day. Two minutes were reserved for an automatic filter exchange. Particles were collected on quartz filters with a diameter of 47 mm. The filters were carried out through a stabilization and weighing procedures strictly according to the requirements of the standard EN 12341:2014. In this campaign, PMS5003 produced coefficient of determination $R^2 = 0.9827$ with reference measurements (for daily average concentrations of $PM_{2.5}$) and performed better than more expensive OPC-N2 (Alphasense, London, UK) sensor [8]. The absolute values are overpredicted by PMS5003, mostly due to the effect of the hygroscopic growth of aerosols as we explained in detail [8]. However, this behavior is suitable for corrections, especially if take into account good linearity of the sensor and no observable time drift in the long-term use [8].

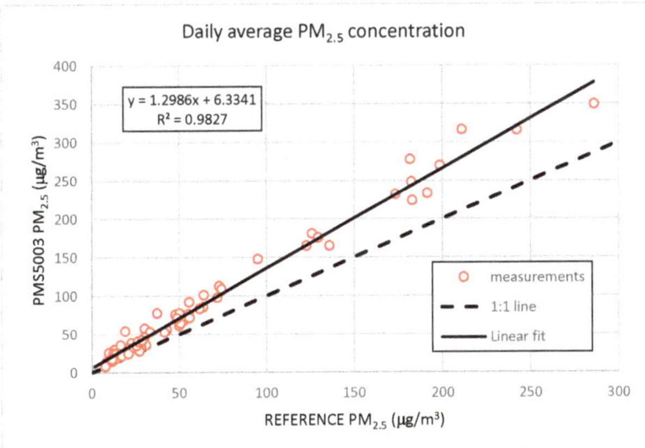

Figure 3. Evaluation of PMS5003 sensor: reference gravimetric measurements.

PMS5003 is one of the most analyzed particulate matter sensors. Numerous studies are dedicated to laboratory and field tests of PMS5003. It was tested in [9] using laboratory and field tests where a high bias of PurpleAir (PMS5003) was observed. In [10], PurpleAir (PMS5003) was analysed for 16 months in Charlotte, North Carolina, USA, against BAM-1022, and a high bias of PurpleAir (PMS5003) that increases with humidity was reported. A high mean bias of (PMS5003) was reported in [11] as well.

The PMS5003 sensor uses Mie scattering theory to determine PM concentrations. This theory provides the solution of the Maxwell equations for the scattering of plane waves on spherical particles [12]. Laboratory tests of [13,14] are particularly important for understanding the limitations of PMS5003. According to these tests, two major limitations of PMS5003 are as follows:

1. It cannot detect properly coarse particles (especially those larger than 2.5 µm) because most particles miss the focal point of laser beam. When this happens for large particle, it is incorrectly sized.
2. Strong wind can obstruct sensor's aspiration, depending on the wind direction relative to the sensor inlet.

The reason why PMS5003 works so well in Sarajevo, when urban pollution is strong, is that none of these two conditions apply: during temperature inversion periods, there is no wind under the inversion layer, and mass spectrometry of the particles shows that contribution of coarse particles is very small. In our campaign [8], we simultaneously measured $PM_{2.5}$ and PM_{10} concentrations using reference gravimetric method, from 2 December 2019 until 12 March 2020, and calculated that $PM_{2.5}$ makes 87% of PM_{10} mass. These are favorable conditions for the Plantower sensor. However, when the spectrum of PM contained a large number of coarse particles, such as the Aralkum desert sand observed in Sarajevo on 27 March 2020, the PMS5003 performed poorly (Alphasense OPC-N2 was much better in this scenario) [8]. However, these desert sand episodes are not frequent in Sarajevo, while the pollution from combustion dominates most of the time, so the overall conclusion is that PMS5003 is indeed the optimal sensor for this project.

2.2. Quality Control and Calibration of Sensors

We acquire sensors in batches, usually of 10 sensors at a time. There are three steps of the quality control and calibration procedure. The first step is to compare sensors for uniformity. All sensors should show the same trends, but there is a slight difference in their readings (acknowledged by the manufacturer in data sheet). Every batch of sensors is tested in our laboratory for 24 h under controlled conditions (Figure 4). The sensors that deviate in readings more than 8% from the average value of the batch are discarded. Then, we proceed to the second step: outdoor tests at the Faculty (Figure 5). Sensors are tested against the research-grade instrument Grimm 11-D (Durag, Hamburg, Germany). The duration of this phase is 7 days. An initial set of calibration coefficients is determined for each sensor in this step. Finally, sensors are tested against the reference gravimetric method, using a dual air sampler, with $PM_{2.5}$ and PM_{10} inlets, for an additional 10 days. The sampling period in the third phase is 24 h. Calibration coefficients are adjusted for the final deployment. One sensor from each batch stays in the laboratory for long-term observations, while remaining sensors are deployed to their final locations. The reference gravimetric method is most accurate, but it is a stationary system with long sampling periods (24 h). However, Grimm 11-D is a portable instrument, which can run on battery for 10 h with time resolution of 6 s, and it is used to verify readings at any location. In these three steps, we eliminate any erroneous sensors from the MAQS network.

As we have explained in [8], we propose the correction of raw readings from the sensor using steady (seasonal) calibration coefficients, determined for each sensor individually. These corrections can be linear for concentrations of $PM_{2.5}$ up to 300 $\mu g/m^3$. Above 300 $\mu g/m^3$ (which is rare situation even in Sarajevo), we observed non-linear effects, so the quadratic corrections fit better. In general, real-time corrections in the MAQS system use the second-order equation $y = Ax^2 + Bx + C$, with coefficients A, B and C determined in step 3 of the above-mentioned procedure. These coefficients are steady within a season (winter, summer spring, and autumn) but not permanent during the lifetime of a sensor. We do not recommend calibrations using floating correction coefficients (for example, based on instantaneous value of ambient air humidity), because such a procedure inserts noise into results since the effective hygroscopic growth coefficient depends on the composition of particles. Furthermore, the internal structure of the sensor and self-heating effect represent additional uncertainties in such an approach [8]. It is known that PMS5003 readings show dependence on relative humidity but less than predicted by simple hygroscopic growth theory [15].

It would be useful to test enclosures with some kind of humidity regulation, which is our future task. We already demonstrated the effectiveness of an in-house developed diffusion dryer, but it was applied to a different class of instrument—portable aerosol spectrometer Grimm 11-D (Durag, Hamburg, Germany) [8]. Promising results of the development of dryers for low-cost sensors were reported in [16,17].

In the MAQS system, calibration is implemented at the database level. When a new measurement arrives, it is stored in the main table (without corrections). At the same time, the calibration trigger calls a special function which corrects the measurement based on the parameters from the calibration table. Corrections with three parameters (linear and quadratic polynomial) are supported in real time (apparently any correction can be applied in post processing). The corrected value is calculated and saved in the appropriate table. Hourly and daily averages are also recalculated. In this way, users have calibrated measurements in real time.

Figure 4. Calibration of MAQS sensors: batch of 10 sensors that are being calibrated in our laboratory for particulate matter spectroscopy and gravimetry.

Each MAQS sensor has PMS5003 inside, while other sensors are optional. Table 2 shows physical quantities which have been measured using the MAQS system. Enclosures for MAQS sensors are designed and produced in-house. The basic enclosure is produced using a 3D printer (Figure 4), while advanced enclosure requires machining as well. Measurements of optional quantities can be calibrated the same way as PM concentrations (they have their own entries in the MAQS calibration table).

Table 2. Supported external sensors.

Physical Quantity	Supported Sensors
Position and time	All U-blox GNSS receivers
Barometric pressure	Bosch BME and BMP series
Wind speed	in-house developed cup anemometer
Carbon dioxide concentration	Amphenol Telaire 6703, Plantower DS-CO2-20
Temperature	Sensirion SHT 30 and 40 series
Relative humidity	Sensirion SHT 30 and 40 series

Every MAQS sensor measures time (using GNSS receiver or Internet time servers), PM_1, $PM_{2.5}$, PM_{10}, and 6 size bins from PMS5003 sensor as frequent as possible (once per second), and results are averaged over a period of 60 s and then stored in the database. The same averaging technique applies if there are optional sensors from Table 2.

Figure 5. Dual air sampler and Stevenson screen with MAQS sensors inside.

2.3. Comparison with Governmental Monitor

We have installed an MAQS sensor on the University building (Faculty of Electrical Engineering) in the immediate vicinity of the US Embassy in Sarajevo, which hosts an air quality station with a beta attenuation monitor BAM-1020 (MetOne, Grants Pass, OR, USA). Long-term observations from 21 February 2019 to 4 May 2020 (438 days) are summarized in Figure 6. The MAQS sensor performed well in the long run, with R^2 coefficients 0.919, 0.980, and 0.998 for hourly, daily, and monthly average value, respectively.

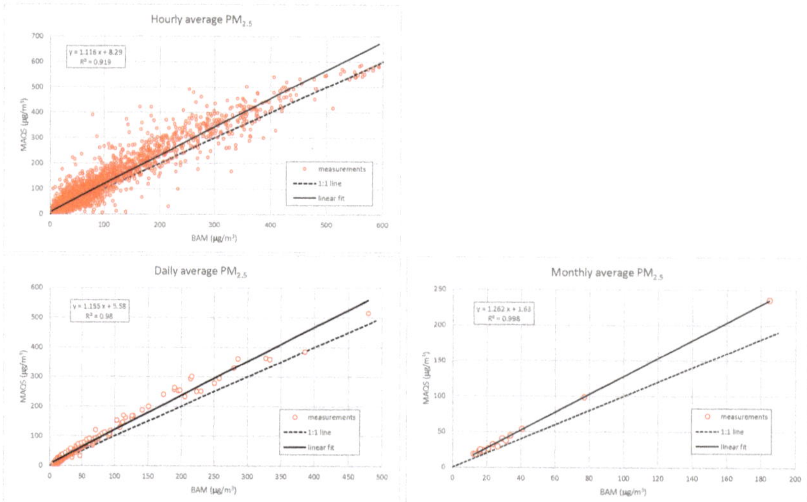

Figure 6. Long-term comparison of MAQS and US EPA station with BAM-1020.

2.4. Limitations of MAQS

MAQS is based on low-cost sensors, and it inherits both advantages and disadvantages of such technology. The PMS5003 sensor performed better than expected during the period of strong urban pollution caused by burning of fossil fuels and biomass. However, observations in spring when the pollution was mild and caused by various components, including the transported dust from afar, significantly decreased the coefficient of determination when MAQS was compared with reference instruments (gravimetric).

Figure 7 shows measurements of $PM_{2.5}$ and PM_{10} concentrations from 13 March 2020 to 4 May 2020. In comparison with the reference gravimetric instrument, the MAQS sensor produced R^2 coefficients of 0.730 and 0.718 for $PM_{2.5}$ and PM_{10}, respectively.

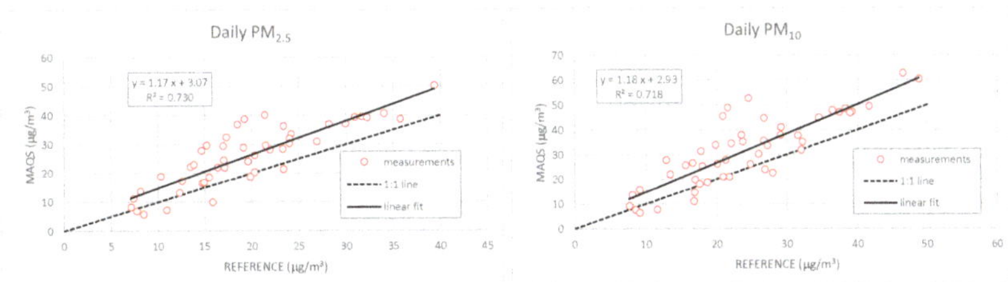

Figure 7. The MAQS sensor evaluation during the mild air pollution period from 13 March 2020 to 4 May 2020.

Research-grade mass spectrometer Grimm 11-D registered atypical particle distribution on 27 March 2020. It was Aralkum desert dust transported from afar, where coarse particles dominate in the spectrum, as Figure 8 shows. The PMS5003 sensor inside the MAQS node was not able to detect coarse particles due to reasons explained in [14]. Under such particle distribution, the MAQS sensor cannot determine the PM_{10} fraction accurately (Figure 8).

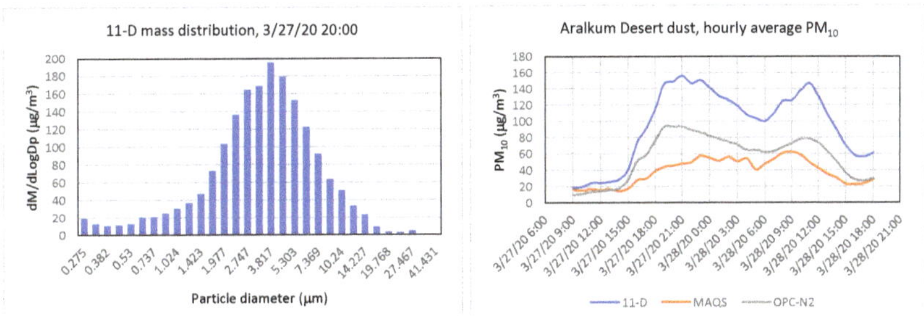

Figure 8. Atypical particle composition: Aralkum desert dust. Comparison of Grimm 11-D, MAQS, and Alphasense OPC-N2 readings.

2.5. Related Works

MAQS is not the only project of this type. For example, the PurpleAir network is often discussed [18]. It is a similar project which employs two Plantower PMS5003 sensors in nodes, which are connected to the Internet cloud (ThingSpeak) using Wi-Fi. PurpleAir is tested against governmental and research stations in USA [15,19] and Greece [11,20]. They reported reasonable correlations in various comparisons with reference or equivalent stations with R^2 values in the range from 0.56 to 0.98 and high bias of the sensor which was corrected using the multilinear regression.

Mobile sensor nodes called ExpoLIS were developed in Portugal [21]. Sensor nodes were installed in buses to map air quality corridors across the city. Particulate matter sensor was OPC-N3 (Alphasense, London, UK). Reported results from the comparison between the OPC-N3 and beta attenuation monitor MP101M (Envea, Poissy, France) were not satisfactory: R^2 coefficient was 0.04 for $PM_{2.5}$ and 0.12 for PM_{10}. After they applied corrections (multilinear regression based on Kohler theory), R^2 coefficient increased to 0.17 for $PM_{2.5}$ and 0.47 for PM_{10}.

Another project for sensor nodes called AIRBOX was implemented in Italy [22]. They measured several parameters including PM_{10} using the PPD20V sensor (Shinyei, Kobe, Japan). In comparison with the monitoring station of the local environmental agency, they reported a coefficient of correlation R (not R^2) in the range from 0.80 to 0.95.

Taking into account above-mentioned projects, we can say that MAQS is a unique long-term project with its own cloud server, vertical measurements for temperature inversion research, live pollution map (based on modified IDW algorithm), and non-linear real-time corrections.

3. Results and Discussion

The first batch of 10 MAQS sensors was installed in the year 2018 across Sarajevo Canton. In cooperation with our partners, the second batch of sensors was distributed over several locations in Bosnia and Herzegovina (outside of Sarajevo Canton) and three locations abroad. Currently, we are focused on Sarajevo, but MAQS is a global system already and supports world-wide locations. We will give three examples in the following subsections.

3.1. World-Wide Locations

Prior to the coronavirus pandemic, MAQS sensors were installed in Calgary (Canada), Zurich (Switzerland), and Akure (Nigeria). Figure 9 shows PM_{10} measurements in Zurich from 30 December 2018 to 1 February 2019. We can see strong air pollution, with variable intensity (depending on local meteorological conditions).

Figure 9. Hourly average values of PM_{10} concentration in Zurich from 30 December 2018 to 1 February 2019.

A very interesting situation was observed in Calgary, as Figure 10 shows. The Canadian province Alberta was affected by wildfires, particularly on 31 May 2019. The MAQS sensor in Calgary detected very high PM_{10} concentrations on that day.

Figure 10. Hourly average values of PM_{10} concentration in Calgary from 21 May 2019 to 7 June 2019.

3.2. Pollution Map

In order to generate the pollution map, we need to estimate air pollution in every single point of the city. With limited number of sensors, we have to use a certain interpolation algorithm for that purpose. Inverse distance weighting (IDW) is commonly used in geosciences for such interpolations. We have adopted IDW here with some modifications. If we have N sensors accros the city, they give values of desired parameter $z_1, z_2, \ldots, z_N$ at the location of each sensor. The value of $PM_{2.5}$ at some arbitrary point M is estimated using the formula

$$z_M = \frac{\sum_{i=1}^{N} \frac{z_i}{d_i^p}}{\sum_{i=1}^{N} \frac{1}{d_i^p}}, \quad (1)$$

where d_i is the distance from sensor i to the point M, and p is the power parameter. The choice of power parameter is not trivial. For a small power parameter, such as $p \leq 2$, interpolated values may be dominated by points that are far away. When p is large, the generated map looks like a mosaic of tiles with a nearly constant interpolated value. Our choice for the live pollution map is $p = 3$. An additional modification was implemented: only nearest neighbors within sphere of radius R were taken into account for interpolation. This is combined with our spatial search algorithm for efficient implementation, which makes the map viewable on mobile devices with limited processing speed.

The live pollution map is presented on web page https://www.maqs.info (accessed on 30 April 2024) using OpenLayers library [23]. Figure 11 shows the screenshot of the live map from 9 December 2019, where we can see interesting spatial variation in the $PM_{2.5}$ concentration. Mild eastern wind cleared air pollution from the eastern part of Sarajevo, but it was not strong enough to reach the western part of the city (this scenario was detected several times using the MAQS live pollution map). Tall buildings (visible in Figure 1) are obstructing ventilation of the city in the case of mild wind. This shows that MAQS can be used as a tool in urban planning, microclimate studies, and future low-emission zones establishment.

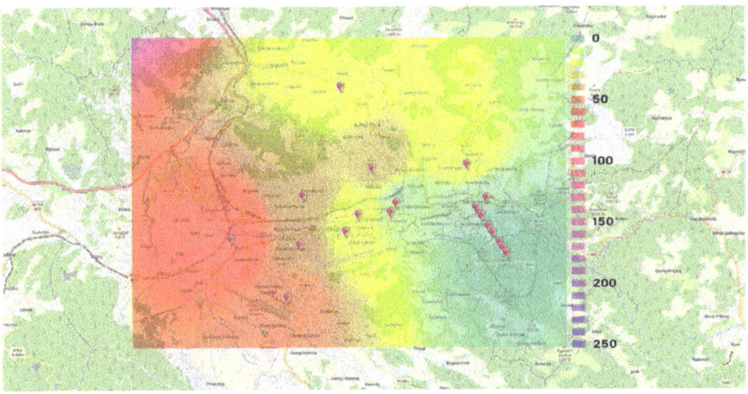

Figure 11. Live pollution map for 9 December 2019: example of spatial variation of $PM_{2.5}$ concentration across the city. Markers on the map are sensor locations.

When the mouse cursors hovers over the map (or the screen is briefly touched on a mobile device), the interpolated value in that point will be displayed as shown in Figure 12.

The map can be animated in two ways: by pressing the 'H' button for hourly animations over the last 24 h or by pressing the 'D' button for daily animations for the last 10 days.

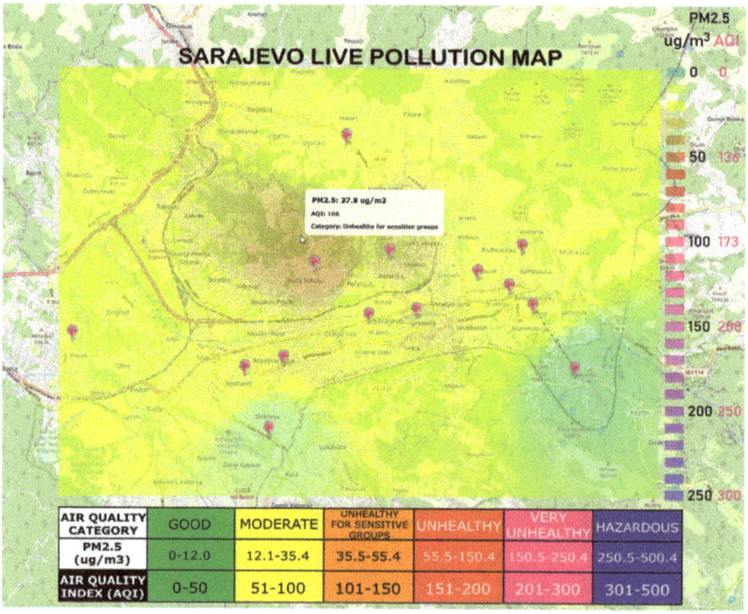

Figure 12. Live pollution map for 28 April 2024: mouse hovering for display of AQI category (calculated using $PM_{2.5}$ only) and $PM_{2.5}$ concentration in every point on the map.

3.3. Vertical Profiles and Temperature Inversion

Figure 13 shows hazardous air pollution in Sarajevo on 6 December 2019. Temperature inversion was strong and stable which resulted in accumulation of pollutants. Only peripheral parts of the City with altitudes greater than 800 m ASL were above the inversion lid.

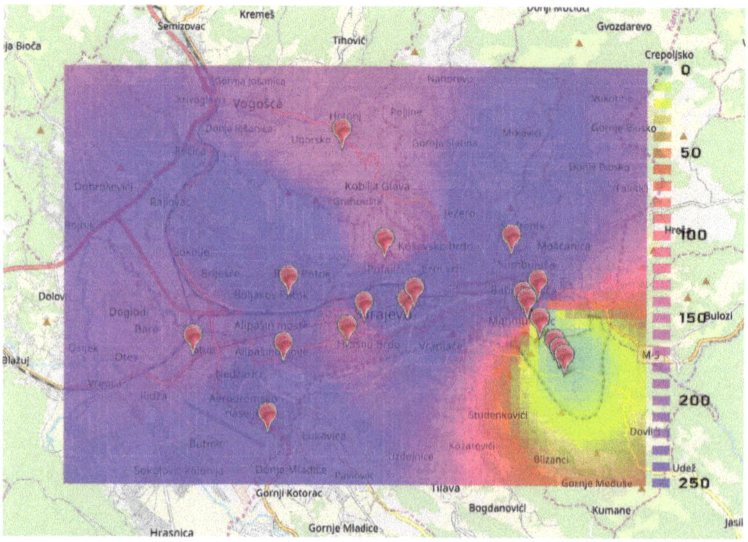

Figure 13. Live pollution map for 6 December 2019: City of Sarajevo under hazardous air pollution.

In order to test the capabilities of the MAQS system to measure vertical profiles, a special campaign was organized during the first 3 weeks of the year 2020, when strong temperature inversion had been expected. We performed several preliminary measurements using drones [3] and concluded that the location of the Sarajevo cable car (which spans from nearby the City Hall to Trebevic mountain, with length of 2.1 km and height difference of 580 m) would be suitable for measurements. In cooperation with the City of Sarajevo, who is the owner of the cable car, we have installed a batch of eight MAQS sensors on the pillars of the cable car, as shown in Figure 14. Each sensor is mounted on top of a pillar, while precise altitude of the sensor is determined using GNSS measurements and listed in Table 3.

MAQS sensors for this campaign were equipped with SHT31-D temperature and humidity sensors (Sensirion, Stäfa, Switzerland) together with mandatory PMS5003 PM sensors (and above-mentioned GNSS receivers). Each SHT31-D sensor was calibrated in our laboratory prior to deployment. There was no Wi-Fi connection over the cable car route, so we had to use our own GSM modems and Wi-Fi repeaters in very challenging conditions. SD cards were installed in each MAQS sensor to prevent data loss in case of Internet connection problems. During this campaign, the data availability was 100% (not a single measurement was lost) which shows the reliability of the MAQS system.

Table 3. Vertical positions of sensors on cable car pillars.

Pillar	Sensor Altitude ASL (m)
S2	585
S3	660
S4	710
S5	810
S6	895
S7	1000
S8	1090
S9	1165

The anticipated temperature inversion in this period appeared to be very strong and stable, which made this campaign successful. Figure 15 shows six characteristic profiles of temperature and $PM_{2.5}$. When there is no temperature inversion (6 January 2020 5:00), $PM_{2.5}$ concentrations were low and well distributed over the city (vertically in the range of measured altitudes). This plot shows a linear decrease in temperature with altitude. Despite low ambient temperatures (well below zero degrees Celsius), wind and natural convection prevented accumulation of pollutants which resulted in excellent air quality at that time.

However, different situation happened two days earlier (4 January 2020 10:00), when temperature profile had characteristic "Z" shape. That morning inversion layer started about 100 m above ground level (AGL) and ended about 500 m AGL, with gradient in the inversion layer of 15.5 K/km. The concentration of $PM_{2.5}$ below the inversion layer reached 250 $\mu g/m^3$. Similar profile was recorded on 9 January 2020 15:00, with a difference often observed in the afternoon: the inversion layer started higher, at about 250 m AGL. The gradient in the inversion layer was even greater, 33.3 K/km, which in combination with cold air in the city resulted in hazardous air pollution with $PM_{2.5}$ exceeding 300 $\mu g/m^3$.

Plot from 7 January 2020 8:00 shows the inversion layer starting from 150 m AGL all the way to top of the mountain, but with a lower gradient of 10.8 K/km. The air temperature in the city of Sarajevo was -10 °C, while the $PM_{2.5}$ concentration was 120 $\mu g/m^3$.

The last two graphs (11 January 2020 2:00 and 16 January 2020 2:00) show similar effects: the inversion layer starting from 150 m AGL with gradients around 13 K/km. Stable and lasting inversion trapped pollutants for longer periods of time, which resulted in $PM_{2.5}$ concentrations of 500 $\mu g/m^3$ even at 2 a.m., when most human activities were suspended.

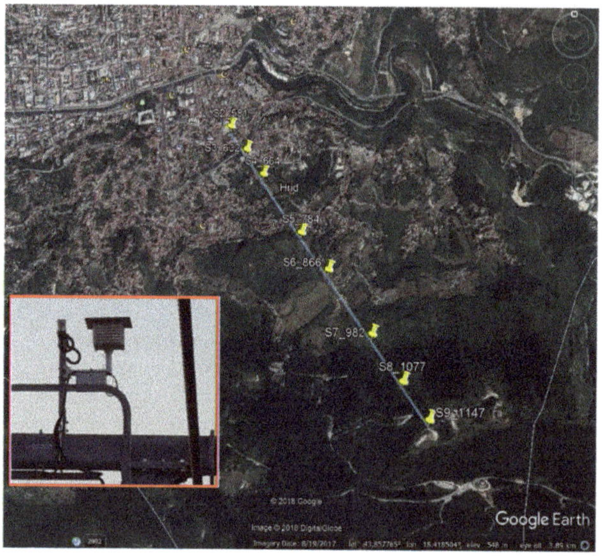

Figure 14. Sarajevo cable car pillars (yellow markers) and terrain from Google Earth. A MAQS sensor is shown in red box.

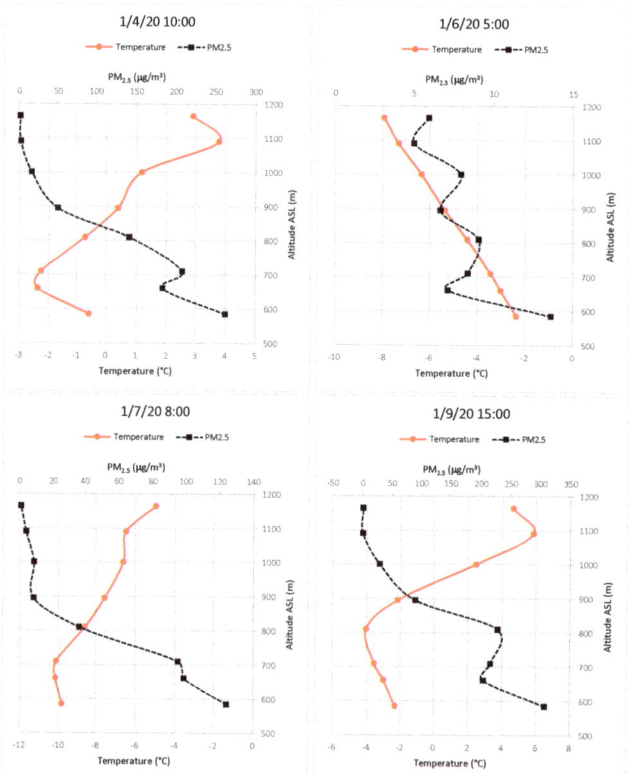

Figure 15. *Cont.*

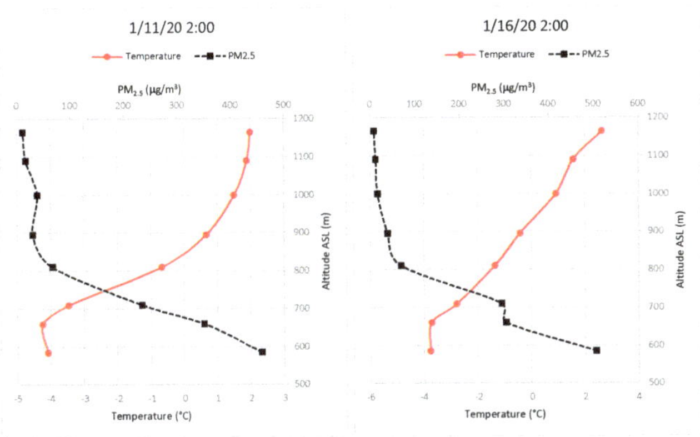

Figure 15. Temperature inversion profiles.

The relative humidity is also stratified along the inversion layer. Figure 16 shows two relative humidity profiles for the morning of 4 January 2020 and afternoon of 9 January 2020. Below the inversion layer (inside the City), the relative humidity was very high (around 90%), while on Trebevic mountain (above inversion layer), it was around 40%.

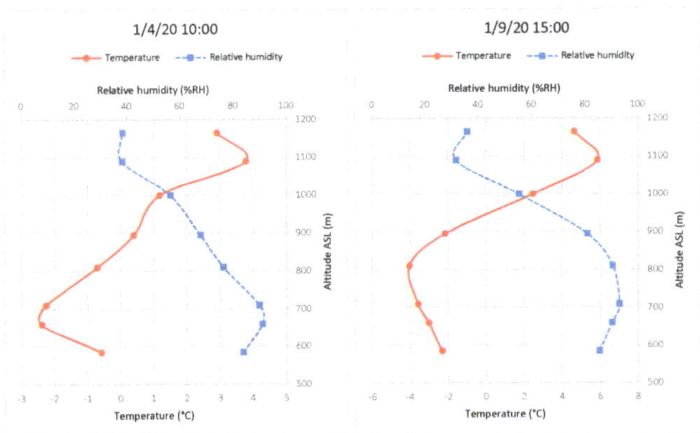

Figure 16. Relative humidity profiles.

4. Conclusions

A new system called MAQS for high resolution spatial and temporal monitoring of air pollution in urban environments has been developed and employed. After comprehensive evaluation, which included laboratory and field tests, an optimal particulate matter sensor was chosen—Plantower PMS5003. The system is mostly developed in-house, while several components are being imported: particulate matter sensor, wireless network adapter and optional external sensors. Advanced code for server, and sensors is written from scratch. An efficient calibration procedure is established and verified.

This system proved its capabilities to augment a sparse governmental network of monitoring stations. It offers high-resolution measurements in space and time, with the ability to generate a live pollution map. Furthermore, it is used to measure vertical profiles of temperature, humidity and particulate matter concentration, which is very important for understanding qualitative and quantitative characteristics of temperature inversion. Using MAQS, we have quantitatively shown that two major parameters of temperature inversion, in the context of air pollution, are height above the ground where the inversion layer starts and temperature gradient in the inversion layer. Unfortunately, both of these parameters are unfavorable in Sarajevo, resulting in extreme episodes of air pollution.

We have collected more than 60 million database records so far. This database is a very valuable asset for the research of air pollution. MAQS can be used in future microclimate studies, urban planning projects, and low-emission zones.

MAQS inherits some limitations from low-cost sensors. It cannot properly measure coarse particles, and if they dominate in the spectrum, measurements of PM_{10} will be inaccurate, as we can see from the Aralkum desert dust example. Furthermore, the current number of installed sensors is relatively small, which leaves some parts of the pollution map far from any sensor.

We plan to increase the number of sensors in the near future, so that we can detect hyperlocal sources of particulate matter emissions and increase the resolution of the live pollution map. In the meantime, we will consider some numerical models for estimating the particulate matter distribution over the area where an observation is missing.

Sensor modifications to include air dryers to prevent high bias of the sensors due to hygroscopic growth of aerosols will be tested in the future.

Funding: This research received no external funding.

Institutional Review Board Statement: Not applicable.

Informed Consent Statement: Not applicable.

Data Availability Statement: Live data are available at https://www.maqs.info (accessed on 30 April 2024). Access to the database with all stored measurements is possible upon request.

Acknowledgments: This work is dedicated in honor of my colleague Boran Pikula. Furthermore, I would like to thank Damir Muslic who contributed the code for the MAQS. I also appreciate the endless discussions with Dzevad Bibic about this project. Assistant Faruk Razic provided several locations for sensor nodes and personally installed several sensors, which is greatly appreciated. Last but not least, I would like to thank Sarajevo Cable Car Company and Adnan Selmanovic for their extraordinary cooperation.

Conflicts of Interest: The author declares no conflicts of interest.

Abbreviations

The following abbreviations are used in this manuscript:

AQI	Air Quality Index
AGL	Above Ground Level
ASL	Above Sea Level
GNSS	Global Navigation Satellite System
IDW	Inverse Distance Weighting
MAQS	Mobile Air Quality System
PM	Particulate Matter
SQL	Structured Query Language
TCP/IP	Transmission Control Protocol/Internet Protocol

References

1. Trako, N.; Masic, F.; Ajanovic, F.; Merdan, S.; Huremovic, J.; Zero, S.; Masic, A.; Gojak-Salimovic, S. Health risk assessment of heavy metals in $PM_{2.5}$ and PM_{10} in Sarajevo air, Bosnia and Herzegovina. *J. Environ. Sci. Heal. Part* **2024**, *58*, 1039–1045. [CrossRef] [PubMed]
2. Awe, Y.A.; Murisic, M.; Kreso Beslagic, E.; Rinnerberger, N.; Oguz Kuntasal, O.; Brody, M.S.; Golub, E.S.; Enriquez, S.; Paranos, D.; Colovic-Daul, M. *Air Quality Management in Bosnia and Herzegovina (English)*; World Bank Group: Washington, DC, USA, 2019. Available online: https://documents.worldbank.org/curated/en/117281576515111584/Air-Quality-Management-in-Bosnia-and-Herzegovina (accessed on 30 April 2024).
3. Masic, A.; Bibic, D.; Pikula, B.; Dzaferovic-Masic, E.; Musemic, R. Experimental study of temperature inversions above urban area using unmanned aerial vehicle. *Therm. Sci.* **2019**, *23*, 3327–3338. [CrossRef]
4. Boost C++ Libraries. Available online: https://www.boost.org/ (accessed on 30 April 2024).
5. PostgreSQL: The World's Most Advanced Open Source Relational Database. Available online: https://www.postgresql.org/ (accessed on 30 April 2024).
6. Arduino. Available online: https://www.arduino.cc/ (accessed on 30 April 2024).

7. ESP-IDF. Available online: https://idf.espressif.com/ (accessed on 30 April 2024).
8. Masic, A.; Bibic, D.; Pikula, B.; Blazevic, A.; Huremovic, J.; Zero, S. Evaluation of optical particulate matter sensors under realistic conditions of strong and mild urban pollution. *Atmos. Meas. Tech.* **2020**, *13*, 6427–6443. [CrossRef]
9. Tryner, J.; L'Orange, C.; Mehaffy, J.; Miller-Lionberg, D.; Hofstetter, J.C.; Wilson, A.; Volckens, J. Laboratory evaluation of low-cost PurpleAir PM monitors and in-field correction using collocated portable filter samplers. *Atmos. Environ.* **2020**, *220*, 117067. [CrossRef]
10. Magi, B.I.; Cupini, C.; Francis, J.; Green, M.; Hauser, C. Evaluation of $PM_{2.5}$ measured in an urban setting using a low-cost optical particle counter and a Federal Equivalent Method Beta Attenuation Monitor. *Aerosol Sci. Technol.* **2020**, *54*, 147–159. [CrossRef]
11. Kosmopoulos, G.; Salamalikis, V.; Pandis, S.N.; Yannopoulos, P.; Bloutsos, A.A.; Kazantzidis, A. Low-cost sensors for measuring airborne particulate matter: Field evaluation and calibration at a South-Eastern European site. *Sci. Total Environ.* **2020**, *748*, 141396. [CrossRef] [PubMed]
12. Mie, G. Beitrage zur Optik truber Medien, speziell kolloidaler Metallosungen (contributions to the optics of diffuse media, especially colloid metal solutions. *Ann. Phys.* **1908**, *25*, 377–445. [CrossRef]
13. Kuula, J.; Makela, T.; Aurela, M.; Teinila, K.; Varjonen, S.; Gonzalez, O.; Timonen, H. Laboratory evaluation of particle-size selectivity of optical low-cost particulate matter sensors. *Atmos. Meas. Tech.* **2020**, *13*, 2413–2423. [CrossRef]
14. Ouimette, J.; Arnott, W.P.; Laven, P.; Whitwell, R.; Radhakrishnan, N.; Dhaniyala, S.; Sandink, M.; Tryner, J. Fundamentals of low-cost aerosol sensor design and operation. *Aerosol Sci. Technol.* **2024**, *58*, 1–15. [CrossRef]
15. Barkjohn, K.K.; Gantt, B.; Clements, A.L. Development and application of a United States-wide correction for PM2.5 data collected with the PurpleAir sensor. *Atmos. Meas. Tech.* **2021**, *14*, 4617–4637. [CrossRef] [PubMed]
16. Chacon-Mateos, M.; Laquai, B.; Vogt, U.; Stubenrauch, C. Evaluation of a low-cost dryer for a low-cost optical particle counter. *Atmos. Meas. Tech.* **2022**, *15*, 7395–7410. [CrossRef]
17. Nothhelfer, M.; Sperber, O.; Todea, A.M.; Schunke, B.; Romazanowa, O.; Schumacher, S.; Bathen, D.; Asbach, C. Effect of an Aerosol Dryer on Ambient PM Measurements with SDS011 Low Cost Sensors during a Two-year Period in Duisburg, Germany. *Aerosol Air Qual. Res.* **2023**, *23*, 230080. [CrossRef]
18. Couzo, E.; Valencia, A.; Gittis, P. Evaluation and Correction of PurpleAir Temperature and Relative Humidity Measurements. *Atmosphere* **2024**, *15*, 415. [CrossRef]
19. Barkjohn, K.K.; Holder, A.L.; Frederick, S.G.; Clements, A.L. Correction and Accuracy of PurpleAir PM2.5 Measurements for Extreme Wildfire Smoke. *Sensors* **2022**, *22*, 9669. [CrossRef] [PubMed]
20. Stavroulas, I.; Grivas, G.; Michalopoulos, P.; Liakakou, E.; Bougiatioti, A.; Kalkavouras, P.; Fameli, K.M.; Hatzianastassiou, N.; Mihalopoulos, N.; Gerasopoulos, E. Field Evaluation of Low-Cost PM Sensors (Purple Air PA-II) Under Variable Urban Air Quality Conditions, in Greece. *Atmosphere* **2020**, *11*, 926. [CrossRef]
21. Correia, C.; Martins, V.; Matroca, B.; Santana, P.; Mariano, P.; Almeida, A.; Almeida, S.M. A Low-Cost Sensor System Installed in Buses to Monitor Air Quality in Cities. *Int. J. Environ. Res. Public Health* **2023**, *20*, 4073. [CrossRef] [PubMed]
22. Penza, M.; Pfister, V.; Suriano, D.; Dipinto, S.; Prato, M.; Cassano, G. Application of Low-Cost Sensors in Stationary and Mobile Nodes for Urban Air Quality Index Monitoring. *Eng. Proc.* **2023**, *48*, 62. [CrossRef]
23. OpenLayers Library. Available online: https://openlayers.org/ (accessed on 30 April 2024).

atmosphere

Article

Using a Low-Cost Sensor to Estimate Fine Particulate Matter: A Case Study in Samutprakarn, Thailand

Supichaya Roddee [1], Supachai Changphuek [1], Supet Jirakajohnkool [1], Panatda Tochaiyaphum [1], Worradorn Phairuang [2,3], Thaneeya Chetiyanukornkul [4] and Yaowatat Boongla [1,*]

[1] Department of Sustainable Development Technology, Faculty of Science and Technology, Rangsit Campus, Thammasat University, Pathumtani 12121, Thailand; supichaya.rodd@dome.tu.ac.th (S.R.); supachai.chang@dome.tu.ac.th (S.C.); supet@sci.tu.ac.th (S.J.); catty20@tu.ac.th (P.T.)

[2] Faculty of Geosciences and Civil Engineering, Institute of Science and Engineering, Kanazawa University, Kanazawa 920-1192, Japan; pworradorn@se.kanazawa-u.ac.jp

[3] Department of Geography, Faculty of Social Sciences, Chiang Mai University, Muang, Chiang Mai 50200, Thailand

[4] Department of Biology, Faculty of Science, Chiang Mai University, Muang, Chiang Mai 50200, Thailand; thaneeya.s@cmu.ac.th

* Correspondence: yaowatat@tu.ac.th; Tel.: +66-924-630-064; Fax: +66-2564-4485

Abstract: This study evaluates low-cost sensors (LCSs) for measuring coarse and fine particulate matter (PM) to clarify and measure air pollution. LCSs monitored PM_{10}, $PM_{2.5}$ (fine particulates), and $PM_{1.0}$ concentrations at four sites in Samutprakarn, Thailand from December 2021 to April 2022. Average daily PM_{10}, $PM_{2.5}$, and $PM_{1.0}$ concentrations at the monitoring locations were 53–79, 34–45, and 31–43 $\mu g/m^3$, respectively. In December 2021, the monitoring station had a daily $PM_{2.5}$ value above 100 $\mu g/m^3$, indicating haze occurrences. However, the monitoring site's daily PM_{10} and $PM_{1.0}$ concentrations did not surpass Thailand's ambient air quality threshold. We also measured and calibrated comparative particulate matter concentrations from LCSs and a tapered element oscillating microbalance (TEOM) monitor (Pollution Control Department (PCD) standard analytical method). $PM_{2.5}$ concentrations from the LCSs were lower than TEOM, but the difference was not statistically significant. The $PM_{2.5}$ monitoring station provided near-real-time air quality data for health risk reduction, especially when PM levels were high. Based on this study, authorities and local agencies may consider improving air quality regulation in Samutprakan, focusing on suburban $PM_{2.5}$ air pollution.

Keywords: fine particles; low-cost sensor; PM sensor; Samutprakan; Thailand

1. Introduction

Particulate matter (PM) refers to a complex mixture of small solid particles and liquid droplets that are suspended in the air and vary in size, shape, surface area, chemical composition, solubility, and origin. Particles in the air with an aerodynamic diameter up to 10 μm are labeled as PM_{10}, while particles in the air with an aerodynamic diameter between 2.5 and 10 μm are referred to as $PM_{10\text{-}2.5}$ or coarse particles. Additionally, particles with aerodynamic diameters up to 2.5 μm are termed fine particle matter ($PM_{2.5}$). Ultrafine particles with a diameter below 1 micron are called $PM_{1.0}$ [1–3]. Both anthropogenic (man-made) and natural PM are known [2,3]. Several sources have shown that PM_{10} sources include industrial activities, transportation, fuel combustion, and construction [4–6], and $PM_{2.5}$ sources include traffic/vehicular [7,8], residential wood burnings, and forest fires [9–11]. Hazardous health effects from PM have been extensively studied [12,13]. Epidemiological evidence has been provided for the human respiratory system [13–15]. Due to their size [16,17], short-term and long-term exposures [18,19] of ambient $PM_{2.5}$ cause more adverse health effects and toxicity than PM_{10} [20–22]. Ambient $PM_{2.5}$ can easily move through the human body inhalation pathway. Before the COVID-19 lockdown, many countries were considered high-risk areas and needed to address the human health effects of the $PM_{2.5}$ concentration distribution. During and after the COVID-19 lockdown, the concept was to reduce air quality (AQ) due to the strict control of the virus transmission, and this potentially led to an air quality change in high-risk areas [23–25]. In addition, an improvement in air quality resulted from policies that controlled and monitored ambient air (AA) monitoring [26–29]. Thailand has serious air quality problems in several areas in every dry season (haze episodes from December to February). Excessive average levels of particulate matter ($PM_{2.5}$ > 50 μg/m^3 is the ambient air standard value in Thailand) are found in the air in many areas of Bangkok [30–33] and Thailand [34,35]. To assess the harmful health impact of ambient air pollution, the air quality index equation was proposed for calculating AQI values. The Thailand Air Quality Index is based on five criteria and each color has meaning in terms of the air pollution level. The index range is from 0 to above 201. The AQI range from 0 to 25 is considered very good air quality and appropriate for outdoor activities and tourism (Excellent: color blue). The AQI range from 26 to 51 is considered good air quality, and outdoor activities and tourism are possible (Satisfactory: color green). For the AQI range from 51 to 100 (Moderate: color yellow), the general public is able to engage in outdoor activities, but sensitive groups are recommended to undertake limited outdoor activity if symptoms such as coughing, difficulty in breathing, and/or eye irritation occur. In the AQI range from 101 to 200 (Very Unhealthy: color orange), the general public should monitor their health. If symptoms such as coughing, difficult breathing, and/or eye irritation occur, outdoor activities should be limited and/or personal protective equipment should be used as needed. For sensitive groups, reduced/minimized outdoor activities and/or use of personal protective equipment is recommended. If symptoms such as coughing, difficulty in breathing, eye irritation, chest pains, headaches, irregular heartbeat, nausea and/or exhaustion occur, medical assistance should be sought. If AQI values are above 201 (Hazardous: color red), outdoor activities should be avoided by all. Additionally, areas with poor air quality should be avoided and/or personal protective equipment should be used as needed. If any symptoms occur, medical assistance should be sought.

The air quality monitoring station in Thailand was provided by the Pollution Control Department (PCD). The PCD has created air quality monitoring networks to measure air pollutants such as PM_{10}, $PM_{2.5}$, and noxious gases (CO, NO_2, SO_2, and O_3). Samutprakarn province is a city near Bangkok (BKK), Thailand. Samutprakarn is an area that has mixed air emission sources, associated with many types of surrounding industry, fuel combustion, high population density, and vehicular sources [36]. The PCD has provided reference equipment [37–39] to measure air pollutants in Thailand, including the gravimetric method for PM_{10} and the federal equivalent method (FEM) for $PM_{2.5}$, as well as the tapered element oscillating microbalance (TEOM) [40–42] and the beta attenuation monitor (BAM). However, these instruments are quite expensive, and they are only deployed at central monitoring sites in each province. One monitoring station cannot record the spatial distribution of PM concentrations. PCD started to monitor ambient air in Samutprakarn province with TEOM. Authorities and local agencies of Samutprakarn province have developed policies for improving air quality in the Samutprakarn region, and air monitoring instruments are used during haze seasons and in the non-haze seasons. However, such instruments could not be distributed to all areas due to the high cost and need for trained personnel. Currently, air pollutants such as PM_{10} and $PM_{2.5}$ are instead measured by several low-cost instruments and equipment.

LCSs have been used extensively to characterize air pollution in the last few years. LCSs have been widely used in monitoring PM_{10} and $PM_{2.5}$ mass concentrations. In general, the LCS is a device that uses one or more sensors with other components to detect, monitor and report on specific air pollutants, such as particulate matter (PM), carbon dioxide (CO_2), and/or environmental factors like temperature and RH. Several reports in Thailand and in other countries have reported PM_{10} and $PM_{2.5}$ levels from many sensor devices [43–46]. Calibration of LCSs is critical due to aerosols' physical and chemical properties and weather conditions. Calibration of the LCS in the field typically involves comparing it to a reference instrument. Chunitiphisan et al. [44] measured particle matter, temperature, and RH with an Unmanned Aerial Vehicle (UAV) and a sensor (Plantower PMS 3003) in Northern Thailand (Nan province). Both PM_{10} and $PM_{2.5}$ levels were found to be lower than the data from reference devices (TEOM); measured PM_{10} values were 10.25 μg/m^3 using the Plantower PMS 3003 (light scattering technique) and 22.69 μg/m^3 using the TEOM, and $PM_{2.5}$ was 8.84 using the Plantower PMS3003 and 11.84 μg/m^3 using the TEOM. The linear correlation of the particulate matter between their sensor and reference devices was 0.5 (PM_{10}) and 0.6 ($PM_{2.5}$). Moreover, a light scattering technique was used to measure PM_{10} and $PM_{2.5}$ outdoors in an area of haze pollution in Chiangmai province, Thailand. During the haze episode, PM_{10} and $PM_{2.5}$ were collected in several areas around Chiangmai over a twenty-four-hour period, and the results were between 12.58 and 149.82 μg/m^3 (PM_{10}) and from 9.63 to 130.37 μg/m^3 ($PM_{2.5}$), respectively. In this study, beta ray detectors were employed to estimate the correlation of PM_{10} and $PM_{2.5}$ levels in their study versus a certified device (TEOM). These observations were in agreement with results in the reported reference device for PM_{10} and $PM_{2.5}$ ($r = 0.8658$). Thus, their sensor has the potential for detecting PM_{10} and $PM_{2.5}$ [45]. Another application of the $PM_{2.5}$ concentration sensor (SN-GCHA1, Panasonic Photo & Lighting Co., Ltd., Osaka, Japan), the light scattering technique, revealed the mass concentration results from real-time indoor measurements (ranging from 20.05–45.85 μg/m^3) and outdoor measurements (ranging from 9.42–56.56 μg/m^3). Additionally, data were also collected for the hourly average temperature (28.8 to 35.2 °C) (temperature sensor: RHHI-112A, Shinyei Technology Co, Ltd., Kobe, Japan) and RH (55.2 to 72.6%) (RH sensor: RHHI-112A, Shinyei Technology Co, Ltd.). In a controlled laboratory environment, Levy Zamora et al. [47] tested three Plantower PMSA003 sensors against eight emission sources: incense oleic acid, NaCl, talcum powder, cooking emissions, and monodispersed polystyrene latex spheres. The accuracy ranged from 13% to more than 90% when compared to reference instruments, demonstrating that PM sources had an impact on LCS performance. The LCS was most accurate for sizes under 1 μm. Kim et al. [48] evaluated LCSs in the laboratory and in the field using two commercial LCSs, Plantower PMS3003 and Plantower PMS7003, which were compared with a reference-grade PM monitor (GRIMM 11-D). The LCSs indicated lower mass concentrations than GRIMM 11-D for laboratory testing while the LCSs showed generally higher PM mass concentrations than GRIMM 11-D in field testing. The result indicated that the outdoor environment had notable impacts on LCS data. Field testing revealed that the LCS error increased with increasing RH levels (>75%). Additionally, meteorological conditions, including wind speed, wind direction, and topographic elevation [49] had an effect on PM concentrations, while combustion emission sources played a significant part in the PM_{10} and $PM_{2.5}$ concentrations. Levy Zamora et al. [47] indicated that temperature and RH significantly influenced LCS performance [47]. Long-term monitoring of $PM_{2.5}$ levels in Hanoi was performed using Panasonic $PM_{2.5}$ sensors (light scattering intensities of single particles technique), and the sensors were calibrated with virtually monodisperse polystyrene latex (PSL) particles. The sensor detects particles with diameters as small as ~0.3 μm and measures $PM_{2.5}$ mass concentrations up to ~600 μg/m^3. The sensors correlated well with a nearby beta attenuation monitor (3.1 km away), with an R^2 value of 0.73 [50]. Ly et al. [51] employed a machine learning technique based on a random forest (RF) algorithm and concentration weight trajectory (CWT) to investigate the impact of long-range transport and meteorological data on $PM_{2.5}$ characteristics and LCS (light scattering intensities of single particles technique) performance. The influence of meteorological and regional transport on winter $PM_{2.5}$ levels was also investigated. In an urban city of Klang Valley, in Malaysia, the $PM_{2.5}$ level was found to be 19.1 μg/m^3 using a detector (AiRBOXSense, Alphasense Ltd., Great Notley, Braintree, UK, the optical particle sensor (OPC-N2)). A GRIMM portable aerosol spectrometer (PAS-1.108) was employed as a reference device to validate $PM_{2.5}$ and PM_{10} levels, with correlation coefficients of 0.71

and 0.83, respectively [52]. The LCS, namely LCS Edimax AirBox AI-1001W V3 (laser sensor), was used to measure the $PM_{2.5}$ levels in indoor and outdoor areas in West Jakarta, Indonesia. This sensor was connected to the Internet for 24 h. The sensor specifications were as follows: concentration range of 0–500 μg/m^3; smallest particle size of 0.3 μg; maximum weight of 210 g; sensor temperature of 0–60 °C ± 1 °C; relative humidity (RH) range of 0–100% ± 5% RH; operation temperature range of −10–60 °C; with one internal antenna. The LCSs were placed at a height of 2.5–3 m, which corresponded to the air quality monitoring station (AQMS) of $PM_{2.5}$. According to this study, $PM_{2.5}$ concentrations are higher outdoors than indoors [53]. A Pocket $PM_{2.5}$ Sensor (LED: light-emitting diode, PD: photodiode, USB: Universal Serial Bus) (Yaguchi Electric Co., Ltd., Miyagi, Japan) was used to measure the $PM_{2.5}$ concentrations in seven townships in Yangon, Myanmar. The Pocket $PM_{2.5}$ sensor showed a strong correlation with the reference monitor [2]. One of the townships showed a high level of $PM_{2.5}$, over 100 μg/m^3. This result indicated that the particular monitoring site had an effect on the $PM_{2.5}$ levels. Also, solar radiation research aids climate models and meteorological forecasts. Meteorological factors, including dawn duration, temperature, and RH, have been used to assess solar radiation at various places. The study showed that higher levels of air pollution were associated with increased solar radiation, which was the greatest during the summer [54–56]. PM concentration using a LCS as a detector showed positive R^2 with a reference device [57,58]. Calibration under actual ambient conditions is necessary before deployment for monitoring $PM_{2.5}$ levels. The LCS should be calibrated for each source in their appropriate environment [58,59].

The LCS has lower accuracy and sensitivity than the reference monitor, and its use is dependent on the type of sensor. However, given their low cost, portability, and ease of installation, LCSs are suited for areas where reference monitors are not established. As a result, we assessed the performance of an LCS (PMS7003, Plantower) to better understand the major elements influencing performance, such as urban emission sources, RH, temperature, and PM_{10}, $PM_{2.5}$, and $PM_{1.0}$ levels. Furthermore, data storage for measurements is vital; we examine wire-free data export to an application, giving quick access to the information. The objective of our study was to develop a LCS device for measuring coarse (PM_{10}) and fine particulate matter ($PM_{2.5}$ and $PM_{1.0}$), along with some environmental data including air temperature and RH. The LCS was calibrated with reference devices from Thailand. The field test was performed at our monitoring sites, which were known to have serious air quality problems. The study location was established in Samutprakarn province, which is an urban environment area. Between December 2021 and April 2022, we collected mass concentrations of PM_{10}, $PM_{2.5}$, and $PM_{1.0}$, as well as temperature and RH data. The tested LCS was operating for a period of five months, The PM results were converted to AQI values using Thailand's AQI equation and then shown as geographic information system (GIS) maps.

2. Materials and Methods

2.1. Monitoring Sites

The monitoring stations are in Thailand's Samutprakarn province, a high-risk location for air pollution. Samutprakarn is a 388 square-kilometer area with a warm temperate, semi-humid continental monsoon climate with two seasons in terms of temperature and RH. The yearly average temperature is 28.8–32.1 °C, and the annual rainfall is 1100–1500 mm. The resident population is 1,344,875 million people [60]. Figure 1A–F shows the installation of four LCSs at four sampling sites for this study. Each sampling site had different characteristics: the first site was in a residential area (DFH1, Latitude, 13.607976, Longitude, 100.744788, Figure 1C); the second site was located in a semi-residential area in the Bangpli district, Samutprakarn province, surrounded by a mixture of many combustion emission sources (cooking, vehicles, construction) (DFH2, Latitude, 13.601481, Longitude, 100.770638, Figure 1D); the third site (DFH3, Latitude, 13.596204, Longitude, 100.639773, Figure 1E) was a closed to residential area in Pak Nam, Meuang, Samutprakarn province; and the last site (DFH4, Latitude, 13.571715, Longitude, 100.786608, Figure 1F) was set up near a commercial area in the Bang Sao Thong district, Samutprakarn province, that had twenty-four-hour a day transportation (Table 1). A Pollution Control Department (PCD)-approved instrument collected daily PM_{10} and $PM_{2.5}$ readings for comparison with our LCSs. The PCD stations measured the air quality in Pak Nam, Meuang, Samutprakarn province (18T, residential area, Figures 1A and S1); this was near the monitoring location of our LCS (DFH3) and in Bang Sao Thong district, Samutprakarn province (19T, semi-residential area, Figures 1B and S2); this was near the monitoring location of our LCS (DFH4). A Tapered Element Oscillating Microbalance (TEOM) (Table S2) was used to gather ambient air samples from the 18T and 19T stations [61]. The PM_{10} and $PM_{2.5}$ mass concentration data were completely delivered to these two PCD stations in Samutprakarn. The LCSs have been placed >1.5 m. The distances between 18T and DFH1 were 16.09 km, DFH2 18.89 km, DFH3 4.70 km, and DFH4 20.90 km. Additionally, the distances between 19T and DFH1 were 6.05 km, DFH2 3.72 km, DFH3 16.12 km, and DFH4 0.13 km. The list of acronyms used in study was presented in Table S1.

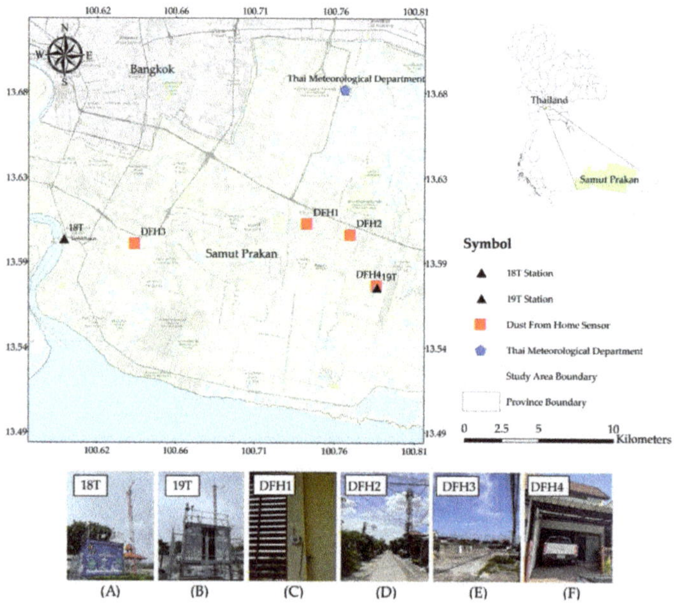

Figure 1. The selected monitoring sites are in Samutprakarn province, Thailand.

Table 1. Monitoring sites of real-time PM_{10}, $PM_{2.5}$, and $PM_{1.0}$ concentrations (24 h).

Monitoring Site Code	Specify of Site	Parameter	Instrument
DFH1	Residential area	PM_{10}, $PM_{2.5}$, $PM_{1.0}$, RH, Temp.	LCS
DFH2	Semi-residential area	PM_{10}, $PM_{2.5}$, $PM_{1.0}$, RH, Temp.	LCS
DFH3	Residential	PM_{10}, $PM_{2.5}$, $PM_{1.0}$, RH, Temp.	LCS
DFH4	Commercial area	PM_{10}, $PM_{2.5}$, $PM_{1.0}$, RH, Temp.	LCS
PCD station (18T)	Residential area	PM_{10}, $PM_{2.5}$	TEOM
PCD station (19T)	Semi-residential area	PM_{10}, $PM_{2.5}$	TEOM

2.2. Details of Sensor Device

In this study, a LCS (Plantower Laser Dust Sensor PMS7003) was tested at several monitoring sites, from December 2021 to April 2022. The Plantower Laser Dust Sensor PMS7003 is a low-cost, commercially available sensor that costs between THB 800–1100 (USD 25–35). The advantages of this sensor are that it can be controlled remotely via a

Raspberry Pi and is small enough to fit into a mobile or wearable enclosure, while previous research has shown that, compared with other sensors such as Plantower PMS5003 and Alphasense OPC-N2, PMS7003 tends to show a significant correlation with the reference instrument and good reproducibility [47,48,62]. Therefore, our study employed PMS7003. To assess the performance of our LCS for the detection of the PM_{10}, $PM_{2.5}$, and $PM_{1.0}$ concentrations, the RH and temperature, and various test data, were recorded. All the devices used were obtained from an electronics store and collocated; they included the Plantower laser dust sensor PMS7003 (a particle detector that uses the light scattering principle), a NodeMCU ESP8266 (V2) Wifi board, a fan (DC 12V, 8 cm), a liquid crystal monitor display (I2C LCD2004A LCD (Blue Screen), a 20 × 4 LCD with back-light), a power switching supply (DC 12V, 12 × 12 mm), a USB adapter (5V, 1A), a micro USB (B) programming cable (30 cm), a digital temperature and RH sensor (DHT22 AM3202), plastic cover boxes, and a jumper (F2F). The data gathered were reported via the Internet of Things (IoT) and was periodically uploaded to remote cloud storage such as Thingspeak. A numerical and graphical value near the real-time monitor was automatically obtained.

2.3. Temperature and RH Sensor

The impact of meteorological conditions on the PM_{10}, $PM_{2.5}$, and $PM_{1.0}$ levels was evaluated using the LCSs and data from Thailand's Meteorological Department (TMD) [63]. TMD station is shown in Figure 1. The distances between TMD and DFH1 were 9.04 km, DFH2 9.43 km, DFH3 17.04 km, and DFH4 12.89 km. The LCS (DHT22, AM3202) measured the environmental parameters, such as air temperature (°C) and RH (%). The air temperature monitoring sensor had an air temperature range of 0 °C to +60 °C (accuracy ± 0.2 °C, resolution 0.1 °C), and the RH monitoring sensor had a range of 0 to 60% (accuracy ± 2%, resolution 0.1%). Four LCSs were placed at different locations (residential, semi-residential, and commercial areas) in various environments as mentioned in Table 1. These four locations had varying meteorological conditions and pollution levels. However, no differences were observed in the temperature and RH values between the TMD and our LCSs over the monitoring period.

2.4. PM Mass Concentration

The LCS (Plantower Laser Dust Sensor PMS7003) was used to measure the concentrations of PM_{10}, $PM_{2.5}$, and $PM_{1.0}$, which are widely found in urban environments. The concentration range was 1–999.90 $\mu g/m^3$. This LCS is a digital and universal particle concentration sensor that can be used to determine the number of particles present in the surrounding air. For this sensor, the laser scattering principle is used (irradiation of particles in ambient air), then light scattered to a specific angle is collected, and lastly the scattered light signal change with time curve is obtained. Finally, the microprocessor is used to determine the equivalent particle diameter and the number of particles with varied sizes per unit volume.

2.5. Air Quality Index (AQI)

The air quality index (AQI) was utilized in accordance with Thailand's air quality index standard, as released by the PCD [61]. The Thai government uses the AQI value to communicate the quality of the air to the general public. Additionally, the AQI is divided into five categories, ranging from 0 to 201, with each category employing colors to symbolize the level of danger. The air pollutants include PM_{10}, $PM_{2.5}$, O_3, SO_2, NOx, and CO. Each of these contaminants has its own set of air quality standards that are utilized to calculate the overall AQI. Equation (1) is used to calculate pollutant levels [61]. The total AQI values were then entered into geographical information systems (GIS) and plotted on a map. The geographical interpolation of the AQI is demonstrated using GIS. The first equation is as follows:

$$I = \frac{\boldsymbol{Ij - Ii}}{\boldsymbol{Xj - Xi}}\,(X - Xi) + Ii \tag{1}$$

where I = air quality index (AQI) for pollutant; X = pollutant concentration; Xi = the concentration breakpoint that is $\leq X$; Xj = the concentration breakpoint that is $\geq X$; Ii = the index breakpoint corresponding to Ii; Ij = the index breakpoint corresponding to Ij.

2.6. Statistical Analysis

This experiment was completed using the Statistical Package for Social Science (SPSS) software version 28, and a *t*-test (*t*-test used to test difference between the means of two variables) was utilized to detect the difference between the concentrations from the LCSs and TEOM. The average, minimum, and maximum values, standard error, mean difference, median, standard deviation, variance, and observation results were generated (Table 2). A one-way analysis of variance (ANOVA) and a post-hoc test were used to compare the concentrations of PM_{10}, $PM_{2.5}$, and $PM_{1.0}$ at each monitoring site. A statistically significant level of *p*-value = 0.05 was used to determine the dust, temperature, and RH levels at each LCS monitor site and whether there are statistical differences between each sampling point.

Table 2. PM Concentrations of each sensor in the study location.

Parameters	Particle Size-Range		
	PM_{10} µg/m^3	$PM_{2.5}$ µg/m^3	$PM_{1.0}$ µg/m^3
DFH-1	59	41	33
DFH-2	58	39	32
DFH-3	60	40	34
DFH-4	52	31	30
PCD (18T)	62	38	NA
PCD (19T)	56	31	NA
Mean	58	36	33
S.D.	3	4	2
Max	62	41	34
Min	52	31	30
t-stat			
DFH-1 vs. 18T	−0.9	1.6	NA *
DFH-2 vs. 18T	1.5 [a]	−0.5	NA *
DFH-3 vs. 18T	0.5	−0.9	NA *
DFH-4 vs. 18T	3.3	3.3	NA *
DFH-1 vs. 19T	−1.4	−6.1	NA *
DFH-2 vs. 19T	−0.8	−0.5	NA *
DFH-3 vs. 19T	−1.5	−4.8	NA *
DFH-4 vs. 19T	1.5	−0.1 [a]	NA *

* NA: not available, [a] statistical analysis with significant difference level at 0.05.

2.7. Geographic Information System (GIS)

In this study, ArcMap version 10.8 software was used and employed the clip tool to make a map that indicates where the measurement instrument is. The buffer command was used to manipulate map data for the study area. The data were then displayed using the colors based on the Thai AQI Index, and the coordinates of the PM measuring devices were specified for display purposes. Reference devices of the Pollution Control Department at two stations were used for comparison data. For better visual comprehension, the findings are arranged geographically by color and Thai AQI value based on severity.

3. Result and Discussion

3.1. Calibration and Validation of Sensor

The combination of LCS enables measurement of a range of parameters such as temperature (°C), RH (%), and coarse and fine particle distributions with the Plantower Laser Dust Sensor PMS7003, as shown in Figure 2a,b. The LCS array for each monitoring site (DFH1-DFH4) was tested, and the PM concentrations were collected before starting the long-term monitoring. The particulate matter (PM) sensors were calibrated against an instrument approved by the PCD Station at Samutprakarn [3]. The LCSs for PM_{10}, $PM_{2.5}$, and $PM_{1.0}$ were set up at the monitoring sites, and their results were compared with a TEOM monitor (18T, 19T). The PM concentrations from the LCSs and TEOM are shown in Figure 3. The highest mass concentration in all LCSs was PM_{10}, followed by $PM_{2.5}$ and $PM_{1.0}$, respectively. Figure 3a shows that the DFH1 sensor recorded an average PM_{10} concentration of 60 $\mu g/m^3$ and $PM_{2.5}$ concentrations of 41 $\mu g/m^3$ (Figure 3b), which were lower than the mass concentrations of 18T (PM_{10} = 79, $PM_{2.5}$ = 46 $\mu g/m^3$) and 19T (PM_{10} = 63, $PM_{2.5}$ = 36 $\mu g/m^3$). The LCS DFH2 had a lower average concentration than the TEOM monitor (18T, 19T). The average mass concentration of PM_{10} was 54 $\mu g/m^3$ and for $PM_{2.5}$ it was 34 $\mu g/m^3$. The average mass concentration of PM_{10} with DFH3 was similar to 18T (79 $\mu g/m^3$), but 19T showed a low concentration. The average $PM_{2.5}$ mass concentrations were 45 $\mu g/m^3$ (DFH3), 46 $\mu g/m^3$ (18T), and 36 $\mu g/m^3$ (19T). DFH4 had an average PM_{10} mass concentration of 62 $\mu g/m^3$, followed by 36 $\mu g/m^3$ for $PM_{2.5}$. The measurement value of both reference devices (18T, 19T) was greater than the LCS in DFH4.

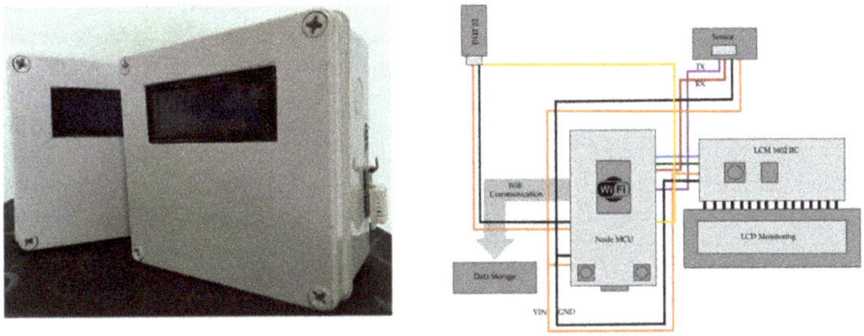

(a) (b)

Figure 2. The LCS used in this study (PMS7003), (**a**) outside box, (**b**) LCS diagram.

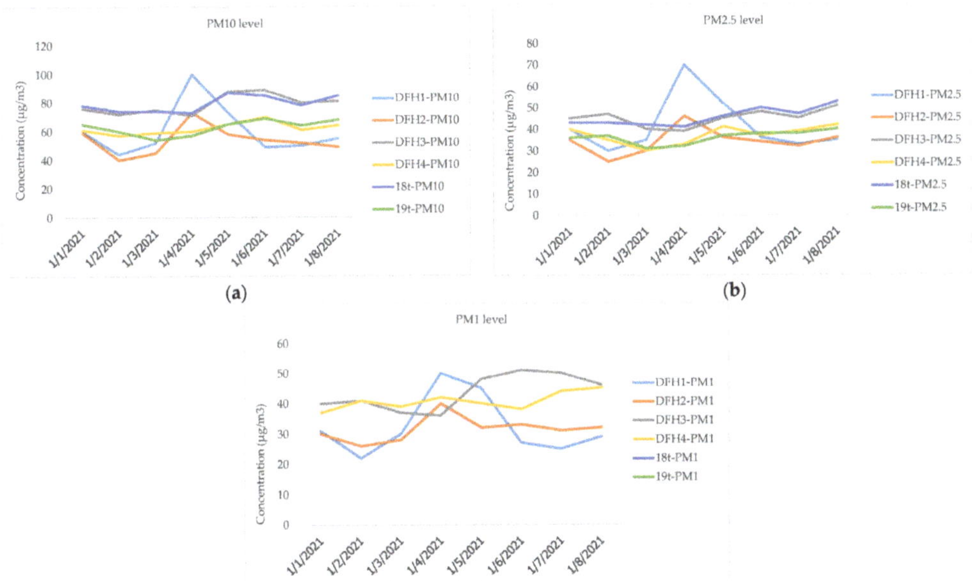

(a) (b) (c)

Figure 3. The comparison of PM concentrations from LCSs and TEOM: (**a**) PM_{10}, (**b**) $PM_{2.5}$, and (**c**) $PM_{1.0}$ ($\mu g/m^3$).

The concentration data from our LCSs (Figure 2) showed a positive correlation with the concentration data from the TEOM (Figures 4 and 5). As shown in Figure 4, the PM_{10} of the LCSs and TEOM are correlated for each site, and the PM_{10} of the LCSs and TEOM scatter plots shows a good correlation for DFH3 (R^2 = 0.9(18T), 0.6 (19T)), showing that the DFH3 LCS measures similarly to reference devices. The LCS DFH4 and TEOM scatter plots have a positive R^2 correlation of 0.7 (18T) and 0.6 (19T). However, the data from the LCSs at the DFH1 and DFH2 sites were not correlated with that from the PCD station, likely due to the large distance between the locations of these LCSs and the PCD station. The relationship between $PM_{2.5}$ from the LCSs and TEOM was also plotted. Figure 5 shows a low correlation between $PM_{2.5}$ from LCSs DFH1 and DFH2 and TEOM. The above has already been explained. R^2 = 0.2, 0.0 (18T), 0.2, and 0.1 (19T), respectively. However, the correlation between DFH3 and DFH4 shows that the LCS DFH3 measures similarly to reference devices (R^2 = 0.7 (18T), 0.9 (19T)). Finally, the LCS DFH4 and TEOM scatter plots display positive R^2 correlations of 0.44 (18T) and 0.7 (19T), respectively. DFH3 and DFH4 had a positive correlation with TEOM because they are close to the reference monitor locations. Larger distances between the LCS and TEOM may lead to larger differences in simultaneous readings. In addition to site factors, field testing may reveal distinct correlations between emission sources (residential and transportation). In terms of PMS7003 correlation with reference stations, Bulot et al. [62] found an R^2 ($R^2 > 0.6$) that is extremely close to our investigation (DFH3, DFH4). LCS testing was conducted during 1–8 December 2021. The average concentrations of PM_{10}, $PM_{2.5}$, and $PM_{1.0}$ at the four monitoring sites were 53–79 μg/m^3, 34–45 μg/m^3, 40–51 μg/m^3, respectively. The PCD station was unable to collect both particle sizes (PM_{10} and $PM_{2.5}$) because of an instrument limitation; thus, only the daily averages of the PM_{10} and $PM_{2.5}$ concentrations measured at DFH1 to DFH4 and PCD were compared. The average concentrations of PM_{10} and $PM_{2.5}$ from the 18T station were 79 μg/m^3 and 46 μg/m^3, respectively. The 19T station gave average concentrations of PM_{10} and $PM_{2.5}$ of 63 μg/m^3 and 36 μg/m^3, respectively.

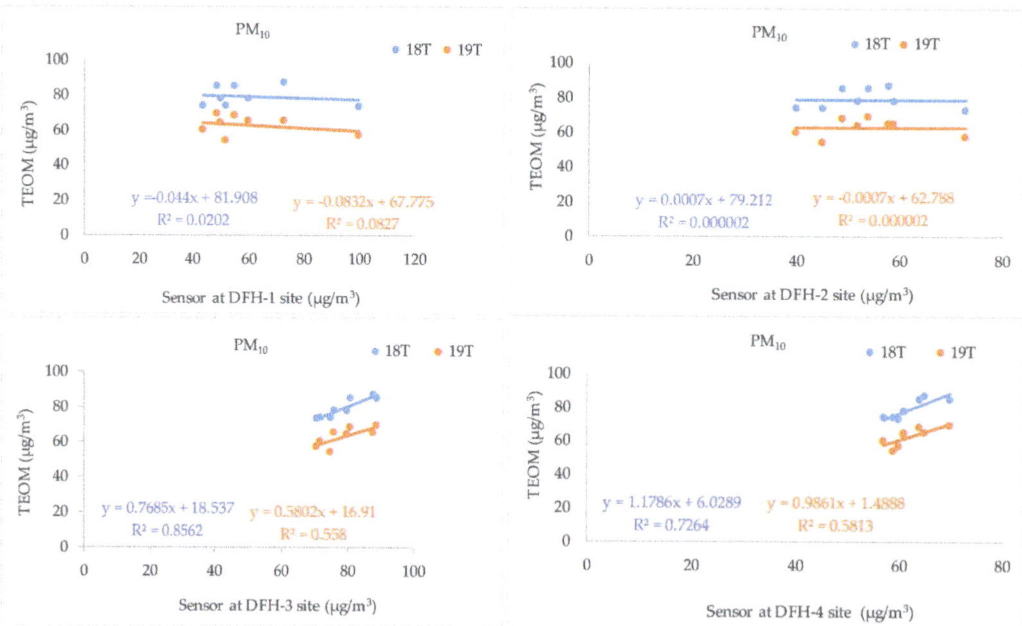

Figure 4. The correlation between LCSs and TEOM values for PM_{10}.

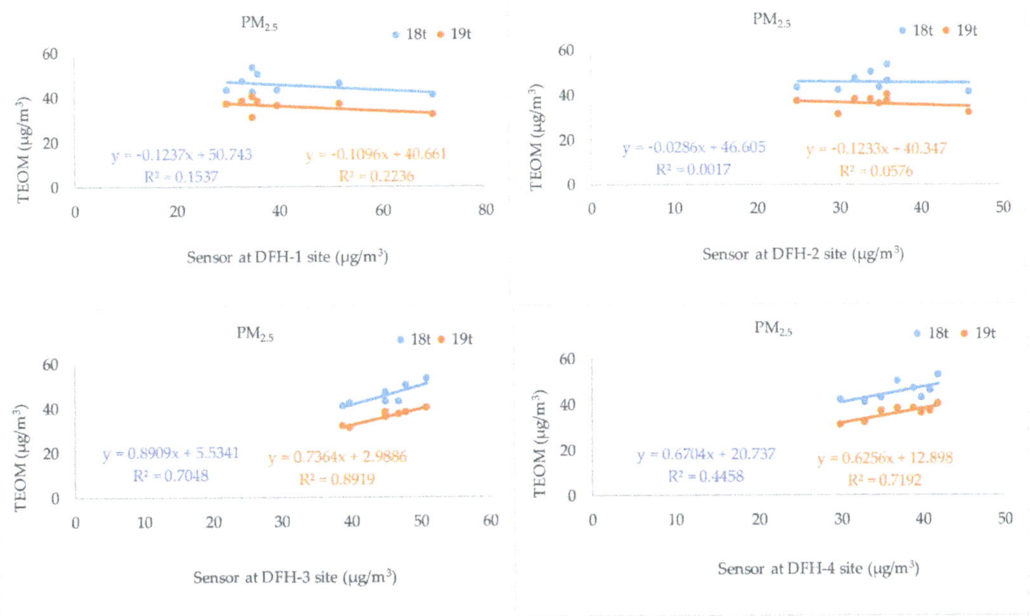

Figure 5. The correlation between LCSs and TEOM values for $PM_{2.5}$.

3.2. Daily Results

The daily average PM_{10}, $PM_{2.5}$, and $PM_{1.0}$ concentrations at the four monitoring sites were measured from 9 December 2021 to 30 April 2022. The results from the coarse and fine-particulate matter concentrations provided values lower than those from TEOM. SPSS analysis was used to determine the difference in the concentration between the LCSs and TEOM. The statistical results are illustrated in Table 2; if the *t*-stat value was less than the *t* critical value, then the hypothesis was acceptable. The mean concentration between the LCSs and TEOM is not statistically significantly different because the value at the significant difference level is more than 0.05 (except DFH-2 vs 18T for PM_{10} and DFH-4 vs 18T for $PM_{2.5}$). The negative number indicates that the average concentration of TEOM is lower than that of the LCSs. The correlation between the LCSs and TEOM values are shown in Figures 4 and 5. The following figures (Figures 6–10) show the time-series trend of the daily average from the important PM contribution at the monitoring sites. The distributions of the PM_{10}, $PM_{2.5}$, and $PM_{1.0}$ concentrations at the monitoring site in Samutprakarn, Thailand, are shown. Each bar represents the average values from the monitoring sites (Figures 6–10). From these results, the PM_{10}, $PM_{2.5}$, and $PM_{1.0}$ concentration data from the LCSs during December 9 to 31, 2021, at DFH1 and DFH3, had highest concentration, as shown in Figure 6a,c. The maximum concentrations of PM_{10} reached 129 $\mu g/m^3$ (22–23 December 2021, green color). The $PM_{2.5}$ concentrations at the DFH1 and DFH3 were 100 and 96 $\mu g/m^3$ (orange color), respectively, followed by the $PM_{1.0}$ concentrations of 75 and 72 $\mu g/m^3$ for DFH1 and DFH3 (22 December 2021, as presented in the blue color), respectively. In January 2022, the daily average concentrations of PM_{10} were the highest at DFH3 (63 $\mu g/m^3$) and DFH4 (58 $\mu g/m^3$) as shown in Figure 7c,d, and the highest $PM_{2.5}$ concentrations were found at DFH1 (42 $\mu g/m^3$) and DFH3 (41 $\mu g/m^3$). Additionally, the $PM_{1.0}$ concentrations had a similar trend at DFH1 (33 $\mu g/m^3$) and DFH3 (35 $\mu g/m^3$). Overall, in February, the detection from the LCSs was lowest at DFH4 (Figure 8d) and highest at DFH1 (Figure 8a). The graph shows a decreasing concentration of the data in March. The lowest concentration for all particle sizes from the LCSs was found at DFH2 (Figure 9b) and DFH3 (Figure 9c). The highest concentration for all particle sizes was found at DFH4 (Figure 9d). Based on the results from April 2022, the PM_{10} concentration was maximum at DFH1 (67 μg/m^3), as shown in Figure 10a, followed by the $PM_{2.5}$ concentration maximum at 48 μg/m^3 (Figure 10a), and the highest concentration of $PM_{1.0}$ was 33 μg/m^3 (Figure 10a).

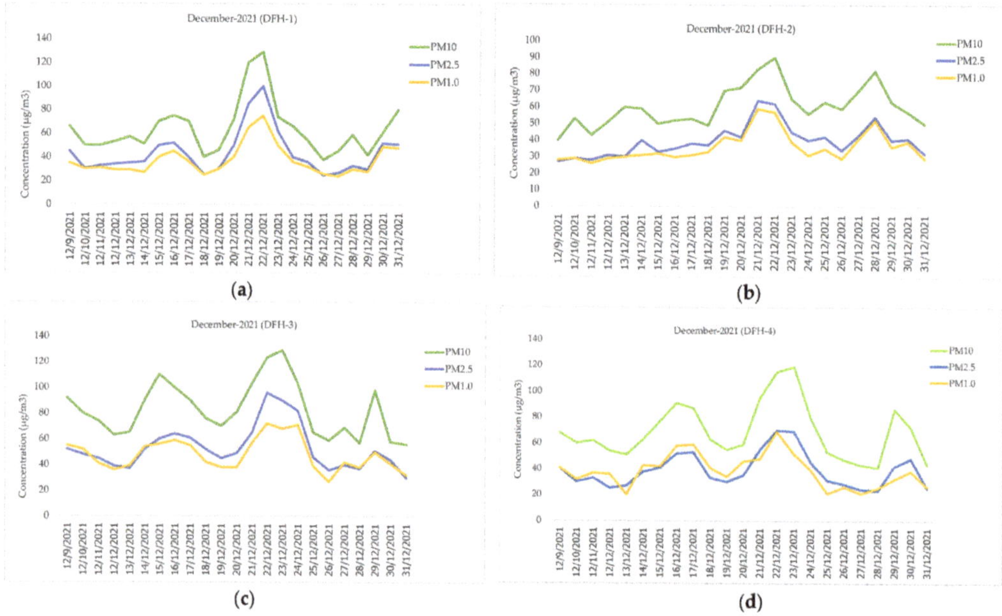

Figure 6. PM_{10}, $PM_{2.5}$, and $PM_{1.0}$ (μg/m^3) measured from LCSs during the period from 9 to 31 December 2021: (**a**) DFH-1, (**b**) DFH-2, (**c**) DFH-3, and (**d**) DFH-4.

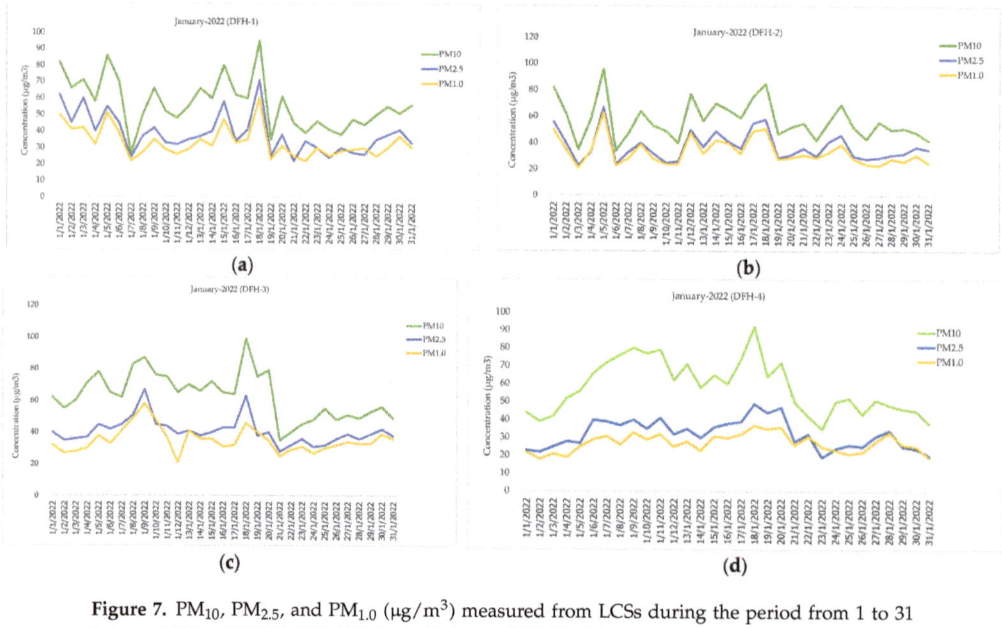

Figure 7. PM_{10}, $PM_{2.5}$, and $PM_{1.0}$ (μg/m^3) measured from LCSs during the period from 1 to 31 January 2022: (**a**) DFH-1, (**b**) DFH-2, (**c**) DFH-3, and (**d**) DFH-4.

Figure 8. PM_{10}, $PM_{2.5}$, and $PM_{1.0}$ ($\mu g/m^3$) measured from LCSs during the period from 1 to 28 February 2022: (**a**) DFH-1, (**b**) DFH-2, (**c**) DFH-3, and (**d**) DFH-4.

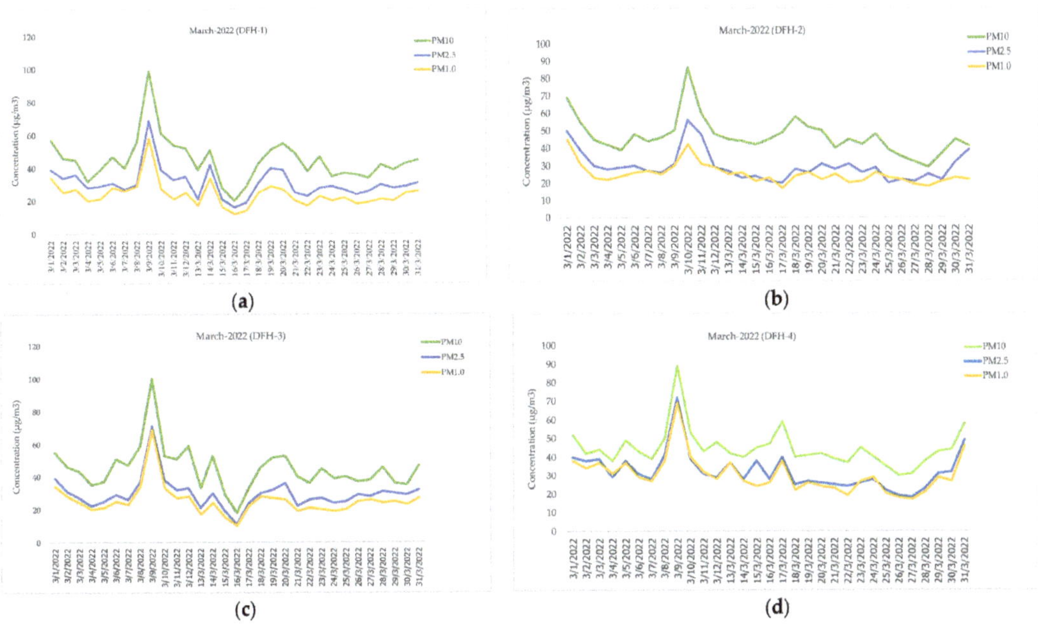

Figure 9. PM_{10}, $PM_{2.5}$, and $PM_{1.0}$ ($\mu g/m^3$) measured from LCSs during the period from 1 to 31 March 2022: (**a**) DFH-1, (**b**) DFH-2, (**c**) DFH-3, and (**d**) DFH-4.

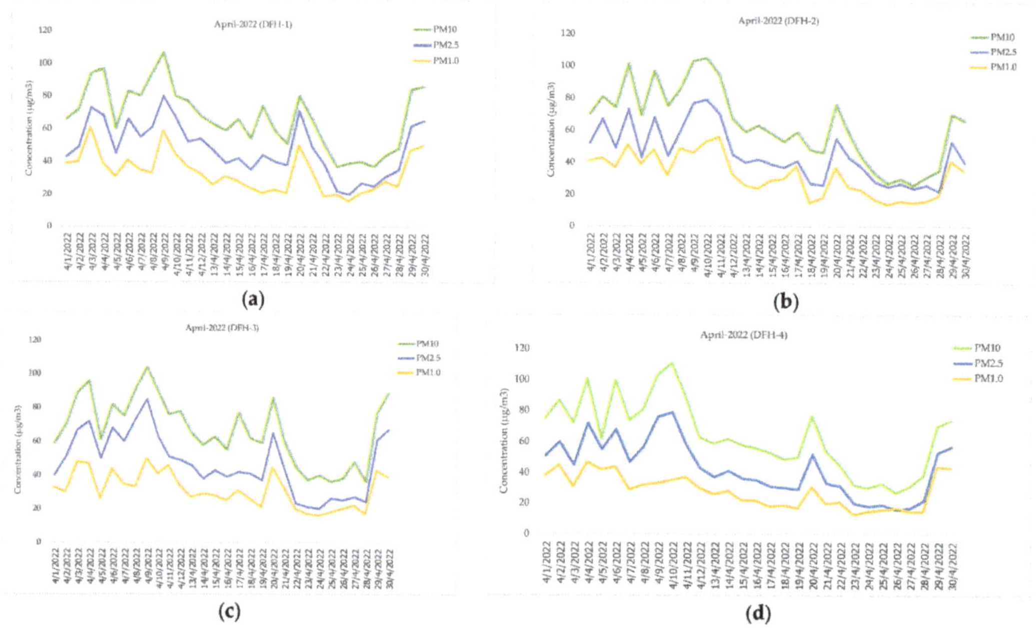

Figure 10. PM_{10}, $PM_{2.5}$, and $PM_{1.0}$ ($\mu g/m^3$) measured from LCSs during the period from 1 to 30 April 2022: (**a**) DFH-1, (**b**) DFH-2, (**c**) DFH-3, and (**d**) DFH-4.

3.3. Highlighted Mass Concentration of Size Distribution

The particle size distributions in the Samutprakarn province are presented in Figures 6–10. The highest mass concentration was PM_{10}, in all monitoring sites, followed by $PM_{2.5}$ and $PM_{1.0}$, respectively. LCSs have been widely used for fine particle analysis in many regions in South-East Asia over the past decade; LCSs provide low-cost, small size capability, easily used instrumentation, and PM mass data. The daily average concentrations of $PM_{2.5}$ at the DFH-1 were 48 ± 16 $\mu g/m^3$ (April 2022) and 44 ± 19 $\mu g/m^3$ (December 2021) as shown in Table 3. The daily average concentrations $PM_{2.5}$ at DFH-2 were 45 ± 17 $\mu g/m^3$ (April 2022) and 42 ± 15 $\mu g/m^3$ (February 2022). The corresponding values at the DFH-3 were 53 ± 17 $\mu g/m^3$ (December 2022) and 47 ± 15 $\mu g/m^3$ (April 2022). The average daily concentrations of $PM_{2.5}$ at the DFH-4 were 43 ± 18 $\mu g/m^3$ (April 2022) and 39 ± 14 $\mu g/m^3$ (December 2021), respectively. The average concentration of $PM_{2.5}$ in Thailand was attributed to a high level in the dry season. Moreover, Chunitiphisan et al. [44] revealed that the ambient $PM_{2.5}$ in Northern Thailand had considerably increased, and the mass level in the wet season was found to be lower than that in the dry season at the sampling place [33,64].

In this work, the study period was designed to be during the haze season in Thailand to estimate the air quality data and sensor performance in different urban environments, and the LCS were placed at four locations in Samutprakarn. The differences in the PM_{10}, $PM_{2.5}$, and $PM_{1.0}$ concentrations from LCS potentially depended on the location performance. The results from the fine particle in this study showed a high concentration in December at a residential area, and its value was higher than some reports in Table 3. Low levels of $PM_{2.5}$ were found in the DustBoy sensor when compared with the results from this study [45]. However, research reports from Myanmar and Vietnam showed higher results than our study. The $PM_{2.5}$ concentrations in Myanmar in Yangon city were higher in morning (91 ± 37 $\mu g/m^3$) and lower in evening (60 ± 22 $\mu g/m^3$) [2]; moreover, the $PM_{2.5}$ concentrations in Mandalay city were higher in the summer (94 ± 10 $\mu g/m^3$) than in winter (53 ± 2 $\mu g/m^3$) [48]. The sensors from Myanmar were calibrated before testing (Pocket $PM_{2.5}$ Sensor and AS-LUNG-O). The $PM_{2.5}$ concentrations reported from Vietnam were the average values. At the first site, Panasonic $PM_{2.5}$ sensors were used in Hanoi and Thai Nguyen Province, and the $PM_{2.5}$ concentrations were measured hourly for three sites and were determined to be 58, 55, and 54 µg/m^3 [51]. At the second location, low levels of $PM_{2.5}$ were observed in Ho Chi Minh City, Vietnam (Plantower PMS 3003); the average concentration data of $PM_{2.5}$ obtained from the two sensors were 34 µg/m^3 for sensor 1 and 34 µg/m^3 for sensor 2 [65].

Table 3. Level of $PM_{2.5}$ from LCS at different locations in various study.

Studied Location, Country	Characteristic	$PM_{2.5}$ Level (µg/m^3)	Sensor Device	Reference
Samutprakarn, Thailand	Residential Semi-residential Commercial Residential	48 ± 16 45 ± 17 53 ± 17 43 ± 18	PMS7003	This study
Bangkok, capital of Thailand	Urban	36 ± 12	DustBoy	[45]
Nan, Northern Thailand	Urban	<5–37 (flight track)	Plantower PMS 3003 (on Drone)	[44]
Bangkok, capital of Thailand Chiangmai, Northern Thailand Ubon Ratchathani, Northeast Thailand	Urban	27 ± 18 (dry season) 14 ± 11 (wet season) 41 ± 29 (dry season) 11 ± 8 (wet season) 39 ± 27 (dry season) 18 ± 16 (wet season)	PMS7003, Plantower	[64]
Yangon, Myanmar	Urban	Kamayut, Morning 91 ± 37 Evening 60 ± 22	Pocket $PM_{2.5}$ Sensor	[2]
Mandalay, Myanmar	Urban	Summer 94 ± 10 Winter 53 ± 2	AS-LUNG-O	[57]
Taipei, Taiwan	Urban	Location A 17 ± 9 Location B 11 ± 4	AS-LUNG-O	[7]
Taipei, Taiwan	Urban (households)	18 ± 11	AS-LUNG-O	[66–68]
Vietnam Hanoi and Thai Nguyen Province	Urban	Hourly: three sites, 58, 55, and 54	Panasonic $PM_{2.5}$ sensors	[51]
Ho Chi Minh City, Vietnam	Urban	Sensor 1: 34 Sensor 2: 34	Plantower PMS 3003	[65]
Jakarta, Indonesia	Urban	50–65	Edimax AirBox AI-1001W V3	[53]
Košice, Slovak republic	Urban and rural	20–80	SPS30 SEN54	[3]
U.S.–Mexico border	Urban Residential Residential household	9 7 7	BlueSky™ Air Quality Monitors, TSI Inc. Shoreview, MN, USA (Model:8143)	[6]

3.4. Temperature and RH Sensor

In this study, the air temperature and RH were monitored using the LCSs. Air temperature performance of LCSs is shown in Figure 11; good agreement was observed between our sensor and TMD. The RH from the LCSs was consistently lower than the results from TMD. The LCS testing occurred during December 2021 (1–8 December 2021). Figure 11a shows the air temperature data. The trend of air temperature from all LCSs was similar to the recoded values from the TMD. The results from the temperature sensor at the DFH-1 (26 °C) site were similar to the DFH-2 site (27 °C). The average air temperature at the DFH-3 and the DFH-4 site showed a general temperature of ~26 °C. Additionally, the RH from our sensor (DFH-1 = 59%; DFH-2 = 60%; DFH-3 = 64%; DFH-4 = 63%) showed lower values than TMD (84%) for all sites. Our sensor devices for air quality data contained temperature and RH sensors. The DHT22 sensor is used for temperature and RH measurement [44]. Figures 12–16 display the study results of temperature and RH over the studied time intervals using LCS. The average air temperature of the LCS ranged between 26 °C and 31 °C, with a RH of 65 to 78%. Bulot et al. [62] observed a strong correlation between PMS7003 and PMS5003 and the background station when the RH was similar (76–98% RH) to the reference device. Moreover, the air temperature and RH can influence PM concentrations. Ly et al. [51] reported that the $PM_{2.5}$ level was associated with meteorological factors, such as wind speed, wind direction, air temperature, RH, and atmospheric pressure. Air temperature had a strong effect on $PM_{2.5}$ in the winter and had a weak effect on $PM_{2.5}$ in summer [33]. Dejchanchaiwong et al. [64] found that low PM concentrations and high RH levels had a substantial impact on LCS performance. Specifically, during the wet season, LCS showed higher relative inaccuracy than during the dry season. In our results, PM mass concentrations were raised at low air temperatures utilizing LCS from December 2021 to February 2022, as shown in Figures 12 and 13. It was found that weather parameters in this period could cause sufficient variations in temperature, like temperature inversion, to reduce the height of the atmospheric boundary layer and dust diffusion, therefore increasing dust concentration. There is also extensive biomass burning from agricultural burning in preparation for the rice planting season in the winter, along with the long-range transport of air pollutants from neighboring countries such as Laos, Vietnam, and Myanmar [61,62]. In comparison, the lowest RH levels were reported in December 2021 and January 2022.

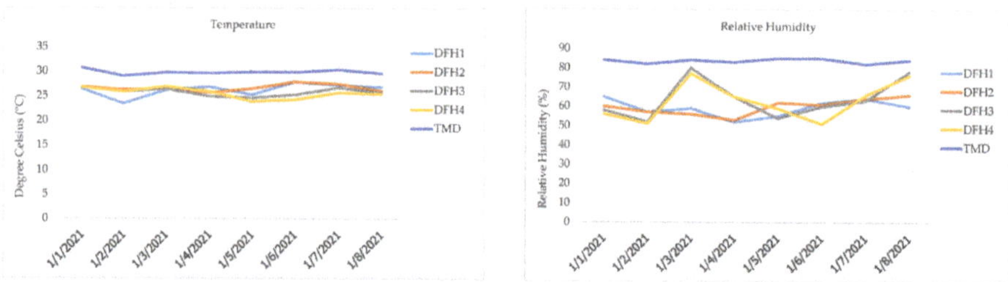

(a) (b)

Figure 11. The comparison of weather from LCSs and TMD: (**a**) air temperature and (**b**) RH.

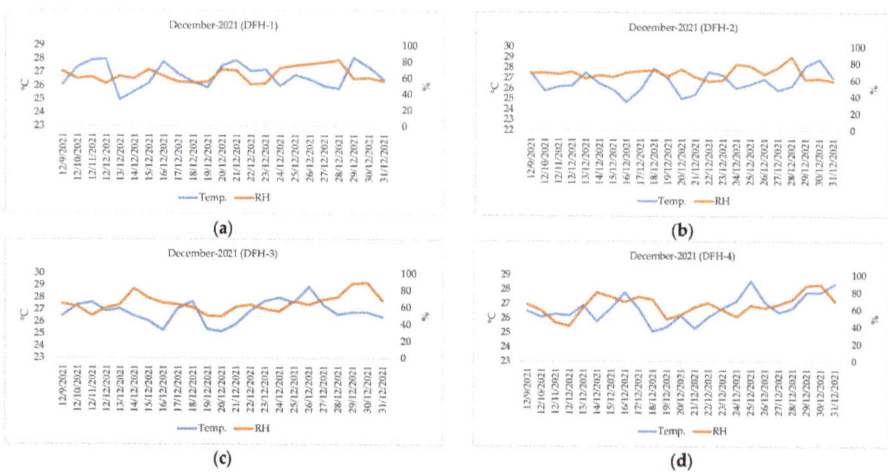

Figure 12. Air temperature (°C) and RH (%) measured from LCSs during the period from 9 to 31 December 2021: (**a**) DFH-1, (**b**) DFH-2, (**c**) DFH-3, and (**d**) DFH-4.

Figure 13. Air temperature (°C) and RH (%) measured from LCSs during the period from 1 to 31 January 2022: (**a**) DFH-1, (**b**) DFH-2, (**c**) DFH-3, and (**d**) DFH-4.

Figure 14. Air temperature (°C) and RH (%) measured from LCSs during the period from 1 to 28 February 2022: (**a**) DFH-1, (**b**) DFH-2, (**c**) DFH-3, and (**d**) DFH-4.

Figure 15. Air temperature (°C) and RH (%) measured from LCSs during the period from 1 to 31 March 2022: (**a**) DFH-1, (**b**) DFH-2, (**c**) DFH-3, and (**d**) DFH-4.

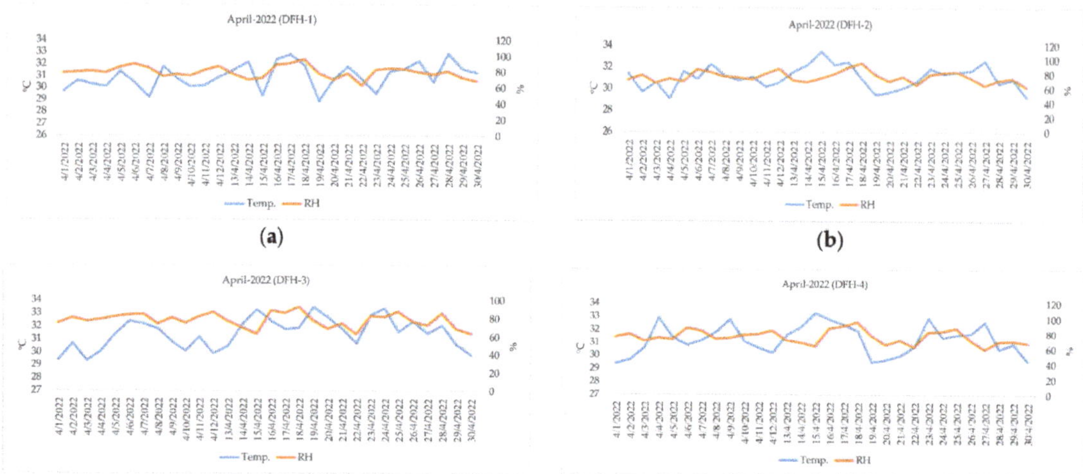

(**a**) (**b**) (**c**) (**d**)

Figure 16. Air temperature (°C) and RH (%) measured from LCSs during the period from 1 to 30 April 2022: (**a**) DFH-1, (**b**) DFH-2, (**c**) DFH-3, and (**d**) DFH-4.

3.5. Air Quality Index (AQI)

The air quality index (AQI) data used to report pollution concentration in Thailand was calculated from the LCSs in all sites; their values and color code of each month are shown in Figure 17. Daily PM_{10} concentrations (calculated from an average over 24-h) at all LCSs were selected to calculate the AQI equation because PM_{10} had higher values than the other particle sizes. Based on our results, PM_{10} played a significant role in the AQI (Figure 17a–e). Since the $PM_{2.5}$ and $PM_{1.0}$ levels were lower than the PM_{10} level, our results could not indicate their hazard level. Different pollutants have different toxicities and health risks; consequently, the AQI is predicted from estimates of the pollutants CO, NO_2, O_3, SO_2, PM_{10}, and $PM_{2.5}$ and is calculated from the average concentrations of the six pollutants. Notably, the fine particles are thought to have a greater influence on the human body because they can penetrate deeper into the human respiratory system. The complete data were imported to GIS and then created in map form. The PM mass concentration was higher in December to February than March–April. As a result, the PM mass concentrations for AQI were estimated using the Thailand AQI standard equation. A pattern of geographic contribution and large variation was shown for the various monitoring sites. The AQI values of Samutprakarn province were visualized via PM mapping. This research revealed that the months of December through February were yellow and green, respectively, while March and April were blue and green at our monitoring sites. The AQI values showed good to moderate results. During our study period, the results demonstrated satisfactory air quality from January to April 2022, as indicated by blue (Figure 17d) and green (Figure 17e), respectively. However, the DFH3 showed higher levels of pollutants, and the pollutants were likely from an anthropogenic contribution since this site was a residential area, as shown in Figure 17a).

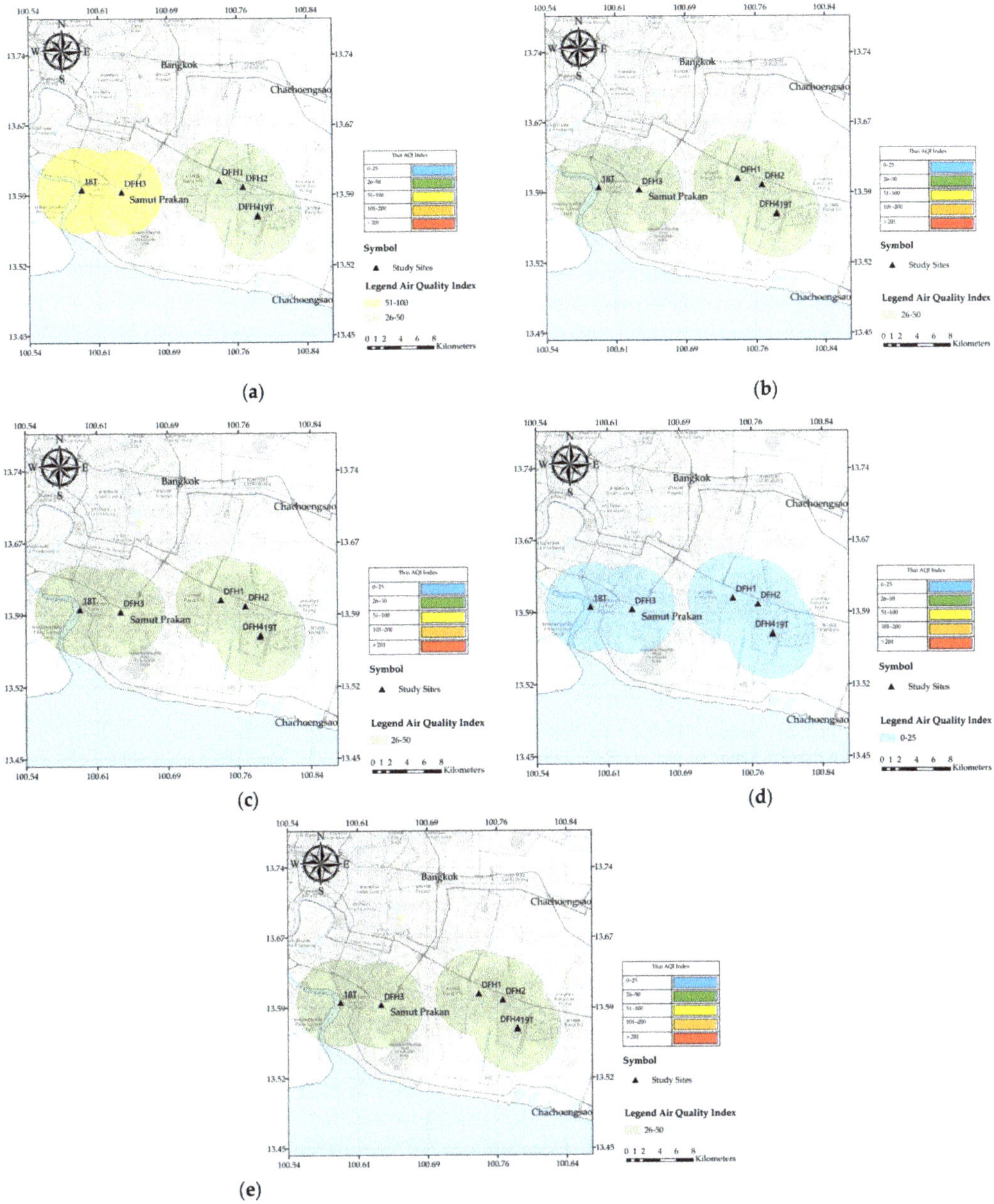

Figure 17. AQI mapping: (**a**) December 2021; (**b**) January 2022; (**c**) February 2022; (**d**) March 2022; (**e**) April 2022.

4. Conclusions

The concept of LCS for monitoring particles in urban ambient air that includes the size range PM_{10}, $PM_{2.5}$, and $PM_{1.0}$ was presented in this paper. The LCSs at all four (DFH1, DFH2, DFH3, DFH4) monitoring sites in urban areas detected PM_{10}, $PM_{2.5}$, and $PM_{1.0}$ during the monitoring periods. Coarse (PM_{10}) and fine particles ($PM_{2.5}$, $PM_{1.0}$) were collected from the LCSs and compared to a TEOM monitor; a TEOM is a reference instrument approved by the Pollution Control Department, Thailand. We have provided evidence that the LCSs (PMS7003) can detect PM_{10}, $PM_{2.5}$, and also $PM_{1.0}$. Most PM concentrations from the LCSs were lower than those from the TEOM. The LCSs were employed continually in an urban area of Samutprakarn Province. Data were collected over a monitoring period of five months (of course a full year would be ideal). Furthermore, temperature and RH sensor were also measured using the DHT22 sensor. Temperature and RH trends from all LCSs showed less sensitivity and accuracy than the TMD.

The observation results revealed a high level of PMs at DFH3, a residential site, while other locations had a lower level of PMs. These results potentially indicated that the level of PMs was affected by the site features. We understand that our study has limitations. Only a few minor sites were chosen for ambient air sampling and monitoring using LCSs. Furthermore, weather conditions had a substantial impact on PM levels. The observation results of PM mass concentration were higher in December to February than March to April. As a result, the PM mass concentrations for AQI were estimated using the Thailand AQI standard equation. A pattern of geographic contribution and large variation was shown for the various monitoring sites. The AQI values of Samutprakarn province were visualized via PM mapping. This research revealed that the months of December through February were yellow and green, respectively, while March and April were blue and green at our monitoring sites. The AQI values revealed good to moderate findings, indicating that there is no health impact at the study sites. Finally, our findings proved the ability of our LCSs to quantify ambient particles and their spatial distribution.

Supplementary Materials: The following supporting information can be downloaded at: https://www.mdpi.com/article/10.3390/atmos15030336/s1, Figure S1: TEOM at air monitoring station of Pollution Control Department (PCD) (18T); Figure S2: TEOM at air monitoring station of Pollution Control Department (PCD) (19T); Table S1: List of acronyms used in study; Table S2: Specific of Tapered Element Oscillating Microbalance (TEOM) ambient particulate monitor.

Author Contributions: S.R., conceptualization, data collection, investigation, methodology, formal analysis, visualization, writing—original draft, writing—review and editing; S.C., conceptualization, data collection, methodology, visualization; S.J., investigation, visualization, writing—original draft, writing—review and editing; P.T., data curation, formal analysis, writing—review and editing; W.P., writing—review and editing; T.C., writing—review and editing; Y.B., conceptualization, data collection, investigation, methodology, formal analysis, visualization, writing—original draft, writing—review and editing. All authors have read and agreed to the published version of the manuscript.

Funding: This research received no external funding.

Institutional Review Board Statement: Not applicable.

Informed Consent Statement: Not applicable.

Data Availability Statement: The data presented in this study are available in file of Supplementary Materials.

Acknowledgments: The authors would like to thank the house owner for the use of their property for research purposes. Furthermore, this study was supported equipment by the Sustainable Development Technology, Faculty of Science and Technology, Thammasat University.

Conflicts of Interest: The authors declare no conflicts of interest.

References

1. Su, Y.; Sofowote, U.; Debosz, J.; White, L.; Munoz, A. Multi-year continuous $PM_{2.5}$ measurements with the federal equivalent method SHARP 5030 and comparisons to filter-based and TEOM measurements in Ontario, Canada. *Atmosphere* **2018**, *9*, 191. [CrossRef]
2. Yi, E.E.P.N.; Nway, N.C.; Aung, W.Y.; Thant, Z.; Wai, T.H.; Hlain, K.K.; Maung, C.; Yagishita, M.; Ishigaki, Y.; Win-Shwe, T.T.; et al. Preliminary monitoring of concentration of particulate matter ($PM_{2.5}$) in seven townships of Yangon City, Myanmar. *Environ. Health Prev. Med.* **2018**, *23*, 53. [CrossRef]

3. Kirešová, S.; Guzan, M.; Sobota, B. Using low-cost sensors for measuring and monitoring particulate matter with a focus on fine and ultrafine particles. *Atmosphere* **2023**, *14*, 324. [CrossRef]
4. Zhang, T.; Su, Y.; Debosz, J.; Michael, N.; Noble, M.; Munoz, A.; Xu, X. Continuous measurements and source apportionment of ambient $PM_{2.5}$-bound elements in Windsor, Canada. *Atmosphere* **2022**, *13*, 374. [CrossRef]
5. Mukherjee, A.; Brown, S.G.; McCarthy, M.C.; Pavlovic, N.R.; Stanton, L.G.; Snyder, J.L.; D'Andrea, S.D.; Hafner, H.R. Measuring Spatial and Temporal $PM_{2.5}$ Variations in Sacramento, California, Communities Using a Network of Low-Cost. *Sensors* **2019**, *19*, 4701. [CrossRef]
6. Mendez, E.; Temby, O.; Wladyka, D.; Sepielak, K.; Raysoni, A.U. Using Low-Cost Sensors to Assess $PM_{2.5}$ concentrations at four South Texan cities on the U.S.-Mexico Border. *Atmosphere* **2022**, *13*, 1554. [CrossRef]
7. Wang, W.C.V.; Lin, T.H.; Liu, C.H.; Su, C.W.; Lung, S.C.C. Fusion of environmental sensing on $PM_{2.5}$ and deep learning on vehicle detecting for acquiring roadside $PM_{2.5}$ concentration increments. *Sensors* **2020**, *20*, 4679. [CrossRef] [PubMed]
8. Zamora, M.L.; Rice, J.; Koehler, K. One year evaluation of three Low-cost $PM_{2.5}$ monitors. *Atmos. Environ.* **2020**, *235*, 117615. [CrossRef] [PubMed]
9. Ahmad, M.; Manjantrarat, T.; Rattanawongsa, W.; Muensri, P.; Saenmuangchin, R.; Klamchuen, A.; Aueviriyavit, S.; Sukrak, K.; Kangwansupamonkon, W.; Panyametheekul, S. Chemical composition, sources, and health risk assessment of $PM_{2.5}$ and PM_{10} in urban sites of Bangkok, Thailand. *Int. J. Environ. Res. Public Health* **2022**, *19*, 14281. [CrossRef] [PubMed]
10. Kanchanasuta, S.; Sooktawee, S.; Patpai, A.; Vatanasomboon, P. Temporal variations and potential source areas of fine particulate matter in Bangkok, Thailand. *Air Soil Water Res.* **2020**, *13*, 1–10. [CrossRef]
11. Sirithian, D.; Thanatrakolsri, P. Relationships between meteorological and particulate matter concentrations ($PM_{2.5}$ and PM_{10}) during the haze period in urban and rural areas, Northern Thailand. *Air Soil Water Res.* **2022**, *13*, 1–10. [CrossRef]
12. California Air Resources Board. Inhalable Particulate Matter and Health (PM2.5 and PM10). Available online: https://ww2.arb.ca.gov/resources/inhalable-particulate-matter-and-health (accessed on 28 February 2023).
13. Cohen, A.J.; Brauer, M.; Burnett, R.; Anderson, H.R.; Frostad, J.; Estep, K.; Balakrishnan, K.; Brunekreef, B.; Dandona, L.; Dandona, R.; et al. Estimates and 25-year trends of the global burden of disease attributable to ambient air pollution: An analysis of data from the global burden of diseases study 2015. *Lancet* **2017**, *389*, 1907–1918. [CrossRef]
14. Mueller, W.; Vardoulakis, S.; Steinle, S.; Loh, M.; Johnston, H.J.; Precha, N.; Kliengchuay, W.; Sahanavin, N.; Nakhapakorn, K.; Sillaparassamee, R.; et al. A health impact assessment of long-term exposure to particulate air pollution in Thailand. *Environ. Res. Lett.* **2021**, *16*, 055018. [CrossRef]
15. Xing, F.Y.F.; Xu, Y.H.; Shi, M.H.; Lian, Y.X. The impact of $PM_{2.5}$ on the human respiratory system. Review Article on Pollution Haze. *J. Thorac. Dis.* **2016**, *8*, E69–E74.
16. Sakunkoo, P.; Thonglua, T.; Sangkham, S.; Jirapornkul, C.; Limmongkon, R.; Daduang, S.; Tessiri, T.; Rayubkul, J.; Thongtip, S.; Maneenin, N.; et al. Human health risk assessment of $PM_{2.5}$-bound heavy metal of anthropogenic sources in the Khon Kaen Province of Northeast Thailand. *Heliyon* **2022**, *8*, e09572. [CrossRef]
17. Tultrairatana, S.; Phansuea, P. Symptoms related to air pollution, mask-wearing and associated factors: A cross-sectional study among OPD pollution clinic patients in Bangkok, Thailand. *J. Health Res.* **2022**, *36*, 1058–1067. [CrossRef]
18. Lin, H.; Liu, T.; Xiao, J.; Zeng, W.; Li, X.; Guo, L.; Xu, Y.; Zhang, Y.; Vaughn, M.G.; Nelson, E.J.; et al. Quantifying short-term and long-term health benefits of attaining ambient fine particulate pollution standards in Guangzhou, China. *Atmos. Environ.* **2016**, *137*, 38–44. [CrossRef]
19. Guo, Y.; Li, S.; Tawatsupa, B.; Punnasiri, K.; Jaakkola, J.J.K.; Williams, G. The association between air pollution and mortality in Thailand. *Sci. Rep.* **2014**, *4*, 5509. [CrossRef]
20. Phosri, A.; Sihabut, T.; Jaikanlaya, C. Temporal variations of short-term effects of particulate matter on hospital admissions in the most densely populated city in Thailand. *Sci. Total Environ.* **2020**, *742*, 140651. [CrossRef]
21. Punsompong, P.; Pani, S.K.; Wang, S.H.; Pham, T.H.B. Assessment of biomass-burning types and transport over Thailand and the associated health risks. *Atmos. Environ.* **2021**, *247*, 118176. [CrossRef]
22. Paoin, K.; Ueda, K.; Ingviya, T.; Buya, S.; Phosri, A.; Seposo, X.T.; Seubsman, S.; Kelly, M.; Sleigh, A.; Honda, A.; et al. Long-term air pollution exposure and self-reported morbidity: A longitudinal analysis from the Thai cohort study (TCS). *Environ. Res.* **2021**, *192*, 110330. [CrossRef] [PubMed]
23. Dejchanchaiwong, R.; Tekasakul, P. Effects of coronavirus induced City Lockdown on $PM_{2.5}$ and gaseous pollutant concentrations in Bangkok. *Aerosol Air Qual. Res.* **2020**, *21*, 200418. [CrossRef]
24. Chu, B.; Zhang, S.; Liu, J.; Ma, Q.; He, H. Significant concurrent decrease in $PM_{2.5}$ and NO_2 concentrations in China during COVID-19 epidemic. *J. Environ. Sci* **2021**, *99*, 346–353. [CrossRef] [PubMed]
25. Niu, H.; Zhang, C.; Hu, W.; Hu, T.; Wu, C.; Hu, S.; Silva, L.F.O.; Gao, N.; Bao, X.; Fan, J. Air quality changes during COVID-19 Lockdown in an Industrial City in North China: Post-Pandemic Proposals for Air Quality Improvement. *Sustainability* **2022**, *14*, 11531. [CrossRef]
26. Zheng, H.; Kong, S.; Chen, N.; Yan, Y.; Liu, D.; Zhu, B.; Xu, K.; Cao, W.; Ding, Q.; Lan, B.; et al. Significant changes in the chemical compositions and sources of $PM_{2.5}$ in Wuhan since the city lockdown as COVID-19. *Sci. Total Environ.* **2020**, *739*, 140000. [CrossRef] [PubMed]
27. Kanniah, K.D.; Zaman, N.A.F.K.; Kaskaoutis, D.G.; Latif, M.T. COVID-19's impact on the atmospheric environment in the Southeast Asia region. *Sci. Total Environ.* **2020**, *736*, 139658. [CrossRef]

28. He, G.; Pan, Y.; Tanaka, T. The short-term impacts of COVID-19 lockdown on urban air pollution in China. *Nat. Sustain.* **2020**, *3*, 1005–1011. [CrossRef]
29. Yang, Q.; Wang, B.; Wang, Y.; Yuan, Q.; Jin, C.; Wang, J.; Li, S.; Li, M.; Li, T.; Liu, S.; et al. Global air quality change during COVID-19: A synthetic analysis of satellite, reanalysis and ground station data. *Environ. Res. Lett.* **2021**, *16*, 074052. [CrossRef]
30. Suleiman, A.; Tight, M.R.; Quinn, A.D. Assessment and prediction of the impact of road transport on ambient concentrations of particulate matter PM_{10}. Transp. *Res. Part D Transp. Environ.* **2016**, *49*, 301–312. [CrossRef]
31. Supasri, T.; Intra, P.; Jomjunyong, S.; Sampattagul, S. Evaluation of particulate matter concentration by using a wireless sensor system for continuous monitoring of particulate Air Pollution in Northern of Thailand. *J. Innov. Technol. Res.* **2018**, *2*, 59–78. (In Thai)
32. Bran, S.H.; Macatangay, R.; Surapipith, V.; Chotamonsak, C.; Chantara, S.; Han, Z.; Li, J. Surface $PM_{2.5}$ mass concentrations during the dry season over northern Thailand: Sensitivity to model aerosol chemical schemes and the effects on regional meteorology. *Atmos. Res.* **2022**, *277*, 106303. [CrossRef]
33. Chirasophon, S.; Pochanart, P. The long-term characteristics of PM_{10} and $PM_{2.5}$ in Bangkok, Thailand. *Asian J. Atmos. Environ.* **2020**, *14*, 73–83. [CrossRef]
34. Uttamang, P.; Aneja, V.P.; Hanna, A.F. Assessment of gaseous criteria pollutants in the Bangkok metropolitan region, Thailand. *Atmos. Chem. Phys.* **2018**, *18*, 12581–12593. [CrossRef]
35. Narita, D.; Oanh, N.T.K.; Sato, K. Pollution characteristics and policy actions on fine particulate matter in a growing asian economy: The case of Bangkok metropolitan region. *Atmosphere* **2019**, *10*, 227. [CrossRef]
36. Pollution Control Department (PCD). *Booklet on Thailand State of Pollution 2018*; Pollution Control Department, Ministry of Natural Resources and Environment: Bangkok, Thailand, 2019. Available online: http://www.oic.go.th/FILEWEB/CABINFOCENTER3/DRAWER056/GENERAL/DATA0001/00001462.PDF (accessed on 6 March 2022).
37. Pollution Control Department (PCD), Thailand. Thailand Air Quality Standard. Available online: http://air4thai.pcd.go.th/webV3/#/Home (accessed on 15 January 2024).
38. United States Environmental Protection Agency (U.S. EPA). Standard Operating Procedure for the Continuous Measurement of Particulate Matter. Available online: https://www.epa.gov/sites/default/files/2021-03/documents/905505_teom_sop_draft_final_sept09.pdf (accessed on 15 January 2024).
39. United States Environmental Protection Agency (U.S. EPA). *List of Designated Reference and Equivalent Methods*; U.S. Environmental Protection Agency: Washington, DC, USA, 2017.
40. United States Environmental Protection Agency (U.S. EPA). *Performance Testing Protocols, Metrics, and Target Values for Fine Particulate Matter Air Sensor*; U.S. Environmental Protection Agency: Washington, DC, USA, 2021.
41. Patashnick, H.; Rupprecht, E.G. Continuous PM-10 Measurements using the Tapered Element Oscillating Microbalance. *Air Waste. Manag. Assoc.* **1991**, *41*, 1079–1083. [CrossRef]
42. Cyrys, J.; Dietrich, G.; Kreyling, W.; Tuch, T.; Heinrich, J. PM2.5 measurements in ambient aerosol: Comparison between Harvard impactor (HI) and the tapered element oscillating microbalance (TEOM) system. *Sci. Total Environ.* **2001**, *278*, 191–197. [CrossRef] [PubMed]
43. Kanabkaew, T.; Mekbungwan, P.; Raksakietisak, S.; Kanchanasut, K. Detection of $PM_{2.5}$ plume movement from IoT ground level monitoring data. *Environ. Pollut.* **2019**, *252*, 543–552. [CrossRef]
44. Chunitiphisan, S.; Sirima, P.; Suree, P.; Garavig, T.; Thawat, N. Particulate matter monitoring using inexpensive sensors and internet GIS: A case study in Nan, Thailand. *Eng. J.* **2019**, *22*, 25–37. [CrossRef]
45. Yawootti, A.; Wimonthanasit, P.; ChaithanuSate, K.; Sampattagul, S. Comparison of Particulate Matter Monitoring Using Beta Attenuation Monitor and Light Scattering Method in Bangkok Thailand. In Proceedings of the 33rd International Conference on Circuits/Systems, Computers and Communications (ITC-CSCC), Bangkok, Thailand, 4–7 July 2018; pp. 498–501.
46. Sompornrattanaphan, M.; Thongngarm, T.; Tantilipikorn, P.; Kreetapirom, P.; Foo, J. The Contribution of outdoor fine particulate matter to indoor air quality in Bangkok metropolitan region, Thailand-Are indoor dwellers safe? *Siriraj Med. J.* **2018**, *70*, 4.
47. Levy Zamora, M.; Xiong, F.; Gentner, D.; Kerkez, B.; Kohrman-Glaser, J.; Koehler, K. Field and laboratory evaluations of the low-cost plantower particulate matter sensor. *Environ. Sci. Technol.* **2019**, *53*, 838–849. [CrossRef]
48. Kim, D.; Shin, D.; Hwang, J. Calibration of low-cost sensors for measurement of indoor particulate matter concentrations via laboratory/Field evaluation. *Aerosol Air Qual. Res.* **2023**, *23*, 230097. [CrossRef]
49. Danek, T.; Weglinska, E.; Zareba, M. The influence of meteorological factors and terrain on air pollution concentration and migration: A geostatistical case study from Krakow, Poland. *Sci. Rep.* **2022**, *12*, 11050. [CrossRef] [PubMed]
50. Ly, B.T.; Matsumi, Y.; Nakayama, T.; Sakamoto, Y.; Kajii, Y.; Nghiem, T.D. Characterizing $PM_{2.5}$ in hanoi with new high temporal resolution sensor. *Aerosol Air Qual. Res.* **2018**, *18*, 2487–2497. [CrossRef]
51. Ly, B.-T.; Matsumi, Y.; Vu, T.V.; Sekiguchi, K.; Nguyen, T.T.; Pham, C.T.; Nghiem, T.D.; Ngo, I.-H.; Kurotsuchi, Y.; Nguyen, T.-H.; et al. The effects of meteorological conditions and long-range transport on $PM_{2.5}$ levels in Hanoi revealed from multi-site measurement using compact sensors and machine learning approach. *J. Aerosol Sci.* **2021**, *152*, 105716. [CrossRef]
52. Nadzir1, M.S.M.; Ooi, M.C.G.; Alhasa, K.M.; Bakar, M.A.A.; Mohtar, A.A.A.; Nor, M.F.F.M.; Latif, M.T.; Hamid, H.H.A.; Ali, S.H.M.; Ariff, N.M.; et al. The Impact of movement control order (MCO) during Pandemic COVID-19 on local air quality in an Urban Area of Klang Valley, Malaysia. *Aerosol Air Qual. Res.* **2020**, *20*, 1237–1248. [CrossRef]

53. Yulinawati, H.; Khairani, T.; Siami, L. Analysis of indoor and outdoor particulate ($PM_{2.5}$) at a women and children's hospital in West Jakarta. *IOP Conf. Ser. Earth Environ. Sci.* **2021**, *737*, 012067. [CrossRef]
54. Ismail, A.H.; Dawi, E.A.; Almokdad, N.; Abdelkader, A.; Salem, O. Estimation and Comparison of the Clearness Index using Mathematical Models-Case study in the United Arab Emirates. *Atmosphere* **2023**, *10*, 863–869. [CrossRef]
55. Zhou, X.; Liu, S.; Shan, Y.; Endo, S.; Xie, Y.; Sengupta, M. Influences of Cloud Microphysics on the Components of Solar Irradiance in the WRF-Solar Model. *Atmosphere* **2024**, *15*, 39. [CrossRef]
56. Al Karim Haj Ismail, A. Prediction of global solar radiation from sunrise duration using regression functions. *Kuwait J. Sci.* **2022**, *3*, 1–8.
57. Ohnmar, M.T.H. Strengthening Evidence for Action: Impacts of Open Waste Burning in Mandalay, Asia Pacific Clean Air Partnership (UN Environment), Newsletter Issue 9. 2020. Available online: https://cleanairsolutions.asia/wp-content/uploads/APCAP-Newsletter-Issue-9-November-2020.pdf (accessed on 6 March 2023).
58. Prakash, J.; Choudhary, S.; Raliya, R.; Chadha, T.; Fang, J.; Biswas, P. PM sensors as an indicator of overall air quality: Pre-COVID and COVID periods. *Atmos. Pollut. Res.* **2022**, *13*, 101594. [CrossRef]
59. Jayaratne, R.; Liu, X.; Ahn, K.H.; Asumadu-Sakyi, A.; Fisher, G.; Gao, J.; Mabon, A.; Mazaheri, M.; Mullins, B.; Nyaku, M.; et al. Low-cost $PM_{2.5}$ sensors: An assessment of their suitability for various applications. *Aerosol Air Qual. Res.* **2020**, *20*, 520–532. [CrossRef]
60. Department of Provincial Administration (DOPA). Registration Statistics. Available online: https://www.bora.dopa.go.th/ (accessed on 2 March 2023).
61. Pollution Control Department (PCD), Thailand. Thailand Air Quality Standard. Available online: http://air4thai.pcd.go.th/webV3/#/AQIInfo (accessed on 6 March 2022).
62. Bulot, F.M.J.; Johnston, S.J.; Basford, P.J.; Easton, N.H.C.; Apetroaie-Cristea, M.; Foster, G.L.; Morris, A.K.R.; Cox, S.J.; Loxham, M. Long-term field comparison of multiple low-cost particulate matter sensors in an outdoor urban environment. *Sci. Rep.* **2019**, *9*, 7497. [CrossRef] [PubMed]
63. Meteorological Center, Thai Meteorological Department (TMD). Annual Meteorological Statistics in Samutprakarn Province. Available online: https://www.tmd.go.th/weather/province/samut-prakan (accessed on 2 March 2023).
64. Dejchanchaiwong, R.; Tekasakul, P.; Saejio, A.; Liman, T.; Le, T.C.; Tsai, C.J.; Linn, G.Y.; Morris, J. Seasonal field calibration of low-cost $PM_{2.5}$ sensors in different locations with different sources in Thailand. *Atmosphere* **2023**, *14*, 496. [CrossRef]
65. Nguyen, C.D.T.; To, H.T. Evaluating the applicability of a low-cost sensor for measuring $PM_{2.5}$ concentration in Ho Chi Minh city, Viet Nam. *Sci. Technol. Dev. J.* **2019**, *22*, 343–347. [CrossRef]
66. Lung, S.C.C.; Tsou, M.C.M.; Hu, S.C.; Hsieh, Y.H.; Wang, W.C.V.; Shui, C.K.; Tan, C.H. Concurrent assessment of personal, indoor, and outdoor $PM_{2.5}$ and PM_1 levels and source contributions using novel low-cost sensing devices. *Indoor Air* **2021**, *31*, 755–768. [CrossRef]
67. Boongla, Y.; Chanonmuang, P.; Hata, M.; Furuuchi, M.; Phairuang, W. The characteristics of carbonaceous particles down to the nanoparticle range in Rangsit city in the Bangkok Metropolitan Region, Thailand. *Environ. Pollut.* **2021**, *272*, 115940. [CrossRef]
68. Phairuang, W.; Inerb, M.; Furuuchi, M.; Hata, M.; Tekasakul, S.; Tekasakul, P. Size-fractionated carbonaceous aerosols down to $PM_{0.1}$ in southern Thailand: Local and long-range transport effects. *Environ. Pollut.* **2020**, *260*, 114031. [CrossRef] [PubMed]

atmosphere

MDPI

Article

Low-Cost Sensor Monitoring of Air Quality Indicators during Outdoor Renovation Activities around a Dwelling House

László Bencs

Department of Applied and Nonlinear Optics, Institute for Solid State Physics and Optics, HUN-REN Wigner Research Centre for Physics, P.O. Box 49, H-1525 Budapest, Hungary; bencs.laszlo@wigner.hun-ren.hu; Tel.: +36-1-392-2222 (ext. 1684)

Abstract: A couple of air quality (AQ) parameters were monitored with two types of low-cost sensors (LCSs) before, during and after the garden fence rebuilding of a dwelling house, located at the junction of a main road and a side street in a suburban area of Budapest, Hungary. The AQ variables, recorded concurrently indoors and outdoors, were particulate matter (PM_1, $PM_{2.5}$, PM_{10}) and some gaseous trace pollutants, such as CO_2, formaldehyde (HCHO) and volatile organic compounds (VOCs). Medium-size aerosol ($PM_{2.5\text{-}1}$), coarse particulate ($PM_{10\text{-}2.5}$) and indoor-to-outdoor (I/O) ratios were calculated. The I/O ratios showed that indoor fine and medium-size PM was mostly of outdoor origin; its increased levels were observed during the renovation. The related pollution events were characterized by peaks as high as 100, 95 and 37 μg/m^3 for PM_1, $PM_{2.5\text{-}1}$ and $PM_{10\text{-}2.5}$, respectively. Besides the renovation, some indoor sources (e.g., gas-stove cooking) also contributed to the in-house PM_1, $PM_{2.5\text{-}1}$ and $PM_{10\text{-}2.5}$ levels, which peaked as high as 160, 255 and 220 μg/m^3, respectively. In addition, these sources enhanced the indoor levels of CO_2, HCHO and, rarely, VOCs. Increased and highly fluctuating VOC levels were observed in the outdoor air (average: 0.012 mg/m^3), mainly due to the use of paints and thinners during the reconstruction, though the use of a nearby wood stove for heating was an occasional contributing factor. The acquired results show the influence of the fence renovation-related activities on the indoor air quality in terms of aerosols and gaseous components, though to a low extent. The utilization of high-resolution LCS-assisted monitoring of gases and PM_x helped to reveal the changes in several AQ parameters and to assign some dominant emission sources.

Keywords: indoor and outdoor atmosphere; air quality metrics; low-cost environmental monitor; air pollution; garden fence; refurbishment; house dust; ventilation rate; building construction

1. Introduction

Anthropogenic air pollutants have a considerable influence on atmospheric climate [1] and the health of the human population in terms of proven negative effects, such as asthma and COPD [2]. They also have an impact on the preservation of historic buildings and cultural heritage items on display [3]. Thus, accurate monitoring and setting up proper measures of air quality (AQ), as well as the assessment and future predictions of related health effects for the population, are important, especially for locations exposed to a heavy burden of anthropogenic pollution such as, for instance, densely travelled roads [4], industrial areas [5] and sites with building construction/renovation activities, where high levels of dust and various gases/aerosols are emitted into the ambient air.

Concerning the latter topics, studies mostly related to building renovation activities and the accompanying indoor air quality (IAQ) changes have been reported in the literature [6–24]. Several of the research works discussed a worsened IAQ due to refurbishment activities and the so-called sick building syndrome, for instance, in connection with educational buildings [6–11], hospitals [11,12], dwelling houses [11,13–18] and office buildings [11]. The related AQ problems are manifested in terms of increased indoor pollutant levels of CO [6,7], CO_2 [6,7,13,15,16,19], $PM_{2.5}$ [6,16–18], PM_4 [15], PM_{10} [7,9,17,23], ultrafine PM [12,18], total suspended PM [20], formaldehyde (HCHO) [7,14,16,19], BTEX [15], volatile organic compounds (VOCs) [7,8,11,13–17,19], semi-volatile organic compounds (SVOCs) [14], black carbon (BC) [18], water-soluble PM constituents (nitrate, sulphate, chloride, ammonium) [20], toxic metals (Pb [9,20], Cd, Cu, and Zn [9]), asbestos particles [9] and fungi [20]. A couple of the compiled pollution cases are connected to inefficient air exchange in the study area, particularly indoors, before, during and after the renovation, e.g., Refs. [6,13], indicating a higher risk of negative health effects on humans [6–8]. In addition, the accumulation of toxic metals in suspended particulate matter was observed during the renovation of the interior [9,10]. Preschool children were reported to be highly sensitive to household renovation [21,22], while the enhanced level of outdoor air pollution could cause the early development of pneumonia [22] and childhood asthma [23]. Another study reported a high indoor-to-outdoor (I/O) ratio of HCHO in the air, and the wooden furniture as the main indoor source [8]. Increasing the air ventilation rate with mechanical systems and/or using low-emission furnishing was the recommended solution to mend this problem [7]. Microorganisms and carbonaceous aerosols are additional indicators of the deteriorated IAQ, for instance, in kindergartens [24]. In addition, during interior renovation, fungi suspended with the ambient aerosols may also be present indoors, though their contribution was greater before renovation [12]. Whereas ultrafine and fine PM was present at 1.6–3.3-fold higher concentrations during the renovation of the building [12]. Valdoulakis et al. [25] extensively reviewed several key air pollutants affecting the indoor air quality in homes, including their sources and the related health effects on humans.

Research on low-cost sensors (LCSs) applied for indoor and outdoor air monitoring has gained increasing attention in the past decade [26–28]. This is partially due to their light-weight design and ease of use in field campaigns, mostly operating with battery-powered electricity. LCSs are generally less accurate than laboratory/research-grade devices; therefore, they require more frequent calibration and accuracy checks. Therefore, such measurement systems demand scrutiny of analytical performance and instrumental limitations under field and laboratory conditions, preferably against a calibrated, reference instrument, e.g., [29,30].

As one can conclude from the above literature survey, no study has been reported so far on the application of two types of LCSs for the indoor and outdoor air of individual dwelling houses with a garden, subjected to shorter/longer renovation/rebuilding of external items, such as the garden fence. These kinds of refurbishment activities consist of several construction stages, such as the removal of the old fence and the building of a new one, which incorporates technologies that may release large amounts of pollutants. Consequently, in this study, a couple of atmospheric air pollutants were monitored by the assistance of LCSs to gain better insight into the changes in outdoor and indoor air quality before, during and after the reconstruction activities of the garden fence and to predict its possible health effects on the people living around.

2. Materials and Methods

2.1. Description of the Sampling Site and Related Refurbishment Activities

The sampling site for the present investigation was selected at a dwelling house that is typical for the region, located at the junction of a nearby main road and a side street, in a suburban area of Pesterzsébet, a district in Budapest, Hungary. The main features of the local anthropogenic activities of the sampling area of interest are described in detail in Ref. [29]. Originally (before renovation), the garden of the house was separated from the side street by a gate (width: ≈1.0 m) and a porous (≈50% permeability) fence (≈13.3 m long section), built of limestone brick (style: running bond with gaps, respectively, wall width: 0.12 m and capped brick poles of 0.25 × 0.25 m), dating back to the early 1930s. Part of this fence also served as a wall of a small toolshed (≈1.6 m width), built of brick, located in the corner of the garden (length: ≈4.5 m). This wall was part of the railing on the side of the main road. The toolshed and the original (old) fence were about 1.75 m tall, measured at the streetside. An about 40-year-old chain link fence (height: 1.70 m, length: 18.2 m) and a car gate (width: ≈2.7 m) sealed off most of the garden side lying next to the main road. The dwelling house is situated ≈18.5 m off the main road (the closest lane), at the opposite side of the garden. Along the fence, the garden is planted with a couple of grapevines and various kinds of bushes (e.g., mint, cedar, lavender, rosemary) of similar height (1–1.5 m), in rows, with some gaps at the ends. This arrangement provides a low-height barrier against resuspended street dust and other air pollutants, expected to penetrate the garden from the street. In front of these plants, about 40 pine trees of different heights (ranging at 1–9 m) were grown; they served as an additional shield for filtering the resuspended ambient dust and air pollutants. The house was built in the 1930s, using the same type of brick as the garden fence. The house has plastered walls, which were supplied with energy-efficient thermal insulation in 2017. The entry door of the house to the garden is located at the SE side. This type of building is characteristic for the outskirts of the cities and the countryside too; thus, the presented background data could be applied as general models. An overview drawing of the sampling site is depicted in Figure 1.

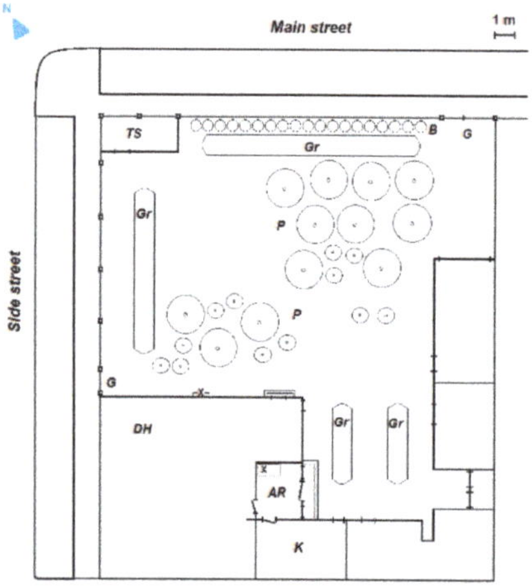

Figure 1. Scheme of the sampling/renovation area with the dwelling house, the garden and the old fence, with capped brick poles (squares); abbreviations: AR—anteroom, B—bushes, DH—dwelling house, G—gate, Gr—grapes, K—kitchen, P—pine trees, TS—toolshed, X—sampling points.

The renewal of the garden fence was performed from 8 August till 11 September 2022, with intermissions on the weekends and on some workdays due to rainy/stormy weather. The masons usually worked on the site from 8 a.m. till 5 p.m., with a couple of short breaks (10–20 min), though in some instances they stayed till 6 p.m. to finalize one-on-one scheduled daily task. The renovation-related activities included the demolition of the former brick/grid fence and the toolshed, the cutting/demounting of the original iron posts and their replacement with new iron poles (e.g., applying an angle grinder with a cutting disc and the use of primer and topcoat paints, with thinners), the use of a temporary fence consisting of iron bars, concrete mixing, the construction of the concrete substructure (total fence plinth height: 0.8 m) along the entire length of the fence (29 m), the laying and concreting of prefabricated fence brick stones, the application of the primer and topcoat paint on the posts and the plinth, as well as setting up one new side gate for the garden. The description of the daily activities of the renovation-related work during the sampling campaign are compiled in Table S1 of the Supplementary File, while the meteorological conditions recorded at a nearby weather station near the sampling site are summarized in Figure S1.

2.2. Instrumentation and Methods

Two different designs of LCSs, i.e., Geekcreit® (China) model PM1.0/PM2.5/PM10, built with a PMS5003 Plantower PM sensor (herein referred to as GPM), and Bohu BH1 Models A3/B3 (Bohu IoT Enterprise, China) based on a Pando Technologies G7 sampler were utilized for the daily monitoring of indoor and outdoor air pollutants (Figure 2). Although both types of sensors were calibrated according to the manufacturer's specification, they were additionally checked against a research-grade apparatus, as described in Refs. [29,30]. The outdoor sensors were deployed on a windowsill of the house, overlooking the garden (NE wall), while the indoor sensors were set up in the anteroom of the house, on a table, about 2.5 m from the entrance door. A living room and the kitchen open into the anteroom. The residents usually open the front door of the house more often in the summer period, due to the hot weather, and for airing the rooms. However, during the daily reconstruction-related activities, it was usually kept closed, similar to the windows of the house, in order to prevent the ingress of dust and aerosols. The ventilation rate of the building was assessed from the exponential decay curves of the ambient CO_2 concentration, recorded in the anteroom during the sampling campaign. This method has already been well described in the literature, e.g., [31]. The calculation resulted in an average ventilation rate of 0.79 h^{-1} for evenings (open windows), which corresponds to an efficient indoor–outdoor air-exchange, while lower rates (0.25–0.45 h^{-1}) were calculated for the daytime and at the end of the campaign (closed windows and doors).

(**a**) (**b**)

Figure 2. Pictures of the GPM and BH1 sensors deployed at the indoor sampling point (anteroom) of the dwelling house (**a**) and the outdoor sampling point (**b**).

The calibration of the BH1 sensors for the monitored gases was performed against analytical grade chemicals (37% HCHO, abs. EtOH, acetone, supplied by Merck) in a smoke chamber, while for CO_2, stoichiometric reactions of commercial chemicals were applied as described in Ref. [32]. The sensitivity for the monitored PM_x species was 1.0 $\mu g/m^3$ for both sensor types. The limit of detection (LOD) of PM_x acquired by the GPM and the BH1 sensors was presented in detail in Refs. [29,30], ranging about 1.5 $\mu g/m^3$. For the gaseous components, the sensitivity of 1.0 ppm CO_2, 0.001 mg/m^3 HCHO and 0.01 mg/m^3 VOCs could be realized with the applied BH1 sensors. The accuracy of the determinations for CO_2, HCHO and VOCs was $\pm$40 ppm, $\pm$0.005 mg/m^3 and $\pm$0.1 mg/m^3, respectively, according to the manufacturers' reports.

2.3. Statistical Methods

For the evaluation of the monitoring data, Microsoft Excel© was applied throughout this study. First, the raw data were filtered for outlier values of PM_x, some of these corresponding to the internal calibration cycles of the LCSs. The monitoring data, collected every second (GPM) and minute (BH1), were averaged to data of 5 min resolution for each LCS. The statistical data of various PM_x and gaseous pollutants were calculated, including minimum, maximum, median, arithmetic mean, first and third quartiles. The average indoor–outdoor (I/O) ratios were calculated for each air pollutant. The Pearson's correlation coefficient (*R*) was calculated for the indoor and outdoor monitoring data to reveal the relationship of the investigated air pollutants and microclimatic variables.

3. Results and Discussion

3.1. Indoor and Outdoor Aerosols, Microclimatic Variables and Their Ratios

3.1.1. Weekly Outdoor Aerosols and Microclimatic Variations

Changes in the weekly average concentrations of various PM_x species and some of the microclimatic conditions were evaluated before (week 1), during (weeks 2–6) and after (weeks 7–8) the fence renovation (Figure 3). As seen, low levels of PM_1, $PM_{2.5\text{-}1}$ and $PM_{10\text{-}2.5}$ could be observed for the week before the start of the renovation activity, with averages of 10.3, 3.0 and 0.6 μg/m^3, respectively. The PM_x did not show any increase on the subsequent week, when renovation started (week 2), resulting in values of 8.1, 2.9 and 0.6 μg/m^3, respectively. However, as compared to the corresponding data of week 1, the peak concentrations of PM_1, $PM_{2.5\text{-}1}$ and $PM_{10\text{-}2.5}$ showed a drastic increase of about 3-, 9- and 10-fold, respectively, during week 2, reaching values as high as 67, 95 and 37 μg/m^3, respectively. This effect could be expected in the renovation area, which was mostly due to the demolition of the old brick fence and the toolshed, and the related emission of a larger amount of coarse and fine dust and aerosols. Nevertheless, the sharp pollution peaks contributed to the weekly average values only to a low extent.

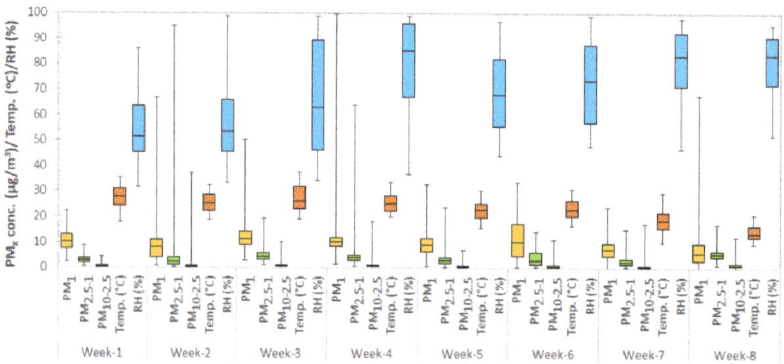

Figure 3. Weekly trends of PM_x and microclimatic data evaluated with 5 min resolution, outdoors at the sampling site with the GPM-sensor; n = 2040 for each week.

As it can be seen in Figure 3, the subsequent period of the renovation (weeks 3–6) was also characterized by increased average levels of PM_1, $PM_{2.5\text{-}1}$ and $PM_{10\text{-}2.5}$, falling within the ranges of 8.9–12, 3.1–4.7 and 0.7–1.5 μg/m^3, respectively, with averages of 10.7, 4.1 and 1.1 μg/m^3, respectively. Peak concentrations of PM_1, $PM_{2.5\text{-}1}$ and $PM_{10\text{-}2.5}$ ranged from 33 to 100; 14 to 64; and 7 to 18 μg/m^3, respectively. This time interval of the sampling campaign corresponds to the fence renovation activities (the construction of a substructure, laying of the bricks, using cement/adhesives). Week 7 of the sampling campaign, representing the period after the renovation activities, is characterized by considerably decreased average levels of PM_1, $PM_{2.5\text{-}1}$ and $PM_{10\text{-}2.5}$, i.e., 7.6, 2.7 and 0.7 μg/m^3, respectively. These data indicate that the resuspended and on-site-produced aerosol/dust was present at increased air levels and contributed to the pollution of the outdoor air around the dwelling house. On the other hand, the PM_x data of week 8 already indicates the use of wood heating in nearby dwelling houses, due to the cold autumn weather. That week was characterized by enhanced average $PM_{2.5\text{-}1}$ and $PM_{10\text{-}2.5}$ levels, i.e., 5.6 and 1.6 μg/m^3, respectively.

Before the start of the renovation (week 1), the average RH was about medium high (57%), but its average value started to increase gradually from the onset of the renovation (weeks 2–5). This could be due to the application of higher amounts of water in this area for the purpose of the reconstruction work (e.g., concrete production) and the hot summer weather, which contributed to the continuous evaporation of the water applied, and an increased RH. However, after a decrease (week 5), the RH showed an increase till week 7 (i.e., the period after the renovation) between 69% and 81%. The sampling campaign was featured with a rather modest air temperature (range: 18.5–27.8 °C, average: 24.4 °C), showing a gradual decrease on the final weeks of the renovation (weeks 5–6), due to arrival of slightly colder early autumn weather.

3.1.2. Weekly Indoor Aerosols and Microclimatic Parameters

The weekly mean concentrations of the monitored PM_x species and microclimatic conditions were evaluated similar to those of the outdoor air sampling, i.e., before (week 1), during (weeks 2–6) and after (weeks 7–8) the fence refurbishment (Figure 4). As can been seen, low concentrations of PM_1, $PM_{2.5\text{-}1}$ and $PM_{10\text{-}2.5}$ could be observed for the week before the renovation activity, with averages of 7.7, 3.3 and 1.3 μg/m^3, respectively. Of these variables, the mean value of PM_1 and $PM_{2.5\text{-}1}$ showed a small decrease to 6.2 and 3.2 μg/m^3 in the week when renovation started (week 2), whereas the average of $PM_{10\text{-}2.5}$ developed some increases, i.e., to 1.7 μg/m^3, respectively. Interestingly, this sampling week was characterized by several lower and some large peaks of PM_1, $PM_{2.5\text{-}1}$ and $PM_{10\text{-}2.5}$, reaching as high as 160, 255 and 220 μg/m^3, respectively, as compared to the former values (week 1). These data, along with the increased coarse background PM_x levels and higher weekly variance point towards some influence of the ongoing renovation activities on the indoor air quality.

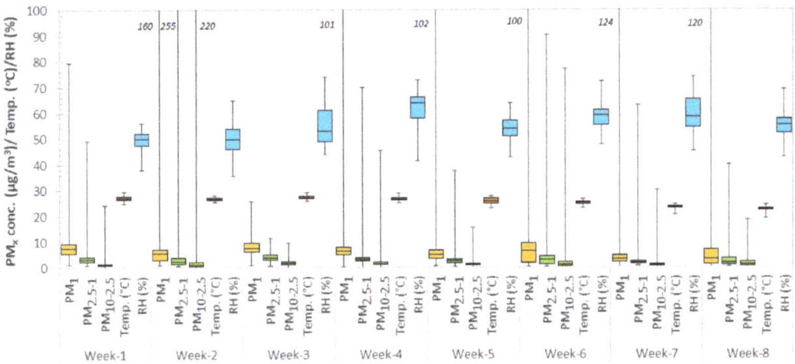

Figure 4. Weekly trends of size-segregated aerosol concentrations and of microclimatic data indoors (anteroom of the dwelling house); n = 2040 for each week/variable.

As seen in Figure 4, the subsequent period of the refurbishment (weeks 3–6) could be characterized by rather fluctuating average levels of PM_1, $PM_{2.5\text{-}1}$ and $PM_{10\text{-}2.5}$, in the ranges of 5.3–8.2 (average: 7.1), 2.6–4.1 (average: 3.6) and 1.2–2.1 (average: 1.8) μg/m^3, respectively. However, week 7 of the sampling campaign, i.e., after the renovation activities, was characterized by significantly dropped average levels of PM_1, $PM_{2.5\text{-}1}$ and $PM_{10\text{-}2.5}$, with 4.3, 2.1 and 1.1 μg/m^3, respectively. For the reasons explained above, week 8 was already characterized by slightly enhanced average aerosol levels, e.g., 5.7, 2.9 and 1.7 μg/m^3, for PM_1, $PM_{2.5\text{-}1}$ and $PM_{10\text{-}2.5}$, respectively.

Hansen et al. [12] reported that suspended PM with lower aerodynamic diameters was present at considerably higher levels during a nearby building demolition in the outdoor air around a hospital than before, but only by low factors: ultrafine particles increased by 1.6-fold, particles ≥ 0.3 μm by 1.6-fold, particles ≥ 0.5 μm by 2.9-fold and particles ≥ 1 μm by 3.3-fold. However, no significant effects of the outdoor construction work on the indoor air quality were reported. On the other hand, the present results point towards an increase in the concentration of the indoor PM_x in the dwelling house during the fence renovation, while their levels dropped quite quickly after finishing these activities.

With regard to the microclimatic data, recorded by means of the LCSs on the first two weeks of the sampling campaign, the average indoor RH level was moderate, with similar values, i.e., ≈49.7%. However, during the subsequent sampling weeks, it developed a slightly increasing and fluctuating trend, e.g., ranging from 54% to 62% (average: 58%). The weekly indoor air temperature showed high values at the beginning of the sampling campaign (average: 27 °C), but started to decrease on week 5 (from 25.9 to 23.4 °C, average: 24.8 °C), in response to the gradual seasonal cooling of the outdoor air.

3.1.3. Weekly Trends of Indoor–Outdoor PM_x Ratios

The weekly average I/O ratios for various PM_x fractions are depicted in Figure 5. As one can see, the I/O ratio of PM_1 was quite low, ranging from 0.6 to 0.8 (average: 0.7), showing that the fine aerosol was mostly originating from the outdoor air. This parameter featured a slightly fluctuating trend, and peaked on week 2, due in part to the rising influence of the fence renovation activity, and the related ingress of a large number of fine particulates indoors. Quite the opposite pattern could be observed for medium and coarse particles. Interestingly, I/O ratios for $PM_{2.5\text{-}1}$ slightly higher than 1 could only be observed on week 1 and week 2 of the sampling campaign, i.e., about 1.1, pointing towards indoor sources of these particulates. On the other hand, starting from week 3, the I/O ratios of $PM_{2.5\text{-}1}$ showed a gradually decreasing trend, with lower values (range: 0.79–0.92, average: 0.86), corresponding to outdoor sources of this aerosol fraction. Interestingly, the I/O ratio of the coarse aerosol ($PM_{10\text{-}2.5}$) reached fairly high values during the whole sampling campaign, ranging from 1.4 to 2.6 (average: 1.9). These findings point towards the internal source of coarse aerosols, partly from cooking activities (related to the use of a gas stove) and/or resuspended dust/aerosol, or even the ingress of finer particles from outdoors and their efficient coagulation/accumulation in the indoor air.

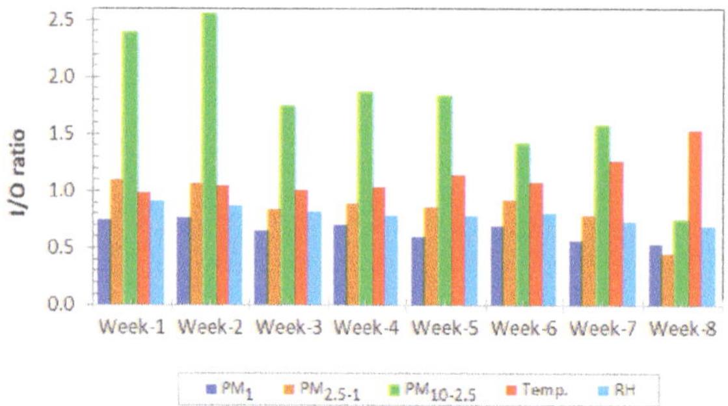

Figure 5. Weekly indoor–outdoor (I/O) ratios of PM_x and microclimatic variables calculated for the sampling campaign; n = 2040 for each week.

The I/O ratios of the air temperature only slightly fluctuated from 1.0 to 1.3 (mean: 1.1). These data did not show any temporal pattern related to the outdoor renovation activities. However, the I/O ratios of the RH values ranged between 0.7 and 0.9 (average: 0.8). During the first two weeks of the campaign, slightly increased values for I/O RH were observed, pointing towards the lowered impact of the water content of the outdoor air. These data point towards the ingress of outdoor air with higher water content into the interior of the dwelling house during the renovation activities.

3.1.4. Weekly Outdoor Levels of Gaseous Components

The variation in the weekly average concentrations of the monitored gaseous species was evaluated before (week 1), during (weeks 2–6) and after (weeks 7–8) the renovation activities (Figure 6). As appears from the data, the weekly average air level of HCHO ranged around 0.01 mg/m^3 before, during and after renovation (weeks 1–8) and demonstrated low variability for each week as well. However, larger HCHO maxima, ranging from 0.2 to 0.95 mg/m^3, were observed on the fence renovation weeks. As with HCHO, the average air concentration of CO_2 showed quite low fluctuation during the sampling weeks, ranging between 517 and 540 ppm (average: 530 ppm). This pollutant was characterized by variance of a medium degree, corresponding to a rather moderate motor vehicle traffic on the nearby main street.

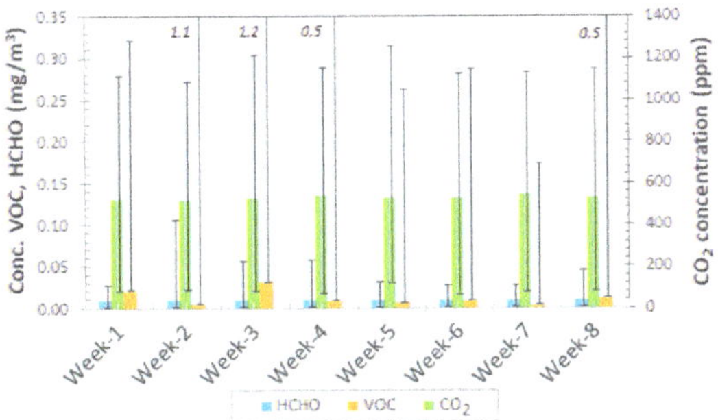

Figure 6. Variation in the weekly average concentrations of the monitored gaseous pollutants outdoors at the sampling site; resolution: 5 min, n = 2020 for each week; positive and negative error bars show maximum and minimum values, respectively.

The average weekly outdoor level of VOCs was about 0.02 mg/m^3 for week 1. It showed a fluctuating pattern onwards in the next sampling weeks, peaking as high as 0.032 mg/m^3 on week 3, but it decreased gradually on weeks 4–7, although a modest increase could be seen on week 8, with high fluctuations. In summary, the VOC levels varied over quite a large scale, with high peaks (e.g., 1.1 mg/m^3) observed on weeks 2–3. This was mostly due to some of the painting/adhesive-related reconstruction activities and the use of thinners for the paints and on the iron support structure of the fence against weathering, as well as the painting on the plinth of the fence (primer and topcoat paints). Interestingly, on week 8, an outstandingly high VOC concentration of 2.5 mg/m^3 could be observed, which was most likely due to wood burning in the neighborhood in the evening hours.

3.1.5. Weekly Indoor Levels of Gaseous Components

The variation in the weekly average concentrations of the gaseous species monitored indoors were assessed too (Figure 7). As it can be seen, the weekly mean CO_2 level indoors was fluctuating between 540 and 645 ppm, with rather high variance for each sampling week. These levels are well below the recommended IAQ values summarized by the ASHRAE Standard (3500 or 5000 ppm) [33]. The indoor source of CO_2 is predominantly cooking activities, related to emissions from combustion-based stoves (gas/coal/biomass stoves, e.g., [34]), and exhalation of occupants [15], each contributing to the indoor accumulation of this pollutant in contexts with sealed doors and windows or nonproper ventilation of the apartment. For instance, Sánka and Földváry [13] reported that the CO_2 concentration ranged between 660 and 2050 ppm (mean: 1205 ppm) before the renovation of a residential building (block of flats). However, after the renovation to introduce energy-saving measures, significantly higher CO_2 levels (range: 880–2770 ppm, mean: 1570 ppm) were recorded in several of the study apartments. This was mainly due to the lowered air-exchange rate of the building, i.e., 0.61 h^{-1} vs. 0.44 h^{-1}, before and after the renovation, respectively. In addition, they reported a similarly enhanced indoor VOC concentration, i.e., 0.57 mg/m^3 and 0.77 mg/m^3, before and after renovation, respectively [13].

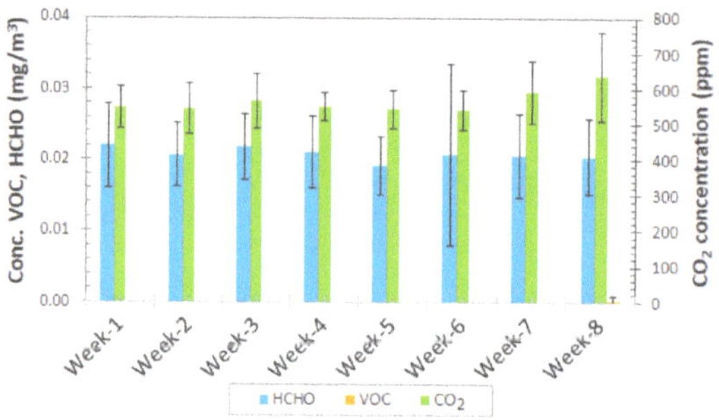

Figure 7. Changes in the weekly average air levels of the monitored gases, evaluated with 5 min resolution, indoors (anteroom); error bars denote fluctuations (SD), n = 2020 for each week.

In this study, interestingly, the weekly average indoor VOC level was negligibly low, being only rarely detectable during the sampling campaign with the presently applied sensors. However, one high peak of VOCs was caught on week 8, reaching a concentration as high as 0.036 mg/m^3, which was related to cooking activities in the morning. This observation is in line with the quite sharply increased CO_2 concentration (range: 640–1080 ppm, average: 860 ppm) and the slightly increased PM_x levels during this pollution episode, with PM_1, $PM_{2.5}$ and PM_{10} peaking at 40, 50 and 57 µg/m^3, respectively. Moreover, indoor emissions of VOCs (range: 0.01–0.03 mg/m^3, average: 0.02 mg/m^3), whose source could not be identified, could also be observed on some occasions in week 8. They were not related to outdoor pollution, since the simultaneous monitoring with the LCSs could not detect VOCs in the outdoor air, but were likely due to some cleaning activities, which release higher levels of organics. The related literature reports on the sources of VOCs arising partly from various chemicals/materials (e.g., adhesives, paints, lacquers) of newly established/renovated buildings and partly from the occupants' belongings and activities [14]. Overall, in the study dwelling house, the level of VOCs was found to be lower than the recommended IAQ value of the WHO (0.5 ppm).

The average weekly indoor HCHO level was rather high during the sampling campaign, ranging from 0.019 to 0.022 mg/m^3, as compared to that observed for the outdoor air (cf. data of Section 3.1.4). In addition, broad fluctuations in the HCHO levels could be seen indoors, for instance, on week 6. On the other hand, it should be noted that this pollutant is primarily of indoor origin; its source is mostly related to adhesives used in wooden boards and furniture [8]. However, Dodson et al. [14] reported that besides occupants' activities, building sources could contribute as well. The interior of the studied dwelling house is also most likely characterized by a mixed pattern of HCHO sources too, due to the predominant presence of furniture made of fiberboard and/or plywood, as well as the use of various cleaning agents and fragrances, in addition to daily cooking. Overall, the median indoor level of HCHO (0.020 mg/m^3) was below the IAQ standard of the WHO (0.1 mg/m^3, [35]). Similar mean/median values were documented for dwelling houses in Austria (0.025 mg/m^3, n = 160) [36], France (0.0196 mg/m^3, n = 567) [35], Arizona (0.021 mg/m^3, n = 181) [37] and Quebec City, Canada (geometric mean: 0.0295 mg/m^3, n = 91) [38]. On the other hand, the present HCHO data are twice higher than those reported for post-occupancy buildings in Boston (0.011 mg/m^3) [14], but half of the average level found in Helsinki (0.041 mg/m^3) [39].

3.1.6. Weekly Indoor-to-Outdoor Ratios for Gaseous Air Pollutants

The weekly average I/O ratios for the monitored gases are depicted on Figure 8. As seen, the weekly average I/O ratio of CO_2 ranged between 1.0 and 1.2, showing an increase in the first three weeks of the sampling period, followed by a decrease (weeks 4–5), and then another increase towards week 8. Similarly, the I/O ratio of HCHO developed a rather similar trend over the campaign weeks, fluctuating between 1.9 and 2.1. As expected, the

VOC levels showed negligibly low I/O ratios, which was due to the low level of these species indoors, resulting in a rare detection frequency. The lower airing rates of the rooms also contributed to the less efficient introduction of fresh air, resulting in an ingress of VOCs too. Two of the monitored gaseous species (CO_2 and HCHO) were mainly of indoor origin in the dwelling house, i.e., related to the usage of the gas stove for daily cooking activities (CO_2, HCHO), the exhaling of residents (CO_2) and/or the presence of plywood and fiberboard furniture (HCHO).

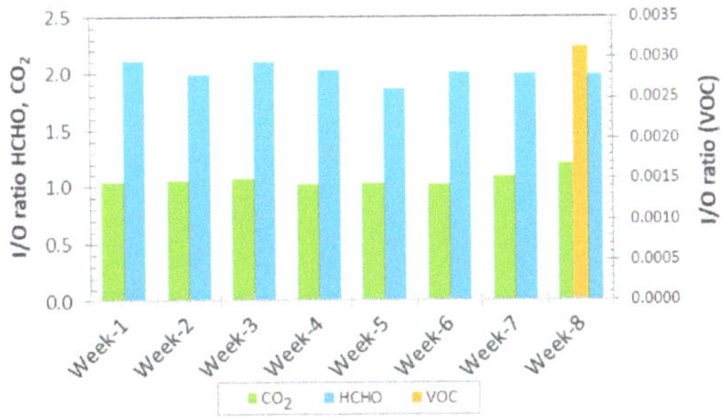

Figure 8. Trends of weekly average indoor-to-outdoor (I/O) ratios of the studied gaseous air pollutants, calculated for the sampling campaign; n = 2040 for each week.

3.2. *Over-Campaign Temporal Evolution of PM_x and Gaseous Pollutants*

3.2.1. Trends of Indoor and Outdoor PM_x

The temporal evolution of the indoor and outdoor PM_x data is depicted in Figure 9a–c. As it appears, the first two weeks of the campaign could be characterized by high levels of outdoor PM_1 and moderate levels of $PM_{2.5\text{-}1}$ and $PM_{10\text{-}2.5}$. At the middle of the week (10 and 12 August), there were some decreases in the overall PM_x content in ambient air. On the other hand, enhanced background levels of PM_x appeared on subsequent days and weeks, a couple of them featuring intensive peaks. A higher variability in the PM_x concentrations could be seen from 6–8 September, indicating worsened air quality. Nevertheless, the overall PM_x concentration began to fall after 9 September, on the last weekend of the refurbishment activities.

The temporal evolution of the indoor PM_x concentration appears to follow that of the outdoor PM_x regarding each aerosol fraction (Figure 9a–c). Nevertheless, the indoor air could be characterized by lower background PM_x levels during the sampling campaign, whereas high peaks occurred more frequently indoors compared to the outdoor air. The high indoor peaks of PM_x were mostly due to the daily cooking activities on the gas stove. As such, one of the most serious types of air pollution events was observed when using cooking oil in an open frying pan and encountering its accidental overheating and burning, which brought a significant contribution to the in-house aerosol levels.

3.2.2. Temporal Evolution of Gaseous Pollutants

The temporal evolution of the monitored gaseous pollutants is depicted in Figure 10. As it appears, the indoor air concentration of CO_2 shows a fluctuating pattern, i.e., around 550–600 ppm on average. Only a few CO_2 maxima above 1000 ppm can be seen, whereas a large number of peaks appear between 800 and 1000 ppm. This air concentration was already considered as the safety level for people staying indoors and exposed to CO_2 levels, as suggested in Ref. [40]. The observed relatively high peaks occurred usually during the last week of the sampling period, when the outdoor temperature dropped, and lower ventilation rates (0.25–0.39 h^{-1}) for the dwelling house were observed due to the less frequent airing of the rooms.

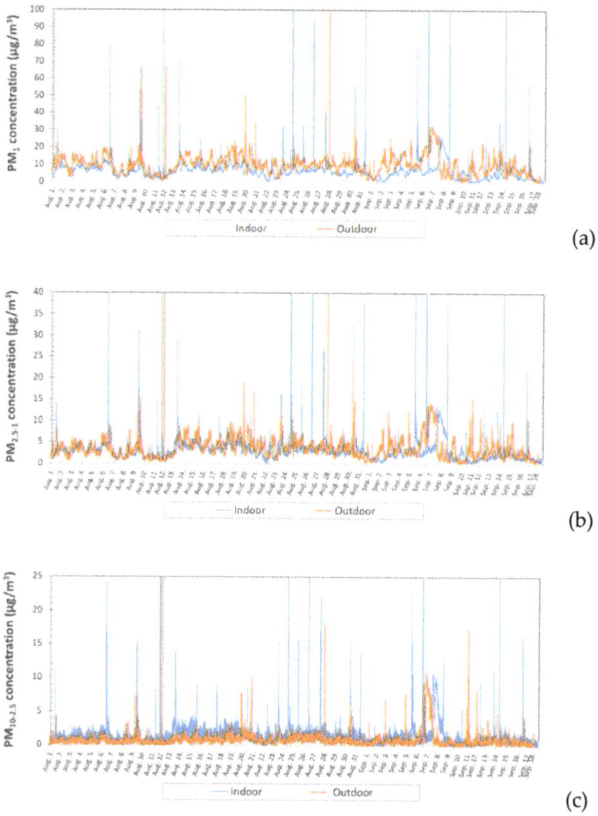

Figure 9. Temporal evolution of the indoor and outdoor PM_x concentrations recorded with the GPM sensor during the sampling campaign; PM_1 (**a**), $PM_{2.5\text{-}1}$ (**b**) and $PM_{10\text{-}2.5}$ (**c**).

The indoor HCHO level was observed to be fluctuating around 0.021 mg/m^3 (range: 0.009–0.49 mg/m^3), with several sharp peaks appearing on each week of the sampling. These peaks were rather due to indoor sources of varying intensities, e.g., furniture, adhesives and cooking activities. On the contrary, the outdoor air level of HCHO was mostly found to be low and rather constant, around 0.01 mg/m^3 on average during the sampling campaign (Figure 10b). Only some occasional peaks of this pollutant were seen, reaching as high as 1.1–2.5 mg/m^3 on weeks 2–4, each related to some fence construction activities, and on week 8, when some nearby wood burning activities occurred, as noted above.

The indoor VOC levels were low during the sampling campaign, mostly undetectable with the applied LCSs (Figure 10). However, some of the days at the end of the sampling campaign (e.g., week 8) were characterized by enhanced VOC concentrations, which reached values as high as 0.036 mg/m^3. These maxima were mostly associated with gas stove-assisted cooking in the morning hours, when the indoor CO_2 levels were also peaking, i.e., about 1000–1080 ppm. The average outdoor concentration of VOCs was 0.012 mg/m^3, while peak concentrations occurred several times during the renovation-related work, especially on weeks 2–3, each related to the use of paints and corresponding thinners, and even on week 8, likely related to the use of wood/coal-fueled stoves for barbecues and/or domestic heating.

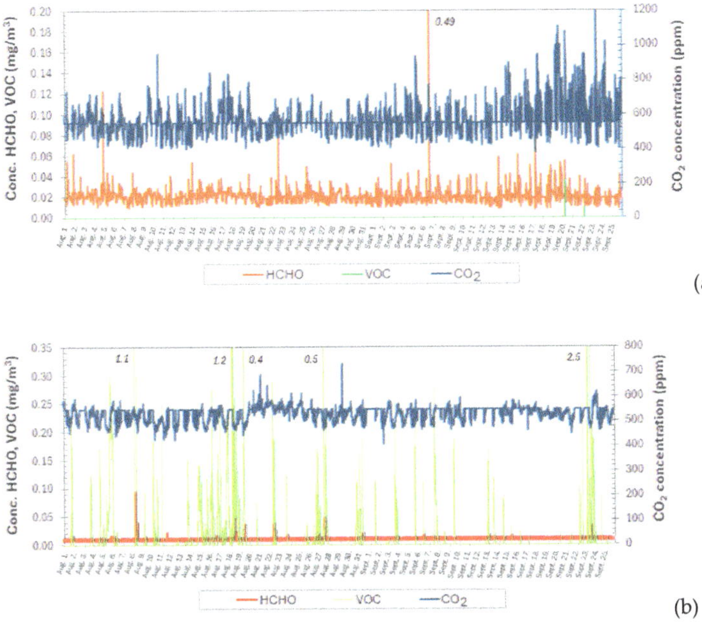

(a)

(b)

Figure 10. Temporal evolution of the concentrations of the monitored gaseous pollutants during the sampling campaign; indoors (**a**) and in the outdoor air (**b**).

3.3. Correlation of Indoor and Outdoor PM_x and Gaseous Pollutants

Pearson's correlation coefficients calculated between indoor and outdoor data sets for the whole campaign showed only very weak or no correlations of the variables. Therefore, the most relevant air pollution episodes of the renovation (weeks 2–3) were evaluated for correlation (Table 1). As it can be seen, the indoor and outdoor PM_1 and $PM_{2.5\text{-}1}$ levels were weakly correlated with each other (R = 0.34–0.43) as expected on the base of the above findings. Much lower/no correlation was attained between the indoor or outdoor data of the finer fractions and $PM_{10\text{-}2.5}$ (R = 0.13–0.22).

Table 1. Indoor–outdoor correlation coefficients (R) for air pollutants and microclimatic variables for weeks 2–3 of the sampling campaign.

Indoor Parameter	Outdoor Variable							
	PM_1	$PM_{2.5\text{-}1}$	$PM_{10\text{-}2.5}$	CO_2	HCHO	VOC	T	RH
PM_1	0.37	0.34	0.16	−0.13	0.04	0.08	0.31	−0.12
$PM_{2.5\text{-}1}$	0.40	0.43	0.13	−0.04	0.03	0.09	0.30	−0.06
$PM_{10\text{-}2.5}$	0.22	0.14	0.15	−0.20	0.04	0.03	0.26	−0.15
CO_2	−0.05	0.13	−0.19	−0.19	0.02	0.07	0.32	−0.22
HCHO	0.07	0.25	−0.20	−0.20	−0.01	0.08	0.32	−0.15
VOC	-	-	-	-	-	-	-	-
T	0.01	0.10	−0.18	−0.67	0.05	0.11	0.85	−0.66
RH	0.50	0.50	0.22	0.45	0.06	0.11	−0.15	0.67

The indoor–outdoor data of HCHO and VOCs did not show any correlation. The indoor CO_2 concentration was strongly negatively correlated with the air temperature ($R = -0.67$), while it showed a weak correlation with RH. However, the indoor and outdoor data of microclimatic variables (T, RH) were strongly correlated, with R values of 0.85 and 0.67, respectively. The indoor air temperature was anticorrelated fairly closely with outdoor

RH, while the outdoor temperature showed weak correlation with the concentrations of most of the studied indoor pollutants (R = 0.30–0.32).

The correlation data of the outdoor variables are compiled in Table S2, while those of indoor data are listed in Table S3. For the outdoor variables, the fine, medium-size and coarse aerosol showed strong correlations with each other (R = 0.76–0.89). The air CO_2 concentration featured some correlation with the RH (R = 0.67), but some negative correlation with the air temperature ($R = -0.53$). Interestingly, VOCs were loosely correlated with HCHO (R = 0.33), suggesting the latter compound is a partial degradation product. For the indoor air, the various PM_x fractions were more strongly correlated (R = 0.84–0.94) than the outdoor PM_x. Only a weak correlation between CO_2 and HCHO was observed, pointing towards the possible source of HCHO from combustion/cooking activities. All these data point towards a low influence of outdoor air pollutants on the indoor AQ parameters.

3.4. Temporal Changes in Pollutants at a Nearby Air Quality Station

The weekly trends of PM_x and CO concentrations, calculated from the hourly raw monitoring data of an official AQ station, are depicted in Figure 11. As it appears, the weekly average $PM_{2.5}$ concentrations ranged between 6.0 and 9.1 μg/m^3 (average: 7.3 μg/m^3). Weeks 1, 2, 5 and 7 could see lower averages, while the other weeks were characterized by slightly higher concentrations. Apart from week 6 and week 8, the within-week variation of $PM_{2.5}$ was low, manifested in the observed low variances. The weekly average $PM_{10\text{-}2.5}$ levels ranged between 3.1 and 8.8 μg/m^3 (average: 5.5 μg/m^3, median: 4.5 μg/m^3). Ostensibly, higher levels were experienced for the summer period (weeks 1–4). This generally high fluctuation was manifested in the enhanced variances (i.e., higher maxima), peaking on week 5 of the sampling campaign. Overall, it can be seen that the temporal evolution of the PM_x levels indoors is in agreement with those observed outdoors at the sampling site, as background pollution.

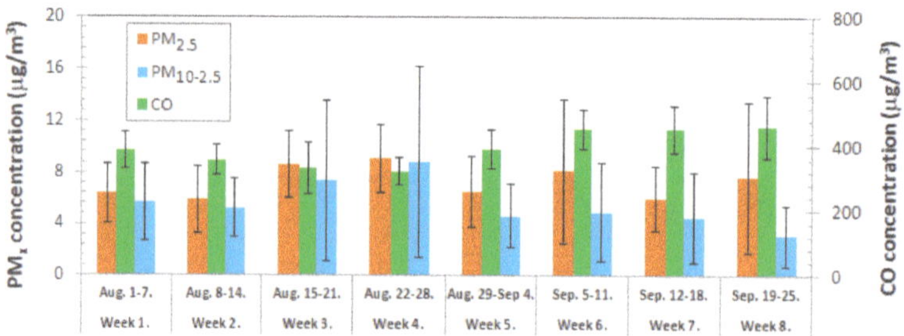

Figure 11. Weekly trends of PM_x and CO levels evaluated from the data of an official air quality station on Gergely street (Kőbánya district, Budapest).

Besides PM_x, air CO levels were evaluated at the AQ station. This gas mostly stems from incomplete, high temperature combustion, such as biomass burning and vehicle exhaust emissions, and is an indicator of the extent of emissions for these types of nearby sources, similar to air CO_2 levels. As seen, the average weekly CO level ranged from 322 to 461 μg/m^3 (average: 395 μg/m^3). Interestingly, its air concentration was small during the summer period (weeks 1–4), whereas a slightly rising level was observed during the autumn weeks, likely due to increased traffic nearby. A rather low fluctuation in the CO level was obtained (Figure 11). Overall, this indicates the smooth contributions of traffic on summer weeks, whereas around 25% increase in the CO level was obtained on the early autumn weeks of the sampling period.

4. Conclusions

This study reported on the indoor and outdoor levels of fine, medium-sized and coarse PM as well as some gaseous pollutants (CO_2, HCHO, VOC) in a dwelling house, monitored concurrently during the reconstruction activities of the garden fence, performed for several weeks. The weekly average data for size-segregated aerosols indicate that the resuspended and on-site produced aerosol/dust was present at increased concentrations and contributed to the pollution of the outdoor air around the dwelling house. This enhanced aerosol burden was mostly due to the work processes related to the renovation, which involved the resuspension of dust, including the demolition of the old fence, excavation and substructure construction.

An enhancement in indoor PM_x levels was observed during the fence renovation, but only to a small extent. These variables featured a decrease shortly after the finalization of the outdoor renovation activity. The indoor-to-outdoor concentration ratios indicate that the fine and medium-size aerosols were mostly of outdoor origin; their concentrations depended on the type of the renovation activity. However, the coarse aerosols were mostly of in-house origin, partly due to cooking activities and/or likely the coagulation of finer PM.

Among the studied gases, HCHO showed a rather small and constant outdoor concentration pattern, whereas its average indoor level was twice as high. Furthermore, indoor HCHO levels varied over a wide range due to some in-house emission sources, such as cooking on the gas stove, the presence of plywood/fiberboard furniture, fragrance and detergent use, etc. The outdoor CO_2 concentration developed a similar smooth weekly trend, but with increased variability. However, the indoor CO_2 level was rather dependent on the in-house emission sources, such as cooking duration, the number and length of stay of occupants and the room ventilation frequency. On the other hand, the outdoor VOC level changed over a broad range, demonstrating its dependence on reconstruction work, such as the use of paints and thinners for the new fence. Peak concentrations of the outdoor pollutants (e.g., PM_x, VOC) were found to be better indices for the nearby reconstruction activities and related pollution episodes than the weekly average/median values, which showed only slight difference during the sampling period. The utilization of high-resolution low-cost sensors for monitoring gaseous pollutants and suspended particulate matter helped to reveal changes in several air pollutants/AQ parameters in fine temporal resolution and to assign some dominant emission sources.

Apart from CO_2 and VOCs, some correlations of the pollutants were observed regarding indoor and outdoor data. Some of the PM_x data were compared to those observed at a nearby official AQ station, reflecting a similar background air pollution pattern before, during and after the fence renovation activities, as observed in the outdoor air around the studied dwelling house.

Based on the observed PM_x values and the gas pollutant levels, the fence renovation work posed a small health risk to the surrounding residents. It is to be noted that the air permeability of the new fence is lower than that of the old one, which may cause an alteration in the concentration of air pollutants via changing their ingress into the garden and indoors, which can have consequences for the health effects of the residents. This aspect of renovation-related pollution has not yet been exploited; thus, it remains for future research.

Supplementary Materials: The following SMs are available online at https://www.mdpi.com/article/10.3390/atmos15070790/s1, Figure S1: Summary of the meteorological conditions, Table S1: Timeline of the renovation activities, Table S2: Correlation coefficients of the outdoor air pollutants and microclimatic variables, Table S3: Correlation coefficients of the indoor air pollutants and microclimatic variables.

Funding: This research was partially supported by the Hungarian National Research Development and Innovation Fund under grant No. TKP2021-NVA-04.

Institutional Review Board Statement: Not applicable.

Informed Consent Statement: Not applicable.

Data Availability Statement: The research data are not publicly available due to file size limitations. The data presented in this study are available on request from the corresponding author.

Acknowledgments: The author thanks the owners/family members of the dwelling house for providing the opportunity to deploy the indoor and outdoor air sampling units, and for their patience and supportive attitude during the sampling campaign.

Conflicts of Interest: The author declares no conflicts of interest.

References

1. IPCC. *Climate Change 2021: The Physical Science Basis, Sixth Assessment Report*; Cambridge University Press: Cambridge, UK, 2021.
2. Dockery, D.W.; Speizer, F.E.; Stram, D.O.; Ware, J.H.; Spengler, J.D.; Ferris, B.G., Jr. Effects of inhalable particles on respiratory health of children. *Am. Rev. Respir. Dis.* **1989**, *139*, 587–594. [CrossRef] [PubMed]
3. Spolnik, Z.; Bencs, L.; Worobiec, A.; Kontozova, V.; Van Grieken, R. Application of EDXRF and thin window EPMA for the investigation of the influence of hot air heating on the generation and deposition of particulate matter. *Microchim. Acta* **2005**, *149*, 79–85. [CrossRef]
4. Stranger, M.; Krata, A.; Kontozova-Deutsch, V.; Bencs, L.; Deutsch, F.; Worobiec, A.; Naveau, I.; Roekens, E.; Van Grieken, R. Monitoring of NO_2 in the ambient air with passive samplers before and after a road reconstruction event. *Microchem. J.* **2008**, *90*, 93–98. [CrossRef]
5. Bencs, L.; Ravindra, K.; de Hoog, J.; Spolnik, Z.; Bleux, N.; Berghmans, P.; Deutsch, F.; Roekens, E.; Van Grieken, R. Appraisal of measurement methods, chemical composition and sources of fine atmospheric particles over six different areas of Northern Belgium. *Environ. Pollut.* **2010**, *158*, 3421–3430. [CrossRef] [PubMed]
6. Zaeh, S.E.; Koehler, K.; Eakin, M.N.; Wohn, C.; Diibor, I.; Eckmann, T.; Wu, T.D.; Clemons-Erby, D.; Gummerson, C.E.; Green, T.; et al. Indoor air quality prior to and following school building renovation in a Mid-Atlantic school district. *Int. J. Environ. Res. Public Health* **2021**, *18*, 12149. [CrossRef] [PubMed]
7. Yang, W.; Sohn, J.; Kim, J.; Son, B.; Park, J. Indoor air quality investigation according to age of the school buildings in Korea. *J. Environ. Manag.* **2009**, *90*, 348–354. [CrossRef] [PubMed]
8. Ismaeel, W.S.E.; Alamoudy, F.O.; Sameh, R. How renovation activities may jeopardise indoor air quality: Accounting for short and long-term symptoms of sick building syndrome in educational buildings. *Archit. Eng. Des. Manag.* **2023**, *19*, 360–377. [CrossRef]
9. Latif, M.T.; Baharudin, N.H.; Velayutham, P.; Awang, N.; Hamdan, H.; Mohamed, R.; Mokhtar, M.B. Composition of heavy metals and airborne fibers in the indoor environment of a building during renovation. *Environ. Monit. Assess.* **2011**, *181*, 479–489. [CrossRef] [PubMed]
10. Latif, M.T.; Na, T.S.; Baharudin, N.H.; Srithawirat, T.; Mohamad, N.; Khan, M.F.; Sulaiman, F.R. Composition of trace metals in indoor dust during and after building renovation. *Environ. Eng. Manag. J.* **2018**, *17*, 1781–1790. [CrossRef]
11. Reitzig, M.; Mohr, S.; Heinzow, B.; Knoppel, H. VOC emissions after building renovations: Traditional and less common indoor air contaminants, potential sources, and reported health complaints. *Indoor Air* **1998**, *8*, 91–102. [CrossRef]
12. Hansen, D.; Blahout, B.; Benner, D.; Popp, W. Environmental sampling of particulate matter and fungal spores during demolition of a building on a hospital area. *J. Hosp. Infect.* **2008**, *70*, 259–264. [CrossRef] [PubMed]
13. Sánka, I.; Földváry, V. Indoor air quality of residential building before and after renovation. *Slovak J. Civ. Eng.* **2017**, *25*, 1–6. [CrossRef]
14. Dodson, R.E.; Udesky, J.O.; Colton, M.D.; McCauley, M.; Camann, D.E.; Yau, A.Y.; Adamkiewicz, G.; Rudel, R.A. Chemical exposures in recently renovated low-income housing: Influence of building materials and occupant activities. *Environ. Int.* **2017**, *109*, 114–127. [CrossRef] [PubMed]
15. Kozielska, B.; Mainka, A.; Zak, M.; Kaleta, D.; Mucha, W. Indoor air quality in residential buildings in Upper Silesia, Poland. *Build. Environ.* **2020**, *177*, 106914. [CrossRef]
16. Derbez, M.; Wyart, G.; Le Ponner, E.; Ramalho, O.; Riberon, J.; Mandin, C. Indoor air quality in energy-efficient dwellings: Levels and sources of pollutants. *Indoor Air* **2018**, *28*, 318–338. [CrossRef] [PubMed]
17. Vilčekova, S.; Apostoloski, I.Z.; Mečiarová, Ľ.; Burdová, E.K.; Kiseľák, J. Investigation of indoor air quality in houses of Macedonia. *Int. J. Environ. Res. Public Health* **2017**, *14*, 37. [CrossRef]
18. Omelekhina, Y.; Nordquist, B.; Alce, G.; Caltenco, H.; Wallenten, P.; Borell, J.; Wierzbicka, A. Effect of energy renovation and occupants' activities on airborne particle concentrations in Swedish rental apartments. *Sci. Total Environ.* **2022**, *806*, 149995. [CrossRef] [PubMed]
19. Awad, J.; Jung, C. Evaluating the indoor air quality after renovation at the greens in Dubai, United Arab Emirates. *Buildings* **2021**, *11*, 353. [CrossRef]
20. Hameed, A.A.A.; Yasser, I.H.; Khoder, I.M. Indoor air quality during renovation actions: A case study. *J. Envion. Monit.* **2004**, *6*, 740–744. [CrossRef]

21. Huang, C.; Zhang, J.L.; Sun, C.J.; Liu, W.; Zhang, Y.P.; Li, B.Z.; Zhao, Z.H.; Deng, Q.H.; Zhang, X.; Qian, H.; et al. Associations between household renovation and rhinitis among preschool children in China: A cross-sectional study. *Indoor Air* **2020**, *30*, 827–840. [CrossRef]
22. Jiang, W.; Lu, C.; Miao, Y.F.; Xiang, Y.G.; Chen, L.; Deng, Q.H. Outdoor particulate air pollution and indoor renovation associated with childhood pneumonia in China. *Atmos. Environ.* **2018**, *174*, 76–81. [CrossRef]
23. Deng, Q.H.; Lu, C.; Ou, C.Y.; Liu, W.W. Effects of early life exposure to outdoor air pollution and indoor renovation on childhood asthma in China. *Build. Environ.* **2015**, *93*, 84–91. [CrossRef]
24. Rejc, T.; Kukec, A.; Bizjak, M.; Godič-Torkar, K. Microbiological and chemical quality of indoor air in kindergartens in Slovenia. *Int. J. Environ. Health Res.* **2020**, *30*, 49–62. [CrossRef] [PubMed]
25. Vardoulakis, S.; Giagloglou, E.; Steinle, S.; Davis, A.; Sleeuwenhoek, A.; Galea, K.S.; Dixon, K.; Crawford, J.O. Indoor Exposure to Selected Air Pollutants in the Home Environment: A Systematic Review. *Int. J. Environ. Res. Public Health* **2020**, *17*, 8972. [CrossRef] [PubMed]
26. Alfano, B.; Barretta, L.; Del Giudice, A.; De Vito, S.; Di Francia, G.; Esposito, E.; Formisano, F.; Massera, E.; Miglietta, M.L.; Polichetti, T. A review of low-cost particulate matter sensors from the developers' perspectives. *Sensors* **2020**, *20*, 6819. [CrossRef]
27. Giordano, M.R.; Malings, C.; Pandis, S.N.; Presto, A.A.; McNeill, V.F.; Westervelt, D.M.; Beekmann, M.; Subramanian, R. From low-cost sensors to high-quality data: A summary of challenges and best practices for effectively calibrating low-cost particulate matter sensors. *J. Aerosol Sci.* **2021**, *158*, 105833. [CrossRef]
28. Rai, A.C.; Kumar, P.; Pilla, F.; Skouloudis, A.N.; Di Sabatino, S.; Ratti, C.; Yasar, A.; Rickerby, D. End-user perspective of low-cost sensors for outdoor air pollution monitoring. *Sci. Total Environ.* **2017**, *607*, 691–705. [CrossRef]
29. Bencs, L.; Plósz, B.; Mmari, A.G.; Szoboszlai, N. Comparative Study on the Use of Some Low-Cost Optical Particulate Sensors for Rapid Assessment of Local Air Quality Changes. *Atmosphere* **2022**, *13*, 1218. [CrossRef]
30. Bencs, L.; Nagy, A. A Smoke Chamber Study on Some Low-Cost Sensors for Monitoring Size-Segregated Aerosol and Microclimatic Parameters. *Atmosphere* **2024**, *15*, 304. [CrossRef]
31. Cui, S.; Cohen, M.; Stabat, P.; Marchio, D. CO_2 tracer gas concentration decay method for measuring air-exchange rate. *Build. Environ.* **2015**, *84*, 162–169. [CrossRef]
32. Rivero, R.A.G.; Hernández, L.E.M.; Schalm, O.; Rodríguez, E.H.; Sánchez, D.A.; Pérez, M.C.M.; Caraballo, V.N.; Jacobs, W.; Laguardia, A.M. A Low-Cost Calibration Method for Temperature, Relative Humidity, and Carbon Dioxide Sensors Used in Air Quality Monitoring Systems. *Atmosphere* **2023**, *14*, 191. [CrossRef]
33. *ANSI/ASHRAE Standard 62.1-2007*; Ventilation for Acceptable Indoor Air Quality. Addendum q to ANSI/ASHRAE Standard 62.1-2007; American Society of Heating, Refrigerating and Air-Conditioning Engineers, Inc.: Atlanta, GA, USA, 2010.
34. Neumann, M.; Dlamini, W.N.; Sallah-Ud-Din, R.; Berekute, A.K.; Siregar, S.; Getnet, M.E.; Maulana, M.; Pan, W.C.; Lung, S.C.C.; Yu, K.P. Assessment of air pollution emitted during cooking using biomass and cleaner fuels in the Shiselweni region of Eswatini (Swaziland). *Clean Technol. Environ. Policy* **2024**. [CrossRef]
35. WHO (World Health Organization). *WHO Guidelines for Indoor Air Quality: Selected Pollutants*, 1st ed.; WHO Regional Office for Europe: Copenhagen, Denmark, 2010; Available online: https://www.who.int/publications/i/item/9789289002134 (accessed on 3 June 2024).
36. Hutter, H.P.; Moshammer, H.; Wallner, P.; Tappler, P.; Kundi, M. Volatile organic compounds: Guidelines from the Austrian working group on indoor air. In *Indoor Air 2005, Proceedings of the 10th International Conference on Indoor Air Quality and Climate, 4–9 September 2005, Beijing, China*; Yang, X., Zhao, B., Zhao, R., Eds.; Volume 1–5, pp. 3519–3522.
37. Gordon, S.M.; Callahan, P.J.; Nishioka, M.G.; Brinkman, M.C.; O'Rourke, M.K.; Lebowitz, M.D.; Moschandreas, D.J. Residential environmental measurements in the National Human Exposure Assessment Survey (NHEXAS) pilot study in Arizona: Preliminary results for pesticides and VOCs. *J. Exposure Anal. Environ. Epidemiol.* **1999**, *9*, 456–470. [CrossRef] [PubMed]
38. Gilbert, N.L.; Gauvin, D.; Guay, M.; Héroux, M.E.; Dupuis, G.; Legris, M.; Chan, C.C.; Dietz, R.N.; Lévesque, B. Housing characteristics and indoor concentrations of nitrogen dioxide and formaldehyde in Quebec City, Canada. *Environ. Res.* **2006**, *102*, 1–8. [CrossRef] [PubMed]
39. Jurvelin, J.; Vartiainen, M.; Jantunen, M.; Pasanen, P. Personal Exposure Levels and Microenvironmental Concentrations of Formaldehyde and Acetaldehyde in the Helsinki Metropolitan Area, Finland. *J. Air Waste Manag. Assoc.* **2001**, *51*, 17–24. [CrossRef]
40. Morawska, L.; Allen, J.; Bahnfleth, W.; Bennett, B.; Bluyssen, P.M.; Boerstra, A.; Buonanno, G.; Cao, J.J.; Dancer, S.J.; Floto, A.; et al. Mandating indoor air quality for public buildings. *Science* **2024**, *383*, 1418–1420. [CrossRef]

MDPI AG
Grosspeteranlage 5
4052 Basel
Switzerland
Tel.: +41 61 683 77 34

Atmosphere Editorial Office
E-mail: atmosphere@mdpi.com
www.mdpi.com/journal/atmosphere

www.ingramcontent.com/pod-product-compliance
Lightning Source LLC
Chambersburg PA
CBHW061547210926
51850CB00009B/2026
9783725856558